Socio-scientific Issues in the Classroom

Contemporary Trends and Issues in Science Education

VOLUME 39

SCOPE

The book series Contemporary Trends and Issues in Science Education provides a forum for innovative trends and issues connected to science education. Scholarship that focuses on advancing new visions, understanding, and is at the forefront of the field is found in this series. Accordingly, authoritative works based on empirical research and writings from disciplines external to science education, including historical, philosophical, psychological and sociological traditions, are represented here.

For other titles published in this series, go to
www.springer.com/series/6512

Troy D. Sadler
Editor

Socio-scientific Issues in the Classroom

Teaching, Learning and Research

Springer

Editor
Troy D. Sadler
University of Florida
Gainesville, Florida, USA
tsadler@coe.ufl.edu

ISSN 1878-0482 e-ISSN 1878-0784
ISBN 978-94-007-1158-7 e-ISBN 978-94-007-1159-4
DOI 10.1007/978-94-007-1159-4
Springer Dordrecht Heidelberg London New York

Library of Congress Control Number: 2011928250

Printed on acid-free paper

Springer is part of Springer Science+Business Media (www.springer.com)

To Andrea Sadler, my best friend and wife.

Foreword

The socio-scientific issues (SSI) innovation belongs to a long history of rational attempts to improve science education for the vast majority of students (grades 6–12) who have not been well served by their school science, as documented by five key failures (Aikenhead, 2006, 2010):

1. There is an alarming and chronic decline of interest and enrolment in secondary science education, in spite of the fact that students generally continue to value science in their world outside of school (Schreiner & Sjøberg, 2007).
2. School science tends to alienate students whose cultural identities differ from the culture of science found in typical science classrooms.
3. Although students do grasp scientific ideas meaningfully (as needed) in out-of-school settings (Albright, Towndrow, Kwek, & Tan, 2008), they generally fail to learn academic science content meaningfully in school. For instance, only 20% of students achieved meaningful learning of the molecule concept over 10 years of school instruction (Löfgren & Helldén, 2009).
4. "Empirical evidence demonstrates how students and many teachers react to being placed in the political position of having to play school games [playing Fatima's rules] to make it appear as if significant science learning has occurred even though it has not" (Aikenhead, 2006, p. 28).
5. Similar to the mass media, conventional school science conveys dishonest and mythical images of science and scientists, such as a positivistic ideology of technical rationality that supports the existence of "*the* scientific method."

In light of such accumulated evidence one might ask: Why would students want to engage with school science in the first place? The question anticipates a mission of SSI school science.

With each new decade of innovation toward a more relevant school science, science education researchers have became more sophisticated and less naïve in both practical and theoretical ways, much to the benefit of those few students lucky enough to have been participants in their projects. However, this evolution has often been challenged by a country's or school system's culture of school science. The importance of this observation is its implication that SSI innovations are first and foremost about *changing the culture of* conventional school science (Aikenhead, 2000; Sadler, 2009).

One way to characterize some features of school science culture is to consider the triad: content (the product of past or current learning), process, and context. This book offers a rich diversity of research into all three cultural components.

Potential success for student learning can best be understood in terms of how each of these three cultural components interacts with students' cultural self-identities (i.e., who they are, where they have been, where they are going, and who they want to become). "We need to consider how learning science can change students' identities by changing their ability to participate in the world" (Brickhouse, 2001, p. 288). In short, to learn science meaningfully is to engage in identity formation (Brown, Reveles, & Kelly, 2005; Calabrese Barton, 1998; Carlone, 2004; Sadler, 2009). "Students who do not feel comfortable taking on a school science identity (i.e., being able to talk, think, and believe like a scientist) represent the vast majority of any student population" (Aikenhead, 2006, p. 108).

In the following, I comment on the relationship between this student majority and SSI innovations with respect to school science content, process, and context.

Content

Some researchers recommend that teachers foster positive school science identities by getting students to talk, think, and believe like scientists (Brown et al., 2005). If students begin to talk, think, and believe like scientists, then others will identify them as competent academic science students. However, identity clashes between students' self-identities and a conventional school science identity can cause many students to feel alienated (Brown & Spang, 2008), and consequently they resist forming an academic science identity, that is, they resist meaningful learning. Rather than becoming scientifically literate, they become scientifically indifferent.

The science content encountered in the culture of conventional school science is invariably academically abstract and decontextualized, which serves such purposes as knowledge accumulation for standardized tests and gate-keeping for university science and engineering departments. Evidence demonstrates the negative impact of this academic science content on students' perceptions of SSI innovations: "Students saw the same activities [SSI innovations] as a simple extension of what ordinarily transpires in science classrooms" (Sadler, 2009, p. 36); a conclusion verified by a very extensive study of students' and teachers' perceptions of what transpires in their science classrooms (Wood, Lawrenz, & Haroldson, 2009) and verified in a review of research into students' identity in science learning (Shanahan, 2009).

To be perceived by students as successful, an innovation needs to change the culture of school science. In the first place, school science must prioritize, throughout its science curriculum, scientific content primarily found *outside* of academic school settings – relevant science in the everyday world (Aikenhead, 2010). Conventional science content in schools and undergraduate university programs differs in cultural ways from the science content observed in everyday science-related occupations, events, and issues. These everyday contexts represent culturally different communities of practice compared with conventional school science (Sadler, 2009). Examples of *relevant school science content* appear in this volume and are emphasized in Aikenhead (2006, 2010) who offers a taxonomy of school science content based on the political question: Who decides what is relevant? Relevant school science includes, for instance, concepts and procedural knowledge for understanding and acting upon scientific evidence – "concepts of evidence" (Duggan & Gott, 2002) – in the world of work and in other everyday situations.

Process

From multiple perspectives, SSI educators have consistently given special attention to learning processes that engage students. In fact the innovative nature of these classroom processes is often perceived by teachers as a major challenge to implementing SSI school science (Aikenhead, 2006). Diverse learning processes revolve around instruction strategies such as decision making, reflective judgment, communities of practice, guided inquiry, citizen participation, moral and ethical reasoning and sensitivity, argumentation and reasoning, critical thinking, case-study analysis, and political action (Sadler, 2009).

Given the importance of potential cultural clashes that prevail in all science classrooms for most students, attention should be given to an emerging cultural perspective on school science instruction. Most students tend to experience school science (grades 6–12) as a foreign culture, but their teachers do not treat it that way. To be successful, these students must, without teacher assistance, learn to cross a cultural border between their own culture and the culture of academic school science (Aikenhead, 2006). An alternative process is for a teacher to treat relevant school science content as potentially foreign to students (epistemologically, ontologically, and axiologically foreign) and to develop cultural responsive ways to help students cross cultural borders more smoothly (Aikenhead, 2006; Scott, Asoko, & Leach, 2007). SSI educators need to hone their culturally responsive processes so students can more easily appropriate the relevant science content dictated by everyday science-related occupations or everyday events and issues – the primary contexts for learning.

Context

Of the three cultural features considered here (content, process, and context), context is the hallmark of SSI school science with its situated learning and context-based instruction (Sadler, 2009). Contexts chosen by teachers and curriculum developers, however, are often at odds with most students' views of relevancy (Campbell et al., 1994), unless: (a) there is a shared view of purpose among teachers and students, (b) there are changes in the nature of the dialogue between teachers and students, and

(c) students compare and contrast their personal understanding with scientific views (Rodrigues, 2006). Each point animates a dimension of school science culture.

Conclusion

When summarizing research in Europe, Osborne and Dillon (2008) lamented:

> The irony of the current situation is that somehow we have managed to transform a school subject which engages nearly all young people in primary schools…into one which the majority find alienating by the time they leave school. (p. 27)

An answer to the pervasive question, "Why would students want to engage with school science in the first place?" requires an exploration into the intricate, culture-anchored, context-laden interplay between, on the one hand, students' self-identities, and on the other, relevant school science content, culturally responsive processes, and educationally appropriate contexts.

Some obvious ways for SSI school science to transform the culture of conventional school science include: (a) taking seriously the fact that many students will even experience learning *relevant* school science content as a cross-culture event; (b) dispelling myths and values associated with any academic science content incorporated into school science; and (c) insisting that educational soundness and relevancy be the main criteria for selecting school science content. Other transformative ideas emerge from chapters in this book.

Unfortunately, *educational soundness* often conflicts with *political realities* such as expediency, institutional expectations, customs, ceremonies, beliefs, routines, loyalties, and the power of certain societal stakeholders (Aikenhead, 2006, 2010). SSI research and development projects will be far more influential if their agendas not only embrace the educational soundness of SSI school science, but as well, produce a politically savvy plan for transforming the school science culture in the school system where the project takes place.

Glen S. Aikenhead

References

Aikenhead, G. S. (2000). Renegotiating the culture of school science. In R. Millar, J. Leach, & J. Osborne (Eds.), *Improving science education: The contribution of research* (pp. 245–264). Birmingham, UK: Open University Press.

Aikenhead, G. S. (2006). *Science education for everyday life: Evidence-based practice*. New York: Teachers College Press.

Aikenhead, G. S. (2010). Towards a cultural view on quality science teaching. In D. Corrigan, J. Dillon, & R. Gunstone (Eds.), *The professional knowledge base of science teaching (in press)*. New York: Springer.

Albright, J., Towndrow, P. A., Kwek, D., & Tan, A-L. (2008). Identity and agency in science education: Reflections from the far side of the world. *Cultural Studies of Science Education, 3*, 145–156.

Brickhouse, N. W. (2001). Embodying science: A feminist perspective on learning. *Journal of Research in Science Teaching, 38*, 282–295.

Brown, B. A., Reveles, J. M., & Kelly, G. J. (2005). Scientific literacy and discursive identity: A theoretical framework for understanding science education. *Science Education, 89*, 779–802.

Brown, B. A., & Spang, E. (2008). Double talk: Synthesizing everyday and science language in the classroom. *Science Education, 92*, 708–732.

Calabrese Barton, A. (1998). Teaching science with homeless children: Pedagogy, representation, and identity. *Journal of Research in Science Teaching, 35*, 379–394.

Campbell, B., Lazonby, J., Millar, R., Nicolson, P., Ramsden, J., & Waddington, D. (1994). Science: The Salters' approach – A case study of the process of large scale curriculum development. *Science Education, 78*, 415–447.

Carlone, H. B. (2004). The cultural production of science in reform-based physics: Girls' access, participation and resistance. *Journal of Research in Science Teaching, 41*, 392–414.

Duggan, S., & Gott, R. (2002). What sort of science education do we really need? *International Journal of Science Education, 24*, 661–679.

Löfgren, L., & Helldén, G. (2009). A longitudinal study showing how students use a molecule concept when explaining everyday situations. *International Journal of Science Education, 31*, 1631–1655.

Rodrigues, S. (2006). Pupil-appropriate contexts in science lessons: The relationship between themes, purpose and dialogue. *Research in Science & Technological Education, 24*, 173–182.

Sadler, T. D. (2009). Situated learning in science education: Socio-scientific issues as contexts for practice. *Studies in Science Education, 45*, 1–42.

Schreiner, C., & Sjøberg, S. (2007). Science education and youth's identity construction: Two incompatible projects? In D. Corrigan, J. Dillon, & R. Gunstone (Eds.), *The re-emergence of values in science education* (pp. 231–247). Rotterdam: Sense Publishers.

Scott, P., Asoko, H., & Leach, J. (2007). Student conceptions and conceptual learning in science. In S. K. Abell & N. G. Lederman (Eds.), *Handbook of research on science education* (pp. 31–56.). Mahwah, NJ: Lawrence Erlbaum.

Shanahan, M-C. (2009). Identity in science learning: Exploring the attention given to agency and structure in studies of identity. *Studies in Science Education, 45*, 43–64.

Wood, N. B., Lawrenz, F., & Haroldson, R. (2009). A judicial presentation of evidence of a student culture of "Dealing." *Journal of Research in Science Teaching, 46*, 421–441.

Contents

About the Authors

Glen Aikenhead is professor emeritus, University of Saskatchewan, Canada. His research and development projects over the years have emphasized relevance of school science, under the slogan Science-Technology-Society for some of that time. This interest culminated in his 2006 book *Science Education for Everyday Life: Evidence Based Practice.* Glen's current work is in cross-cultural Indigenous science education that recognizes Indigenous knowledge as foundational to understanding the physical world, for all students. To this end, he has helped edit science textbooks that include Indigenous knowledge and has co-authored a 2011 book for science teachers and educators *Bridging Cultures: Indigenous and Scientific Ways of Knowing Nature.*

Ruth Amos is a lecturer in science education at the Institute of Education, University of London, UK. She received a B.Sc (Hons) in chemistry from the University of Bristol, UK, an M.A. in Science Education from the Institute of Education (IOE), University of London, and is currently undertaking a Ph.D in science education also at the Institute. She began working life as an analytical chemist at the then Imperial Chemical Industries (ICI, more latterly Astra-Zeneca and Syngenta) and then in the chemical industry for 7 years, and subsequently taught in state high schools in London for 12 years. At the IOE, Amos currently leads and teaches on the pre-service high school science teachers' training course and on the M.A. in Science Education. Her research interests include learning outside the classroom in fieldwork settings, the global dimension in science education, argumentation and decision-making in school science and the use of technologies in assessment for learning.

Scott M. Applebaum earned a B.S. from Baylor University and a D.M.D. from the University of Louisville School of Dentistry. Applebaum now teaches High School at Palm Harbor University High School. He has taken advanced graduate classes at the University of South Florida in socioscientific issues, nature of science, epistemology and cognition, qualitative inquiry and trends in science education. It is through these experiences he has collaborated extensively with Drs. Zeidler and Sadler in numerous research projects related to SSI and NOS, including the one described in this text. His applied innovations in the area of SSI has lead him to transform his pedagogy to an SSI model, and he has inspired many other teachers and graduate students with his teaching, paper presentations at NARST and ASTE, and various research reports in the literature. His exemplary teaching serves as a model to those interested in challenging and advancing students' thinking and interest in science.

Timothy Barko is a doctoral student of science education in the School of Teaching and Learning at the University of Florida. He received a B.A. in mathematics from Messiah College and an M.S.T in zoology from the University of Florida. While conducting research in molecular systematics and evolution, his thoughts turned to the role educators play in either promoting or inhibiting student love for science learning and he decided to change his professional interests to suit. He is currently interested in researching the evolution controversy in American science education and the role teachers play in the successful negotiation of prior beliefs with the corpus of evolutionary thought.

Randy L. Bell is an Associate Professor of Science Education in the Curry School of Education at the University of Virginia. He received a B.S. in botany from Marshall University, an M.S. in forest ecology from Duke University, and a Ph.D. in science education from Oregon State University in 1999. After conducting ecological research in the forests of Oregon and Washington, he earned his Oregon teaching license and taught middle and high school physical science, biology, chemistry, space science, computer applications, and investigative field studies. As a member of the Curry School Faculty, Dr. Bell teaches graduate courses in the science teacher preparation and science education doctorate programs. The author of more than 100 articles, chapters and books, Dr. Bell's research focuses on teaching and learning about the nature of science and using digital technologies to facilitate effective science instruction and conceptual change. Dr. Bell has maintained strong ties to the classroom through a variety of collaborative professional development projects. Most recently, he serves as co-PI on a US Department of Education-funded Investing in Innovation (I3) project designed to provide research-based professional development for Virginia's elementary and secondary science teachers.

Jennie S. Brotman is an Educational Consultant for Teaching Matters, Inc. in New York City. She received a B.A. in biology from the University of Pennsylvania and a Ph.D. in science education from Teachers College, Columbia University, where she studied urban high school students' talk about HIV/AIDS and sexual health decision-making. After working as a research scientist at the University of California, San Francisco, Brotman began her career in education by creating and teaching an inquiry-based science curriculum for a start-up middle school in Oakland, California. She has also taught and researched a science methods course for pre-service and inservice teachers as adjunct faculty for the Barnard College Education Program and has done professional development and curriculum development work in numerous New York City public schools. Brotman's research focuses on how young people grapple with controversial, science-related issues and how schools and teachers might influence students' decision-making processes.

Maurice Man Wai Cheng is a Teaching Consultant in the Faculty of Education, the University of Hong Kong. His main research interest is students' visualization and mental representation of science. His teaching areas include science and chemistry education, learning psychology and quantitative literacy. He has taught in secondary schools, and serves as a member of the official curriculum and assessment committees (science and chemistry subjects) of Hong Kong. He holds a B. Pharm., PGDE (chemistry), M.Ed. (Comparative Education) and is now completing his Ph.D. study at King's College London.

Kristin L. Cook is a senior doctoral student at Indiana University in Curriculum & Instruction with an emphasis on science education. Cook obtained her Masters in Science Education and Bachelors in biology, psychology and animal behavior from Indiana University as well. During her 5 years of teaching high-school biology, Kristin became interested in empowering students of science through the connecting of relevant socio-scientific issues to their daily lives. As such, Kristin's research focuses on opportunities to involve students in the development of scientific knowledge in their local communities. As a teaching assistant and research associate in the doctoral program, Kristin conducts professional development with in-service teachers and teaches courses for the secondary science education program and science courses for pre-service elementary teachers.

Vaille Dawson is a Professor in Secondary Science Education in the Science and Mathematics Education Centre at Curtin University, Western Australia. Prior to commencing an academic career, Dawson worked as a medical researcher and taught secondary school science for 10 years. Her research interests relate to adolescents' understandings, attitudes and decision-making about biotechnology issues and the use of information technology by early-career science teachers. She is co-editor of the textbooks for pre-service secondary and primary science teachers *The Art of Teaching Science* (Allen & Unwin, 2004) and *The Art of Teaching Primary Science* (Allen & Unwin, 2007).

Jennifer L. Eastwood is a post-doctoral associate in science education in the School of Teaching and Learning at the University of Florida. She received a B.A. in biology from Truman State University, an M.S. in anatomy and cell biology in medical sciences at Indiana University, and a Ph.D. in science education at Indiana University. She has taught biology at the high school and middle school levels, college level anatomy courses, and science education courses for pre-service elementary teachers. Eastwood's research has focused on innovative learning contexts in college level biology courses and students' science epistemologies. She is interested in how socioscientific issues can promote ethical, evidence-based reasoning in developing science and medical professionals.

Maria Evagorou is a Lecturer in Science Education in the School of Education at the University of Nicosia (UNic), Cyprus. She received a B.A. in education from the University of Cyprus, an M.A. in science education from the same university, and a Ph.D. in science education from King's College London in 2009. She has worked as an elementary school teacher in Cyprus for 2 years before pursuing graduate studies and as a lecturer at King's College before returning to Cyprus. As a researcher, Evagorou explores young students' argumentation and systems thinking skills within the context of science; elementary school students' talk when they engage in the discussion of socio-scientific issues; the role of new technologies in the teaching of argumentation and systemic thinking; the features of collaborative argumentation; and the design and validation of argumentation assessments. As a UNic faculty member, Evagorou teaches science method courses for the undergraduate program for elementary school teachers.

María Pilar Jiménez-Aleixandre is professor of science education at the University of Santiago de Compostela. A former high-school biology teacher, involved in innovation, she was part of the group building a science education community in Spain. Her research explored conceptual change in evolution and then moved to argumentation in science classrooms, with particular attention to causal explanations and environmental and socio-scientific issues. Her current focus is on the development of students' competence in using evidence to evaluate claims and on how to support teachers for this purpose. She has served on the executive committee of ESERA and currently serves on the editorial board of the *Journal of Research in Science Teaching* and is co-editor of the Issues & Trends section of *Science Education.*

Yael Kali is an Associate Professor of Educational Technology at the Technology Enhanced Education Graduate Department, Faculty of Education, University of Haifa, Israel. Her research interests include technology enhanced learning in science, design principles for technology-based learning environments and web-based learning in higher education. Prof. Kali has been closely involved in the development of the Design Principles Database. Her most recent book is *Designing coherent science education: Implications for curriculum, instruction, and policy* (with Marcia Linn and Jo-Ellen Roseman, 2008). Before joining the University of Haifa, Prof. Kali was a visiting scholar at the Centre for Research on Computer Supported Learning & Cognition (CoCo) in the Faculty of Education and Social Work, University of Sydney. Before that she was a faculty member at the Department of Education in Technology and Science at the Technion – Israel Institute of Technology, and a Co-Principal Investigator at the Technology Enhanced Learning in Science (TELS) Center, headquartered at the University of California, Berkeley.

Michelle L. Klosterman is an Assistant Professor of Science Education at Wake Forest University. She received her B.E. in biomedical engineering from Vanderbilt University, an M.Ed. in science education from Peabody College at Vanderbilt University, and a Ph.D. in science education from the University of Florida. Klosterman taught middle and high school physics and biology in Ohio, Tennessee, and Florida. She is currently the coordinator of the science education program at Wake Forest University. Klosterman teaches undergraduate and master's level courses in science education and supervises secondary science pre-service teachers. Her research focuses on how students and teachers use mass media as an instructional tool. She is interested in how teachers negotiate the transformation of everyday resources into classroom instructional tools and how the critical examination of mass media influences students' understandings of science.

Jacqueline J. Madhok is an Education Researcher with TELS (Technology Enhanced Learning in Science) Center at the University of California, Berkeley. She has an M.A. in science education from Stanford University and received a Ph.D. in SESAME (Graduate Group in Science and Mathematics Education) from University of California, Berkeley in 2006. Before entering the SESAME doctoral program, she worked as a high school teacher of biology and mathematics in Campbell and Fremont, California. She works on assessment and evaluation of TELS instructional materials. Her research focuses on the interaction of students' beliefs about science and science learning and their achievement, as well as diversity, including gender equity.

Stella Magid has a B.Sc. in biology education. Currently she is a M.Sc. student in the Department of Education in Technology and Science at the Technion, Israel Institute of Technology. Her masters' thesis deals with teaching genetics using technology and incorporating activities that promote relevancy and authenticity. Stella worked at MadaTech, the National Museum of Science as a facilitator, teacher and as a curriculum developer. She currently she teaches science and biology in a secondary school in Tel Aviv. In addition, she is a teaching assistant for a biology methods course at the Technion. Stella is interested in teaching science and biology in technology reach and informal environments.

Felicia Moore Mensah is an associate professor of Science Education, Department of Mathematics, Science and Technology, Teachers College, Columbia University, USA. She earned her Ph.D. in science education from The Florida State University, her M.S. in Biology and Secondary Education from North Carolina Agricultural &Technical State University, and her B.S. in Biology from the University of North Carolina-Chapel Hill. She also completed a 2-year postdoctoral fellow at Michigan State University (MSU), while in residence at Teachers College during her first year. Professor Mensah has worked collaboratively and independently as a scholar and her research focuses on three broad but related areas of teacher identity, teacher education/professional development and urban and multicultural education; all areas focus on improving science experiences for P-12 teachers and their students. Primarily a qualitative researcher, Dr. Mensah utilizes multiple theoretical frameworks, such as critical, feminist and poststructural perspectives in her work in teacher education and teacher professional development. Professor Mensah is the author of many journal articles, commentaries, book chapters and professional development materials. She serves on the editorial board of the *Journal of Science Teacher Education, Journal of Elementary Science Education* and *Cultural Studies of Science Education.*

Blanca Puig is a doctoral student in the Department of Science Education at the University of Santiago de Compostela (USC). She is supported by a USC pre-doctoral grant partly funded by the S-TEAM European project. She graduated in biology from this university and then worked in oceanography before moving to science education. Puig is currently working as graduate research assistant under the supervision of Prof. Jiménez-Aleixandre. Her research focuses on evidence evaluation, in particular the influence of determinist social representations of human 'races' on the evaluation of evidence about causal explanations for human performances.

Troy D. Sadler is an Associate Professor of Science Education in the School of Teaching and Learning at the University of Florida. He received a B.S. in biology from the University of Miami, an M.Ed. in science education from the University of Florida, and a Ph.D. in science education from the University of South Florida in 2003. After conducting research in ecological and evolutionary genetics at both the University of Miami and Harvard University, Sadler turned his attention to science teaching. He worked as a high school teacher of biology, biotechnology, and earth science in Gainesville and St. Petersburg, Florida. As a UF faculty member, Sadler teaches courses for a master's level secondary science education program, science methods courses for pre-service elementary teachers and advanced graduate level courses. Sadler's research focuses on how students of science negotiate complex socio-scientific issues and how these issues may be used as contexts for science learning. He is interested in how issues-based learning experiences can support student learning of science and development of practices essential for full participation in modern democratic societies.

Victor Sampson is an Assistant Professor of Science Education in the School of Teacher Education and the FSU-Teach program at The Florida State University. He earned a bachelor's degree in zoology at the University of Washington, a master's degree in secondary education from Seattle University, and a Ph.D. in curriculum and instruction with a concentration in science education from Arizona State University. Dr. Sampson's research focuses on the ways students support and evaluate ideas in the context of science, group and individual meaning making during episodes of argumentation, and the development of new instructional models that promote and support student participation in scientific argumentation.

Whitney M. Schlegel is an Associate Professor of Biology at Indiana University Bloomington. She earned her doctorate in environmental animal physiology and behavior from the University of Illinois, Urbana-Champaign, in 1991 and came to Indiana University in 1992 from the Lincoln Park Zoo in Chicago where she directed research and conservation with exotic species. She served as Director of Undergraduate Curriculum and Assistant Professor of Cellular and Integrative Physiology in the IU School of Medicine until 2005, moving to the Biology Department and College of Arts and Sciences to direct and develop the interdisciplinary undergraduate degree program in human biology. She utilizes the complexities of societal concerns and problems to frame student scientific inquiry and learning in the sciences and employs engaging pedagogies that provide practice with evidence-based reasoning and collaborative problem solving. Her pre-health professional students have inspired her investigations of learning environments and experiences that best foster the habits of mind and ways of knowing inherent to the development of a professional identity, one that allows students to utilize their knowledge of science to better local communities and a global society.

Shirley Simon is a Professor of Education in the Faculty of Children and Learning at the Institute of Education, University of London. She began her career in education as a chemistry teacher in East Sussex schools then completed her M.Sc. in chemical education at the University of Reading. She then worked as Head of Chemistry in schools within the Inner London Education Authority before undertaking a Ph.D. in science education at King's College London. Simon became a researcher and lecturer in science education at King's before moving to the Institute of Education, where she teaches in the secondary science pre-service (PGCE) and masters programs. Simon has been Head of Department within the Faculty, and supervises doctoral students in science education. Her recent research has focused on developing the teaching of argumentation in science departments in London schools. She is interested in teacher professional learning, classroom discourse and students' attitudes to science.

Tali (Revital) Tal is an Associate Professor of Science and Environmental Education at the Department of Education in Technology and Science at the Technion, Israel Institute of Technology. She received a B.Sc. and M.Sc. in biology from Tel Aviv University. Between 1984 and 1995 she worked as a science teacher and environmental education facilitator in both formal and informal education systems. Tal received a Ph.D in science education from the Technion in 1998 and continued with a post-doctoral research appointment in science education at the University of Michigan. Since 2000, she has been a faculty member at the Technion and conducts research on teaching and learning science in everyday contexts by teaching science to all students and involving socioscientific issues in general, and environmental issues in particular. In 2007–2008, Tal was a visiting scholar at the University of California in Berkeley, where she collaborated with the Laurence Hall of Science and with Prof. Marcia Linn at the Graduate School of Education. Dr. Tal's main interest is in supporting students' learning of substantial issues in- and out-of-school. Her research group focuses on environmental education and environmental and scientific literacy in the school system and in informal settings.

Mustafa S. Topcu is an Assistant Professor of Science Education in the Department of Elementary Science Education at Yuzuncu Yil University, Turkey. He received a B.S. in science education from the 19 Mayıs University, Turkey, an M.Sc. in science education from the 9 Eylul University, Turkey, and a Ph.D. in science education from the Middle East Technical University, Turkey in 2008. He worked as an elementary school teacher of science in Izmir, Turkey. After teaching science in elementary school, he worked as a research assistant in the Middle East Technical University. As a Yuzuncu Yil University faculty member, Topcu teaches courses for an undergraduate science education program, and a master's level elementary science education program. His research interests are teachers' beliefs and educational practices, argumentation, socioscientific issues, morality and epistemological beliefs.

Zhi Hong Wan is a Senior Research Assistant in the Department of Science and Environmental Studies at the Hong Kong Institute of Education. He received a Certificate of physics education from the Jingdezhen Institute of Education, an M.Ed in educational psychology from Soochow University and a Ph.D in science education from the University of Hong Kong in 2010. Before starting his research in science education, he had taught middle school physics in Mainland China for 5 years. Zhi Hong's research focuses on how nature of science instruction is conducted and perceived in different social-cultural backgrounds. He is interested in how complex socio-scientific issues can be used as the contexts for teaching nature of science.

Siu Ling Wong is an Associate Professor and Assistant Dean in the Faculty of Education at the University of Hong Kong (HKU). She received a B.Sc. in physics from the University of Hong Kong and received her D.Phil. in physics from the University of Oxford. She continued her research in physics in the area of semiconducting nanostructures as a Croucher Foundation Postdoctoral Research Fellow at the Clarendon Laboratory and a College Lecturer at Exeter College in Oxford. She taught physics and mathematics at Diocesan Girls' School when she returned to Hong Kong in 1997. Her interest in science education has grown since then. She joined HKU as a science educator in 2002. She teaches courses for physics and science teachers and supervises Ed.D. and Ph.D. students. She is the Hong Kong National Research Coordinator (Science) of the Trends in International Mathematics and Science Study (TIMSS). Her current research interests include nature of science, scientific inquiry and teacher professional development. She is most interested in making use of authentic scientific inquiries for teachers' and students' appreciation of nature of science, particularly about the inseparable links between science and the ethical, social, cultural and political environment.

Dana L. Zeidler earned his Ph.D. from Syracuse University. His research program incorporates aspects of argumentation and discourse, moral reasoning and character formation, epistemology and the nature of science into a theoretical framework using socioscientific issues as a means to facilitate scientific literacy. His work has attracted international attention both within and external to the community of science education. He has authored over 60 journal articles and book chapters, presented over 75 conference papers, and has given keynote addresses and international talks in Canada, China, Germany, Korea, The Netherlands, Slovenia, Sweden, Taiwan, Turkey and the United States on these and related topics in science education. He works closely with doctoral students and other leaders in the science education community. He is a Professor and Program Coordinator of Science Education at the University of South Florida, USA. He is also currently the President of the National Association for Research in Science Teaching and serves on the Executive Board for the Association of Science Teacher Education. Dr. Zeidler serves as the Series Editor for *Contemporary Issues and Trends in Science Education Library Series* by Springer Publishers, The Netherlands.

Chapter 1
Situating Socio-scientific Issues in Classrooms as a Means of Achieving Goals of Science Education

Troy D. Sadler

Goals of Science Education

In considering what the science education community ought to be doing in terms of curricular and pedagogical innovations, teacher education, and research, I believe the field must consider two fundamental questions: (1) what should the goals of science education be and (2) how can these goals best be achieved? The question of what the goals of science education ought to be has been a long-standing issue. Debates on the meaning and purpose of scientific literacy have consistently explored this territory. Although the science education community is fairly united in its call to enhance scientific literacy for all students, there is considerably less agreement as to what constitutes scientific literacy. This volume does not take up the challenge of mapping out the landscape of scientific literacy; Doug Roberts' (2007) chapter in the latest *Handbook on Research in Science Education* does an excellent job of summarizing and synthesizing the varied perspectives on this contentious construct. Roberts provides a useful heuristic for characterizing the diversity of views on scientific literary by postulating two *visions* of the construct:

> Vision I gives meaning to SL [Scientific Literacy] by looking inward at the canon of orthodox natural science, that is, the products and processes of science itself. At the extreme, this approach envisions literacy (or, perhaps, thorough knowledgeability) *within science*... Vision II derives its meaning from the character of situations with a scientific component, situations that students are likely to encounter as citizens. At the extreme, this vision can be called *literacy* (again, read *thorough knowledgeability) about science-related situations* in which considerations other than science have an important place at the table. (emphases original; p. 730)

The authors featured in this volume adopt perspectives consistent with Vision II scientific literacy. This perspective suggests that scientifically literate individuals should be able to confront, negotiate, and make decisions in everyday situations that involve science. This perspective on scientific literacy prioritizes science for

T.D. Sadler (✉)
School of Teaching and Learning, University of Florida, Gainesville, FL, USA
e-mail: tsadler@coe.ufl.edu

T.D. Sadler (ed.), *Socio-scientific Issues in the Classroom: Teaching, Learning and Research*, Contemporary Trends and Issues in Science Education 39,
DOI 10.1007/978-94-007-1159-4_1, © Springer Science+Business Media B.V. 2011

all, not just the few who engage (or will engage) in science professionally. Defining scientific literacy in this way has important implications for the first fundamental question posed above: what should the goals of science education be? The perspective necessarily foregrounds the role of science education in supporting the development of all students as citizens prepared to engage thoughtfully in the discourses of modern society. Given the role of science in modern society, scientific ideas, practices, and products are essential aspects of these discourses. Therefore, science education has a significant role to play in this process of preparing citizens. The excerpts presented below offer strong justifications for the position being advocated here:

> What is clear is that ordinary citizens will increasingly be asked to make judgements about matters underpinned by science knowledge or technological capability, but overlaid with much wider considerations. Those without a basic understanding of the ways in which science and technology are impacted by, and impact upon, the physical and the sociopolitical environment will be effectively disempowered and susceptible to being seriously misled in exercising their rights within a democratic, technologically-dependent society. (Hodson, 2003, pp. 650–651)

> … [S]cience education must serve as a foundation for the education of an informed citizenry who participate in the freedoms and powers of a modern, democratic, technological society. With the rapid development of scientific knowledge and the advent of new technologies, all members of society must have an understanding of the implications of that knowledge upon individuals, communities, and the "global village" in which we now live. (Berkowitz & Simmons, 2003, p. 117)

While the basic idea of asserting citizenship goals as a part of science education may not be all that controversial, the implications of positioning citizenship as *the primary goal* of science education may very well be contentious. In claiming that the primary aim of science education should be promoting citizenship, we are classifying several other goals as less-than-primary. Historically, science education has offered primacy to the promotion of understandings of scientific facts, principles, theories, etc. (DeBoer, 1991). Consensus research documents such as *Taking Science to Schools* (Duschl, Schweingruber, & Shouse, 2007) prioritize student engagement in the practices of science, and several politically oriented reports (e.g., Member of the 2005 "Rising Above the Gathering Storm" Committee, 2010) call for enhancing the preparation of the science workforce. I am not arguing that supporting student learning of science ideas, engagement in science practices or preparation for careers in science are not desirable goals, but I am saying that these goals are not as important as supporting students' development as citizens.

Critics of the approaches advocated in this volume may be tempted to argue that the distinctions drawn here between various goals of science education are trivial. The goals themselves are not independent. Engaging in science practices requires understanding of science principles and theories. Better preparing the science workforce requires students to develop stronger understandings of science ideas and abilities to engage in science. Preparing better citizens also requires more sophisticated understandings and practices. However, there are important distinctions to be made between learning science ideas and practices for their own inherent value or for more instrumental purposes such as preparing students for careers or citizenship. For example, if the primary goal of science education is to enhance student understandings of scientific principles, then science educators do not need to concern themselves with dimensions of the more instrumental goals that can present tricky pedagogical challenges. If the primary goal of science education is to support student abilities to engage in scientific practices, then educational opportunities should be designed such that they maximize student engagement in those practices. This may mean that more instrumental goals such as career and/or citizenship preparation would be deemphasized or removed from the curriculum.

Delineating between the primary and secondary goals of science education has important implications for the second fundamental question posed at the outset of this chapter: how should the goals of science education be achieved? Prioritized goals should shape curricular and pedagogical decisions. Many science educators would likely agree that promoting citizenship is a valid goal for science education, but relatively few have made this a primary goal such that development of learning experiences are primarily shaped by this focus. Based on what I observe in the classrooms that I visit and what is represented in state and national standards (at least in the United States), the primary goal of most science instruction is promoting science content. The students with whom I am familiar spend most of their time devoted to listening to lectures related to science content, participating in activities designed to reinforce content, and memorizing scientific formalisms to re-present them in testing situations. While learning about science content in these ways may help learners to better engage in negotiation of and decision-making related to social issues conceptually connected to that science content, it is unlikely that this is the most effective way of supporting student engagement in these issues (Zeidler, Sadler, Simmons, & Howes, 2005). If the primary goal of science education is supporting the development of students to be more informed and to be engaged citizens, then science educators must be more deliberate about achieving this goal.

Situated Learning

The decisions we make related to how learning experiences and instruction are structured in the service of educational goals ought to be shaped by theory on teaching and learning. For me, situated learning has become a powerful framework for understanding and explaining teaching and learning phenomena (Cobb & Bowers, 1999; Greeno, 1998; Lave & Wenger, 1991). This perspective emphasizes the situatedness of learners in specific environments. Environments, or contexts, for learning are formed by participants, including the learners themselves, conceptual and physical resources, as well as the norms that guide participation and discourse. These contexts, in turn, shape the ways in which participants engage in activities and ultimately afford and constrain what participants come to know and be able to do. From the vantage of this theoretical perspective, learning cannot be considered as only those things that transpire in the minds of individuals. Therefore, context becomes a very important factor in learning, not just a backdrop against which learning takes place.

If promoting citizenship is a priority for science educators, then we have to consider the contexts we create for learners to experience science (Sadler, 2009). We cannot assume that exposure to science content in any context will necessarily lead to the preparation of more informed citizens who are better ready to participate in democratic discourses and processes. If our goal is to help students become better able to contribute to debates and decisions about important societal issues with links to science and technology, then we need to create learning contexts such that learners actually confront some of these issues and gain experiences negotiating their inherent complexities. The socio-scientific issues approach to science education does just that.

Socio-scientific Issues

Socio-scientific issues (SSI) are controversial social issues with conceptual and/or procedural links to science (Sadler, 2004). They are open-ended problems without clear-cut solutions; in fact, they tend to have multiple plausible solutions. These solutions can be informed by scientific principles, theories, and data, but the solutions cannot be fully determined by scientific considerations. The issues and potential courses of action associated with the issues are influenced by a variety of social factors including politics, economics, and ethics. SSI may be global in nature such as climate change and the use of genetic technologies or local such as addressing a neighborhood environmental crisis or determining the location of a new power plant.

The SSI movement has arisen within science education with a focus on using these complex issues as contexts for teaching science. If we take the implications of situated learning seriously, then supporting learner exploration of complex, socially relevant issues aligns very well with the goal of promoting citizenship. It is not enough for science educators to teach science content if what we really want to do is help students become better able to negotiate the challenges of science as it is represented in the real issues of society. SSI-based education addresses this challenge explicitly by using the complex issues that highlight the need for SL (Vision II: Roberts, 2007) as contexts for science teaching and learning.

Over the last decade, SSI has become a prominent theme within the science education literature. The SSI movement has built upon other approaches that share the goal of better preparing learners to engage in discourses and decisions related to socially relevant issues associated with science. The most notable of these approaches is science-technology-society (STS; Yager, 1996), but other approaches including science-technology-society-environment (STSE), education for sustainability, and context-based science education also have many features in common with the SSI approach. Discussions of ways in which these different orientations to science teaching overlap and are distinct can be found elsewhere (Tal & Kedmi, 2006; Zeidler et al., 2005).

Not surprisingly, the focus of the literature that has emerged around SSI has been varied and diverse. Several authors have explored the construct conceptually particularly with respect to how SSI-based approaches relate to scientific literacy (e.g., Hodson, 2003; Zeidler & Keefer, 2003). Others have looked at reasoning and decision-making in the context of SSI (e.g., Zeidler, Walker, Ackett, & Simmons, 2002; Zohar & Nemet, 2002) and links between SSI decision-making and certain understandings, like nature of science or content knowledge (e.g., Bell & Lederman, 2003; Lewis & Leach, 2006). Other work has explored issues associated with assessment of SSI-related outcomes (e.g., Eggert & Bogeholz, 2010; Vazquez-Alonso, Manaserro-Mas, & Acevedo-Diaz, 2006). SSI as featured in classrooms has also become an important trend in this literature (e.g., Albe, 2008; Grace, 2009). All of these contributions have important insights to offer the science education community, but as we move toward generating empirical evidence for the question of how best to achieve the goals of science education, I believe the classroom-based studies of SSI implementation and outcomes are particularly significant.

Purpose of This Volume

The perspectives described in the beginning pages of this chapter including the primacy of citizenship goals for science education, situated learning as a theoretical perspective on learning, the significance of SSI as educational contexts, and the importance of classroom-based research have shaped the development of this book project. There are several strong studies of SSI featured in classrooms (many of which are presented in a recent review; Sadler, 2009), but there is limited space within these individual reports to feature discussions of important contextual issues and to look across different kinds of classrooms. The purpose of this volume is to provide such a space. The goal is to bring together researchers who are working in classrooms to explore the effects of SSI-based education in order to create new understandings of how SSI can be productively used by teachers and students. In order to achieve this goal, I recruited a group of authors with current or recently completed classroom-based, SSI research projects. Each chapter features research conducted in these settings, but authors were also encouraged to share details of their work that often fail to be fully represented in journal articles. As such, I requested that authors discuss the following elements of their work in classrooms in addition to discussions of research foci, methods, and results:

- Motivation or origins of the work: Discussion of how and/or why the authors initiated the project they describe.
- Teacher-researcher relationships: Discussion of how the authors worked with classroom teachers and how these relationships were built and sustained including the challenges of doing so.
- Nature of the SSI intervention: Discussion of the design process, curricula, and pedagogy employed in the project.
- Implications for teaching, learning, and research: Discussion of how the project has advanced the understanding of how to effectively use SSI in educational contexts and what research questions have been highlighted by the contribution.

In order to build understandings of how SSI are featured in diverse contexts, I deliberately sought an internationally diverse assemblage of projects. Authors from seven nations across four continents present their work in the volume. This sample of projects by no means represents the full range of diversity within the area of SSI, but it does offer multiple international perspectives on the topic.

As a part of this volume, I also wanted to create an opportunity for scholars to reflect on the work of others and discuss these reflections. The discussion of ways in which individual researchers frame their work, deal with challenges, and react to how other research teams approach these tasks can be helpful for the discussants themselves and potentially educative for others working to implement SSI in classrooms or better understand how to do so. Therefore, each chapter that describes a classroom-based SSI project is followed by a *metalogue*. Metalogues are "written conversations among parties that preserve individual voices while revealing contested areas" (Staller, 2007, p. 137), thereby creating space for the exchange and problematization of ideas. I first saw metalogues featured in an edited volume by Yerrick and Roth (2005) related to classroom discourse communities. This text offered several very strong chapters outlining theoretical and empirical considerations on discourse communities, but I found the discussions following each chapter between the authors and other scholars to be the most enlightening aspects of the volume. Having experienced the value of these kinds of metalogues from the perspective of a reader, I wanted to borrow the strategy as an editor. For each of the main chapters a metalogue was created by at least one author of the original chapter, at least one author from one of the other chapters, and in some cases another scholar with related expertise.

Organization of Chapters

The volume features nine SSI research projects presented in separate chapters with accompanying metalogues. The chapters are not grouped in sections; however, they have been organized deliberately. There are several different ways that the chapters could have been organized including by target audiences, SSI featured, scope of interventions, methodologies, or research foci. The target audiences for the projects range from upper elementary students to college undergraduates. The issues themselves range from environmental issues and climate change to human diseases and health care to social controversies such as biological determinism. The scope of interventions varies from relatively short units delivered in a span of a few weeks to year-long projects and, in one case, a four-year SSI-themed program. The authors present diverse methods including different kinds of primary data sets (e.g., observational data, interviews, tests, and surveys) and different analytic approaches (e.g., pre/post intervention statistical comparisons, case studies, mixed methods designs, and discourse analysis).

Ultimately, I chose to organize the chapters around themes of the research. The first two projects (Tal, Kali, Magid, & Madhok; Sadler, Klosterman, & Topcu) explore how teaching in the context of SSI can support student learning of science content. Tal and colleagues also address using technology to mediate SSI learning experiences and how these experiences can be enhanced through interactions with real individuals engaged in the SSI under consideration, both in and out of school settings. In addition to student learning of science content, Sadler and colleagues explore socio-scientific reasoning, a construct they position as a means of operationalizing and assessing the citizenship dimensions of SSI-based education. Reasoning and decision-making is the dominant research theme for the next two chapters (Eastwood, Schlegel, & Cook; Evagorou). Eastwood and colleagues examine the decision-making practices of students in an innovative SSI-based undergraduate major. These authors also discuss the students' perspectives on learning in the context of SSI. Evagorou studies reasoning and decision-making among upper elementary learners engaged in a technology-supported SSI unit. She also explores ways in which in-class experiences can be supported with out-of-school experiences, a theme initially discussed by Tal et al. (2011).

The next two chapters by Simon & Amos and Puig & Jiménez-Aleixandre extend the focus on reasoning and employ argumentation frameworks to understand learner use of scientific knowledge and evidence in support of the claims made in SSI contexts. Simon & Amos study classroom enactment of a SSI unit embedded in a new national curriculum in the United Kingdom. Puig & Jiménez-Aleixandre explore processes associated with the design of SSI-related learning tasks, the translation and interpretation of these tasks by teachers, and how these tasks support (or fail to support) student construction of scientific explanations in SSI contexts. Wong, Wan, and Cheng direct attention to the learning of nature of science (NOS) through SSI-based instruction. These researchers explore how classroom teachers modify and create SSI-based units to support student learning of NOS. The final two chapters (Zeidler, Applebaum, & Sadler; Dawson) present case studies of long-term enactment of SSI approaches in science classrooms. Zeidler and colleagues explore student and teacher transformations over the course of a year-long case study in which a teacher adopted a SSI framework for his high school science classes. Dawson presents a semester-long case study exploring the impact of a teacher's efforts to raise students' ethical awareness and decision-making skills in the context of issues associated with human reproduction.

Final Thoughts

In the opening sections of this chapter, I made several assertions including that all of the contributors to this text subscribe to a Vision II orientation to SL (Roberts, 2007) and that citizenship goals ought to be *the* primary goal of science education. Although I did not solicit explicit comments on these issues, I am confident, based on the work presented, in asserting that the authors do share consensus in the need to frame SL in terms of abilities to use science in students' everyday lives. I am far less confident in the assertion that citizenship goals ought to be *the* primary goal of science education. If pushed, I do not even subscribe to this assertion in the extreme. However, my fellow contributors and I would agree that citizenship goals ought to be *a* primary goal of science. Earlier in the chapter, I framed the discussion of goal prioritization as an either/or option: a goal is either positioned as the primary goal of science education or it is not. I did so to stress the importance of prioritizing citizenship goals which have historically received rhetorical support but limited or nonexistent support for actual implementation. I think that if we consider the broad range of formal and informal science education opportunities, then multiple goals can be prioritized. However, given the historical marginalization of citizenship goals in science education (Hughes, 2000), the community needs to be very careful in preserving the prioritization of these goals, even if multiple goals are prioritized.

In this introductory chapter, I have attempted to provide a framework for situating the work featured throughout the volume, describe a rationale for the volume, and outline the organizational structure of the volume. The framework, which draws on notions of scientific literacy and situated learning, is admittedly brief. My goal is to provide a starting place for considering the classroom-based, SSI research described in this volume. However, there is an inevitable tension in presenting a common framework for a collection of work as diverse in scope, focus, and geographic origin as presented here. A simple instantiation of the diversity across chapters that reflects some of this underlying tension is the label SSI itself. In some chapters, authors refer to *socio-scientific* issues (with a hyphen); other authors use *socioscientific* issues (without a hyphen). The presence or absence of the hyphen itself is less important than the underlying conceptual framework that the authors' linguistic choice represents. In some cases, it really may not matter whether the hyphen is present or not; however, some authors are deliberate in their inclusion or exclusion of it. I have chosen to preserve the differences as an explicit acknowledgement of potential differences.

While I can operationalize important terms and make some general arguments related to the significance of scientific literacy and citizenship dimensions of science education, I cannot say that all of the authors subscribe to a comprehensive theoretical framework. Therefore, I have presented an overview of the arguments supporting SSI-based education and a fairly general discussion of fundamental concepts (e.g., scientific literacy) and theory (e.g., situated learning) underlying work in this domain. I offer more complete discussions of the conceptual and theoretical frameworks that guide my own work elsewhere (Sadler, 2009; Sadler & Zeidler, 2009). Issues associated with differences in how the various authors conceptualize SSI and frame SSI research emerge in some of the metalogue discussions. I leave it to the readers to identify and contemplate other issues that are surely present but not directly discussed.

References

Albe, V. (2008). When scientific knowedge, daily life experience, epistemological and social considerations intersect: Students' argumentation in group discussion on a socio-scientific issue. *Research in Science Education, 38*, 67–90.

Bell, R. L., & Lederman, N. G. (2003). Understandings of the nature of science and decision making on science and technology based issues. *Science Education, 87*, 352–377.

Berkowitz, M. W., & Simmons, P. (2003). Integrating science education and character education. In D. L. Zeidler (Ed.), *The role of moral reasoning on socioscientific issues and discourse in science education* (pp. 117–138). Dordrecht, the Netherlands: Kluwer.

Cobb, P., & Bowers, J. (1999). Cognitive and situated learning perspectives in theory and practice. *Educational Researcher, 28*(2), 4–15.

DeBoer, G. E. (1991). *A history of ideas in science education: Implications for practice*. New York: Teachers College Press.

Duschl, R. A., Schweingruber, H. A., & Shouse, A. W. (Eds.). (2007). *Taking science to school: Learning and teaching science in grades K-8*. Washington, DC: National Academies Press.

Eggert, S., & Bogeholz, S. (2010). Students' use of decision-making strategies with regard to socioscientific issues: An application of the Rasch partial credit model. *Science Education, 94*, 230–258.

Grace, M. (2009). Developing high quality decision-making discussions about biological conservation in a normal classroom setting. *International Journal of Science Education, 31*, 1464–1489.

Greeno, J. G. (1998). The situativity of knowing, learning, and research. *American Psychologist, 53*, 5–26.

Hodson, D. (2003). Time for action: Science education for an alternative future. *International Journal of Science Education, 25*, 645–670.

Hughes, G. (2000). Marginalization of socioscientific material in science-technology-society science curricula: Some implications for gender inclusivity and curriculum reform. *Journal of Research in Science Teaching, 37*, 426–440.

Lave, J., & Wenger, E. (1991). *Situated learning: Legitimate peripheral participation*. Cambridge, UK: Cambridge University Press.

Lewis, J., & Leach, J. (2006). Discussion of socio-scientific issues: The role of science knowledge. *International Journal of Science Education, 28*, 1267–1287.

Members of the 2005 "Rising Above the Gathering Storm" Committee. (2010). *Rising above the gathering storm, revisited: Rapidly approaching category 5*. Washington, DC: National Academies Press.

Roberts, D. A. (2007). Scientific literacy/science literacy. In S. K. Abell & N. G. Lederman (Eds.), *Handbook of research on science education* (pp. 729–780). Mahwah, NJ: Lawrence Erlbaum Associates.

Sadler, T. D. (2004). Informal reasoning regarding socioscientific issues: A critical review of research. *Journal of Research in Science Teaching, 41*, 513–536.

Sadler, T. D. (2009). Situated learning in science education: Socioscientific issues as contexts for practice. *Studies in Science Education, 45*(1), 1–42.

Sadler, T. D., & Zeidler, D. L. (2009). Scientific literacy, PISA, and socioscientific Discourse: Assessment for progressive aims of science education. *Journal of Research in Science Teaching, 46*, 909–921.

Staller, K. (2007). Inquiries into conversations among authors, editors and referees. *Qualitative Social Work, 6*(2), 137–157.

Tal, T., & Kedmi, Y. (2006). Teaching socioscientific issues: Classroom culture and students' performances. *Cultural Studies in Science, 1*, 615–644.

Vazquez-Alonso, A., Manaserro-Mas, M.-A., & Acevedo-Diaz, J.-A. (2006). An analysis of complex multiple-choice science-technology-society items: Methodological development and preliminary results. *Science Education, 90*, 681–706.

Yager, R. E. (Ed.). (1996). *Science/technology/society as reform in science education*. Albany: State University of New York Press.

Yerrick, R., & Roth, W.-M. (2005). *Establishing scientific discourse communities: Multiple voices of teaching and learning research*. Mahwah: Lawrence Erlbaum Associates.

Zeidler, D. L., & Keefer, M. (2003). The role of moral reasoning and the status of socioscientific issues in science education. In D. L. Zeidler (Ed.), *The role of moral reasoning on socioscientific issues and discourse in science education*. Dordrecht, the Netherlands: Kluwer.

Zeidler, D. L., Sadler, T. D., Simmons, M. L., & Howes, E. V. (2005). Beyond STS: A research-based framework for socioscientific issues education. *Science Education, 89*, 357–377.

Zeidler, D. L., Walker, K. A., Ackett, W. A., & Simmons, M. L. (2002). Tangled up in views: Beliefs in the nature of science and responses to socioscientific dilemmas. *Science Education, 86*, 343–367.

Zohar, A., & Nemet, F. (2002). Fostering students' knowledge and argumentation skills through dilemmas in human genetics. *Journal of Research in Science Teaching, 39*, 35–62.

Chapter 2
Enhancing the Authenticity of a Web-Based Module for Teaching Simple Inheritance

Tali Tal, Yael Kali, Stella Magid, and Jacqueline J. Madhok

Introduction

In this chapter, we view socio-scientific issues (SSI) as contributing to dialogic argumentation (Ash & Wells, 2006; Driver, Newton, & Osborne, 2000; Tal & Kedmi, 2006) and as enhancing the ability to assess scientific information and data (Jiménez-Aleixandre, Rodríguez, & Duschl, 2000; Zohar & Nemet, 2002), which both contribute to scientific literacy of students in middle and lower high school grades (Roth & Calabrese Barton, 2004). Teaching science through socioscientific issues is in line with ideas brought up by the Science-Technology-Society (STS) movement (Aikenhead, 1994; Hodson, 1994, 1998) that continued to develop into ideas about humanistic science teaching and teaching citizen science (Aikenhead, 2005; Calabrese Barton, 2003; Roth & Calabrese Barton, 2004; Tal & Kedmi, 2006). The essence of all these ideas is that the science content should be situated in real, important, and often controversial issues that gain the public's interest. Ratcliffe and Grace (2003) identified the following characteristics in socioscientific issues: they have a basis in science as they are frequently at the frontiers of scientific knowledge; they involve forming opinions, making choices at personal and societal levels; they are frequently reported by media; they deal with incomplete information; they address local, national, and global dimensions; they involve some cost-benefit analysis in which risk interacts with values; they may involve considerations of sustainable development; they involve values and ethical reasoning; they may require some understanding of probability and risk; they are frequently topical with transient life (pp. 2–3).

T. Tal (✉) and S. Magid
Department of Education in Technology and Science, Technion–Israel Institute of Technology, Haifa, Israel
e-mail: rtal@technion.ac.il

Y. Kali
Technology Enhanced Education Graduate Department, Faculty of Education, University of Haifa, Haifa, Israel

J.J. Madhok
Technology Enhanced Learning in Science Center, University of California, Berkeley, CA, USA

T.D. Sadler (ed.), *Socio-scientific Issues in the Classroom: Teaching, Learning and Research*, Contemporary Trends and Issues in Science Education 39,
DOI 10.1007/978-94-007-1159-4_2,

Integrating societal, environmental, and technological aspects into the science curriculum is not a new idea. The Science-Technology-Society (STS) movement of the 1980s advocated not only the inclusion of controversial issues, but using them as organizers for the curriculum (Bingle & Gaskell, 1994; Solomon & Thomas, 1999). However, as Zeidler and his colleagues argue (Zeidler, Sadler, Simmons, & Howes, 2005), in fact, socioscientific issues were presented merely as additions or anchoring stories to the main stream science that remained disciplinary, standard-based, and free of value. They suggested that the Socioscientific Issue (SSI) movement should replace STS, claiming that while STS education typically stresses the impact of decisions in science and technology on society, it avoids deep engagement with ethical issues and does not consider the moral development of students. With this regard, Tal and Kedmi (2006) argued that this criticism is more about the employment of STS than about its core ideas. Scholars who advocate for a more central role that socio-science should play in science teaching believe that issues such as genetically modified food, nuclear energy and nuclear waste, stem cells research, gene therapy, biodiversity, and so forth that enhance public discourse through the mass media should become the context of science teaching for the future citizens. In an attempt to locate socioscientific issues in the curriculum, Ratcliffe and Grace (2003) point to citizenship, scientific literacy and sustainable development as dealing with values, conceptual understanding and skills. They identify the connections between STS goals and environmental education in contributing to scientific literacy, citizenship, and sustainable development, and argue that despite the different foci, "attention to procedural understanding of reasoning and decision making, combined with acknowledgement and elaboration of values is a feature of all three" (p. 35). Ratcliffe and Grace suggest that socioscientific issues can be a means to achieve the ambitious goal of students acting as informed, responsible citizens when confronted with future scientific advancements. Within the large scope of SSI, in this chapter, more emphasis is given to conceptual understanding and citizenship.

With respect to teaching methods, it is widely agreed among STS/SSI/EfS/EE[1] proponents that teaching should be a process of negotiation and inquiry and that elements of authentic involvement of the students in decision-making and action should be included as well (Hodson, 1994; Sadler, Barab, & Scott, 2007; Sadler & Zeidler, 2004). Decision-making is a major element of being a citizen in a democratic society and the way to support youth in making informed decisions is considered as citizenship education and education for sustainable development. Citizens of the twenty-first century need to take a stand in environmental, health, economical, social justice, and many other issues, but the traditional teaching in most schools does not support students in becoming active citizens (Hodson, 2002). As Hodson argues, teachers struggle when they try to present science as a value-laden activity because the topics they teach are usually neutral. Socioscientific issues which are heavily loaded with values are much more appropriate to convey this message.

[1] STS – Science-Technology-Society; SSI – socioscientific issues; EfS – education for sustainability; EE – environmental education.

From International to Local Context

In Israel, since the 1990s, STS has become the framework of science education in the elementary and the junior high school levels. However, in line with Hodson (2002) and Zeidler et al. (2005), the societal and environmental issues remained as enrichment or merely decoration to the core science content. The only place in which socioscientific issues became a legitimate organizer of the entire curriculum was in the "Science and Technology in Society" (STiS) curriculum (MUTAV – in Hebrew), a curriculum for nonscience majors in the high school, which is studied by small number of students, usually in lower academic tracks. Within the context of STiS, various modules were developed around socioscientific issues. In these modules, the designers aimed at developing the students' questioning skill (Dori & Herscovitz, 1999), their argumentation (Dori, Tal, & Tsaushu, 2003; Tal & Kedmi, 2006) and decision- making, through learning about genetic engineering, air quality, ocean wildlife conservation, and so forth. In doing so, the designers of the modules addressed the four levels of sophistication suggested by Hodson (1994) which, in short, are (1) appreciating the societal impact of scientific and technological change; (2) recognizing that decisions about scientific and technological development are taken in pursuit of interests; (3) developing one's own views; and (4) preparing for and taking action. In the junior high school level, despite the flexible framework of the curriculum, and substantial attempts to develop knowledge integration or higher order thinking skills such as system thinking (Ben-Zvi Assaraf & Orion, 2005; Kali, Orion, & Eylon, 2003), or exposing the students to advanced research technologies to improve conceptual understanding (Margel, Eylon, & Schetz, 2004), only few attempts were made to promote thinking by using socioscientific issues for supporting higher order thinking in science and environmental education (Dori & Tal, 2000; Tal & Hochberg, 2003; Zohar, 2004; Zohar & Nemet, 2002).

Curriculum and Context

The use of socioscientific issues for enhancing students' science literacy will be presented here in the context of technology-enhanced learning in small groups using a Web-based module named Simple Inheritance, developed in WISE. The Web-based Inquiry Science Environment (WISE) was designed to enhance science learning, while taking advantage of the innovations that the Internet can bring into the teaching and learning of science. The WISE library includes several dozen modules, most of which are approximately 2 weeks in length and designed by teams of researchers and teachers, in various fields of science for upper elementary, middle, and high school students (Slotta & Linn, 2009). Many of these modules introduce science contents within health, environmental, and social contexts. For instance, in the asthma module, students investigate how asthma affects the human body, and how it is affected by environmental factors such as pollution (Tate, 2008; Tate, Clark, Gallagher, & McLaughlin, 2008). In the global warming module, students explore the causes for global warming using an interactive visualization which models the various factors involved (Varma, Husic, & Linn, 2008). While learning with WISE modules, students learn scientific content in relevant contexts, and develop a variety of thinking skills such as asking questions, identifying and critiquing evidence, making arguments, making hypotheses, and so forth. Interactive visualizations in WISE modules allow the students to explore complex phenomena and processes and integrate knowledge from various resources (Linn, Lee, Tinker, Husic, & Chiu, 2006). In WISE, students can work individually, as well as in small groups. For teachers, WISE allows modifications, additions, and on-going revisions to improve learning (Slotta & Linn, 2009).

The work reported here takes advantage of another capability of WISE – the authoring environment, which allows one to revise, adapt, and refine existing modules. In discussions involving the WISE research team and our research group, the WISE researchers expressed some concerns that had emerged with the Simple Inheritance module. They felt, and we concurred that this particular module needed some targeted revisions in order for it to support the desired learning outcomes. The WISE Simple Inheritance module along with associated test questions and coding rubrics were developed by Benemann (2005) with the support of the Technology Enhanced Learning in Science (TELS) research group. The module begins and ends with a framework story of Eric, a boy who is sick with cystic fibrosis (CF), a disease that affects his ability to hike with his family. The students explore Eric's family history to arrive at the conclusion that CF is an inherited trait. This context allows for further investigation of other inherited traits and learning about simple genetic mechanisms. Despite the engaging context and the anchoring story, of a sick child that launches the learning sequence, we believed that a "real life" context could make a greater contribution to students' learning. We assumed that other opportunities for social interactions to advance learning will further contribute to the students' engagement and learning (Ash, 2002, 2004; Ash & Wells, 2006).

Our endeavor is based on a previous study in which Tal and Hochberg (2003) employed the WISE Malaria project and attempted to strengthen the argumentative dialog in the classroom. Tal and Hochberg added two socioscientific issues to the basic module – one that dealt with the dilemma of eradication of the small pox virus, and the other dealt with a debate about vaccination against the West Nile fever virus. These socioscientific issues were used to support learning as well as assessment goals. For both issues, a whole-class discussion followed web-based learning exercises. In addition, Tal and Hochberg incorporated a sociocultural dimension to the learning process. Three classes, one of students from a middle-high socioeconomic suburban community, another of students from an urban school of mainly immigrants from the former Soviet Union, and the third class of Arab students, all who learned the malaria module at the same time, met for a "socioscientific conference" in which the students presented posters of their learning outcomes about malaria and participated in mixed groups to discuss the societal issues that affect science, health, and the environment. Tal and Hochberg, who employed the assessment scheme suggested by Zohar and Nemet, found that in the West Nile fever issue, which was given at the end of the module, the students' performances in tasks that required complex reasoning were significantly higher than their performance in the small pox virus preproject assignment. They also found that the students addressed more perspectives on the issue and that they better addressed scientific knowledge in supporting their justifications. Following the same line of thought, we believed that enhancing the authenticity of the Simple Inheritance module by adding a meaningful social interaction to the learning process will contribute to students' learning. We postulated that the contribution will be in both the affective and the cognitive domain.

In addition to better contextualizing the module to the Israeli context, we added two components to the original module: The first component, experienced by one class, was a visit to a CF unit in a children's hospital, and the other component was authentic communication through an asynchronous forum (online interaction). This forum allowed students to talk with a young CF patient over a period of a few days. Generally, we were interested in patterns of learning with the adapted WISE module, and more specifically, we were interested in the value of the two additions that aimed at improving the relevance of the module. Our research team consisted of an expert in technology-enhanced learning, an expert in teaching socioscientific issues who studies learning in informal settings, and an experienced science teacher in grades 8–10 (age 14–16). In this chapter, we share our experience and discuss the advantages and limitations of the project.

A Socioscientific Approach in the Design of the Module

The original WISE module begins with the story of Eric, a sick boy who intends to go hiking with his family. Our revised module, which was created in Hebrew, begins with introducing a newspaper ad, which reminds the public about a forthcoming CF donation day. In this ad (see Fig. 2.1) a real girl, Shefa, tells the public about her daily routine: one hour of physiotherapy, three inhalation treatments, 50 pills, controlled physical activity, special high calorie nutrition, and frequent hospitalizations. The ad culminates with the saying "For you it is a donation, but for us this is like air for the next inhale." We would like to note that in Israel, junior high school and high school students are requested to participate in door-to-door fund-raising for certain approved nonprofit organizations such as the CF, diabetes, and breast cancer organizations. In the revised module, after students are presented with the ad in the first activity, they are asked whether they would have volunteered to participate in such a CF fund-raising program. In order to make an informed decision, students are invited to learn more about CF. This opening dilemma is then reiterated as a final activity in the module, as we describe below.

Fig. 2.1 The opening page of the Hebrew CF module – Fund raising

As can be understood from the above description, right at the beginning of the module the students, who worked in groups of three, were requested to make a decision. In various other tasks, students were required to make decisions and provide arguments for their socioscientific decisions. After that, the students are referred to the Israeli CF nonprofit organization where they can watch a short interview with two boys and can get additional information about the disease and its treatment. By this point, the students begin learning about CF by suggesting questions for further learning, sharing their questions with their peers, and choosing together the questions for their investigation. Already in this first activity we encourage socioscientific reasoning (Sadler et al., 2007) and we highlight the need to make informed decisions that are based on social and scientific perspectives.

In the two additions we made, the hospital visit and the online interaction, we emphasized the opportunity to learn about CF patients' real life dilemmas. In the hospital visit, the students met a young female patient who told them about her everyday life and her after-school activities. One anecdote this girl shared with the students, in an attempt to indicate her relatively good condition was that she was not accepted to a "make your dream come true" program for children with major diseases. Her dream was to visit Disney World, but she did not qualify for the program because her condition was not considered as major. The students met a social worker who provided examples to the way she and the staff present the disease to the patients and suggest strategies to cope with it. She also presented them with tensions between the everyday lives of patients and their need to get continuous treatment. Students met a doctor who answered their questions about CF, heredity, and fertility, which had emerged as a topic of particular interest among many of the students. In the online interaction with the CF patient, students had an opportunity to interact with David, an undergraduate student, about his social life, his sports activities, and the way he manages to study engineering, get treatment, and lead a normal life, as he describes it. All these activities were in conjunction with learning the science behind the disease and learning about other inherited traits.

To sum up this section, in this study we explore how socioscientific issues provide learning opportunities in different contexts. A socioscientific issue can be presented in a text format, in an oral discussion, in a TV item or program, in a Web page or through any number of other channels. The use of socioscientific issues is intended to highlight the complex relationships between science and economical, health, environmental, and social issues, and they provide students with an opportunity to deal with real and relevant dilemmas. Our project involved teaching genetics in an everyday context, while engaging students in dealing with dilemmas of patients, parents of patients, and the general public. The students were requested to make decisions about social action (fundraising), about what should be done with such publically-raised funds, and about whether or not to try to prevent birth of sick babies (acting as genetic counselors), while interacting with real patients in person and online.

The Field Trip

Learning in out-of-school environments is common worldwide. Students get to visit science, natural history, and art museums. They visit zoos and have field trips to nature parks. There is much evidence in the research literature that out-of-school learning has many positive impacts on learning outcomes of various sorts (Bamberger & Tal, 2008; Falk & Dierking, 2000; Rennie & McClafferty, 1995, 1996). Students learn scientific content, develop positive attitudes toward science, interact with each other while being engaged in learning meaningful things, and gain opportunities to use all senses to experience phenomena in real-contexts (Dillon et al., 2006). Field trips can help students to visualize and understand controversial issues such as whether a wetland was dried to provide more land for farmers or to consider the positive and negative environmental consequences of farming (Tal, 2004, 2008). Field trips can also be used to enhance discourse and collaboration between groups in conflict. Tal and Alkaher (Tal & Alkaher, 2008, 2010) investigated multicultural environmental activities of Jewish and Arab youth in nature parks in Israel. Eighth graders from different cultures who speak different languages learned about development vs. conservation in a nature park in the region. The socio-environmental conflict had significant associations with the greater national conflict between Israelis and Palestinians. In our view, all these different types of field trips promote meaningful learning by situating learning in authentic contexts, providing hands-on experiences, embedding those experiences in issues, and accounting for the sociocultural dimensions of learning.

In order to carry out the field trip, we contacted a few CF units in Israeli hospitals. Fortunately, we got several positive responses, which allowed us to prepare a school visit that included the following components: (a) meeting a social worker, a nurse, a physiotherapist, and a doctor; (b) visiting the treatment unit, and experiencing some (real) tests and exercises that CF patients need to go through; and (c) meeting a middle school patient and talking with her about her everyday life. Based on principles of how to carry out educational field trips (DeWitt & Storksdieck, 2008; Orion, 1993; Tal & Morag, 2009), the preparation for the field trip included not only the science and health relevant topics. The students watched a short movie which was available at the CF association website that presented the CF unit and the staff. It happened that with no intention, the girl that appeared in the movie was the same girl that the students met at the hospital. Special attention was given, in the preparation, to ethics, and the students discussed what would be appropriate and inappropriate to ask the patients, to avoid unintended but possibly insensitive inquiries. Overall, the field trip lasted for 3 h. Throughout the field trip, Stella, who planned the visit, acted as a mediator. This function was crucial, as hospitals are not arranged for school visits. No one can know in advance what unexpected event could come up, whether the doctor will have enough time to talk with the students, whether the patient will be in the right mood to open-up to the students, how students will react to experiencing the actual tests on their own bodies, and many other possible challenges. As already mentioned, eventually, we were able to carry out all the planned activities, and even the young patient who was very shy at the beginning, eventually was very friendly and talkative and shared with students some of her life experiences. In school, after the visit to the CF unit the students continued working on the module and were challenged to draw the family tree of the CF patient they met. After this range of activities, the students resumed the more general tasks of the module.

Online Interaction with a Patient

Since computers have been widely introduced into schools in the 1980s, extensive evidence has accumulated showing that technology-based learning environments, when appropriately designed, can have a great impact on student learning of science (Pea & Collins, 2008). As internet access became more abundant in schools, much energy has been put in research and design of Web-based learning environments (e.g. Slotta and Linn, 2009). One important added value of these environments is their capability to allow students to break the boundaries of the classroom, and extend their interaction to include, in addition to their peers and teachers, people around the world who can widen their horizons regarding science topics they study in class (Roschelle, Pea, Hoadley, Gordin, & Means, 2000). These interactions, of course, require careful preparation.

In the current study we decided to take advantage of the affordances of online environments in order to develop another version of the module that would serve as an alternative to the field trip. In many countries, including Israel, field trips in general, and a sensitive fieldtrip such as the one described above, are not easy to carry out (Dillon et al., 2006). To make the unit applicable for other places in the world – in which limitations such as lack of financial support (for transportation), difficulties in collaboration with a nearby hospital, or incapability of hospitals to allow such visits – we made a design decision to provide an alternative authentic experience to students. The online interaction version of the module included a forum, in which students had an opportunity to interact with a real patient. In a pilot study of this version of the module, students interacted indirectly with a nine year old boy. Due to his age, the interaction was indirect – the mother interacted mainly with the teacher (Stella), to answer the students' questions. In an attempt to avoid possible inappropriate or unethical exposure of a youth, we found a young adult – an undergraduate student whom we refer to as David – who was willing to collaborate with us. David was willing to participate in an asynchronous discussion with students within the WISE module, which lasted for about two weeks. He provided a brief personal background in the first posting in the forum and invited students to ask him questions. Each student group wrote a question in the forum. David made an effort to answer questions from each and every group. The questions that the students asked David demonstrate the various aspects they were concerned with, which include social, health, and scientific aspects. Examples of such questions are: "How does coping with the disease affect your life?"How much time per day are you occupied with treatment?" "Whether, and in what ways does the disease affect your social life?" "How did the disease develop when you turned into your teens?" "How many people in your family are sick?" "Are there any sick people or carriers in your family?" "Do you have concerns about passing the disease to your kids?" Some students continued with more questions, and David answered some of those as well. He told students about his sickness, his life history, his family, and his everyday experiences. We stressed that finding the right person for this work was challenging, and eventually, it was the CF unit personnel at the hospital who connected us with him. We also want to note that David suffered from depressions, due to his condition, and that he declared that interacting with the students was a therapeutic activity for him.

The Study

The study comprised three phases: a pilot study and two phases of the main study. The participants were 8–10 graders from a 6-year secondary school (grades 7–12) in Tel Aviv. This school serves a heterogeneous population of low to high socioeconomic status. Altogether, one eighth grade, one ninth grade, and two tenth grade classes participated in the study. Typically, simple inheritance is taught in Israel in the ninth grade, but in some schools it is taught in tenth grade.

In the *pilot study*, we used a version of the module in which we adapted the original WISE Simple Inheritance module to the Israeli context. This included changing the framework story of the module and the associated learning tasks. The adaptation was based on design guidelines for educational technologies found in the Design Principles Data Base (Kali, 2006; Kali & Linn, 2007), and specifically, a design principle which calls to "connect science to personally relevant contexts" (Kali, Fortus, & Ronen-Fuhrmann, 2008) was used. Stella was the teacher of a cohort of 41 students from one ninth-grade class, who participated in the pilot study. With respect to data collection, at that stage we collected descriptive data in the form of students' work as expressed in the "notes" function of the WISE module. We also documented students' reactions while working with the module, and we observed their work in small groups of 2–3 students. In addition, we used this phase to test our scoring rubrics and to examine and revise the open-ended reflection questionnaire and the Likert type feedback survey. Participating as both teacher and researcher, at this stage, allowed Stella, who knew her students very well, to identify issues that required design revisions, and to distinguish them from other issues related to the specific learning context. We refined the design of the adapted version of the Simple Inheritance module to improve usability issues. We also made some modifications to the scoring rubrics to make them more reliable.

The research questions that we pursued in the main study were:

1. What were the learning characteristics of the students who learned simple inheritance using the adapted Simple Inheritance module?
2. How did the two enhancements (the hospital visit and the online interaction with a patient) contribute to [a] the students' interest in genetics? [b] the understanding of scientific ideas in genetics?
3. Was there a difference between the two enhancements with respect to their contribution to the increase in the relevance of the module?

Following the pilot study described above, the main study included two stages: (a) enactment of the revised Simple Inheritance module with one class of 28 eighth graders (taught by Stella) to answer research question 1, and (b) enactment of the two additional versions of the module (basic + hospital visit and basic + online interaction), with two classes of tenth grade students (about 30 students each) to answer research questions 2 and 3.

In the next stage, two other teachers, guided by Stella, taught two tenth grade classes of about 30 students each, in which students had not studied genetics earlier. In this quasi-experimental stage, each class used the adapted module with one addition: either the hospital visit or the online interaction with the CF patient, David. The two classes were similar to one another in terms of student ability levels and female-male ratio; both were also heterogeneous with respect to student socioeconomic status. The additions were randomly assigned to the two classes.

Unlike in the USA, where teaching the module takes about two weeks, in Israel, due to fewer science classes per week and to holidays, it took the teachers more than a month to complete the same number of lessons. The additional activities required more time – 3 h for the field trip plus a preparation activity of about one class period, and about two sessions for the online interaction.

Data Collection and Analysis

Data collection included:

(a) A science-knowledge integration test, which was administered 1 week after students completed their learning with the module. The knowledge integration test that was designed to measure students' explanations was developed by the original developers in the WISE project in the USA (Benemann, 2005; Linn et al., 2006; Liu, Lee, Hofstetter, & Linn, 2008). Duncan (2007) revised the test from which we used three open-ended questions that examine students' integrated understandings of the principles of simple inheritance (see Appendix 1). To this test, we added another complex question that required students to apply their knowledge to a typical situation in Israel, in which many families of CF patients are uncertain about whether their ancestors had the disease as many large families were exterminated during the Holocaust.

(b) A feedback questionnaire that included two parts: six Likert type questions with four possible answers and two open-ended reflection questions (see appendix 2).
(c) The answers of the students to the written tasks in the module.
(d) Observation data collected throughout the four enactments of the adapted Simple Inheritance module.
(e) Evidence from students' work in the module; for example, to assess student engagement, we used the question about their tendency to participate in fund-raising for CF.

The knowledge integration framework was used to develop a rubric with a 0–5 point scale to assess student responses (on the science-knowledge integration test) in order to identify the number of incorrect, partial, and complete connections that students make (Liu et al., 2008). Levels 0–2 are considered low level scores: Score 0 indicates no response was given. Score 1 indicates that even though something is written, the response is off task. Responses that contain incorrect or irrelevant ideas or connections receive a score of 2. Levels 3–5 are considered higher level responses: A score of 3 means that students have relevant correct ideas, but fail to make connections between them. A score of 4 means the student response contains one basic scientifically valid connection between two ideas. A score of 5 is the highest score and must contain multiple valid connections between 2 or more scientifically correct ideas. The scoring levels were refined by careful analysis of student responses, so that they were distinct enough to differentiate students' reasoning, but at the same time capture all possible student ideas.

Differences between students' outcomes in the two conditions (field trip and online interaction with a patient) were calculated using a *T*-test procedure. As we could not make a normal distribution assumption, we compared between students' attitudes toward learning with the field trip vs. the online interaction by employing Mann–Whitney *U* test. This test is an alternative to the independent group *t*-test, when the assumption of normality or equality of variance is not met. Like other nonparametric tests, the Mann–Whitney test uses the ranks of the data rather than their raw values to calculate the statistic. In order to analyze the students' responses to the open-ended questions in the module, we were influenced by the work of scholars who studied student argumentation in the context of socioscientific issues (Hodson, 1994; Jiménez-Aleixandre et al., 2000; McNeill & Krajcik, 2007; Sadler, 2004; Sadler & Zeidler, 2004; Zohar & Nemet, 2002). We looked at students' claims and their justifications. For example, for a family tree task, in which the

Table 2.1 Scoring rubric for the family tree task (max = 4)

Claim	0 Inaccurate claim (CF is not a genetic disease)	1 Accurate claim (CF is inherited)	
Evidence	0 Uses the family tree diagram incorrectly	1 Referring to correct details in the family tree diagram	
Reasoning	0 No attempt to explain the relationship between claim and evidence	1 Insufficient explanation	2 Provides accurate reasoning that ties the evidence to the claim

students were requested to predict which family members will carry the CF gene, we used the rubric presented in Table 2.1. In this task the students had to present a claim with respect to heredity of CF. This claim was supposed to use the evidence, which was their own drawing of the family tree based on given textual data. In their justification, they had to tie the claim and evidence.

A few examples for scoring students' answers are:

"CF can be genetic disease since one of the ancestors of the family was sick" [claim-1; evidence-0 (inaccurate tree); justification-1 (partial)]

"CF is indeed genetic because in the family tree we found that descendants in the family have the disease in different generations"
[claim-1; evidence-1 (referring to correct family tree); justification-1 (insufficient, does not refer to both sides of the family)]

"Yes, CF is inherited, but we don't know in which generation it develops" [claim-1; evidence (tree)-1; justification-0]

"Yes, CF is inherited because you can see other two sick family members in both side of the family"
[claim-1; evidence-1 (correct tree); justification-2 (refers to sick people on both sides of the family)]

It is important to note that students created the family trees based on textual information in order to generate evidence for supporting claims regarding CF. Given that they never saw such a diagram prior to this task, the task was quite sophisticated.

Outcomes

Interest and Engagement

In answer of our first research question, we found that the vast majority of the eighth grade students expressed interest and enjoyment regarding the WISE Simple Inheritance module referring to their comfort in using technology. This was indicated in several ways. In the open-ended reflection question of the feedback questionnaire, many students noted that they preferred the "notes" function in the online module than writing in physical notebooks, which they usually use in biology lessons. In addition, several students addressed the new ideas they learned. One student, for example, wrote: "Working with the web-based module has enriched me with new knowledge. Now I know things and concepts I did not know before." Another student indicated: "I think computer-based learning is good, since kids are taught in a way they are familiar with and it is more creative and fun. I think it helps kids open their minds." Other examples from students' responses in the feedback questionnaire, indicate students' increased interest following their learning of the module: "we want to learn [look] more closely on information about CF, we want to understand more specifically, why this disease is more problematic than other [diseases]"; "all this probability thing and the looking on our ear lobes was interesting." Only one student stated in the open-ended question that the module was not interesting.

In our observations, we found extensive evidence for increased interest in genetics among the students. While learning from the module, many students asked the teachers for recommendations of websites dealing with CF in addition to those provided in the module. A few students who did not find satisfying answers in the module approached "BaShaar" – a nonprofit scientists' organization for the Israeli society, that has an "ask a scientist" forum in its website.

Another piece of evidence for students' deep engagement came from a task we added to the original module in an attempt to increase relevancy and encourage reasoning activities. In a short paragraph, we described a young couple expecting a baby. This couple found out that they both carry the gene for CF, which means they have a 50% chance of having a sick child. The students were asked to "Imagine that you are a genetic counselor, what you would recommend to this couple?" After a short whole-class face-to-face discussion, students were required to write their recommendations. We observed the students enthusiastically negotiating and debating this task. The variety of student answers indicated they understood the sensitivity involved. There were students who argued that the genetic counselor should only give the scientific and health information, with no recommendation regarding a particular decision. One group suggested that the counselor should help the couple better prepare themselves for the situation: "They should learn about CF, for any case, so they won't be surprised and in order to face all the challenges." Another group suggested examining the fetus: "it's 50%, so it's a chance the baby will be healthy, but if they know it's a sick baby, we would recommend an abortion." A different group was convinced that the counselor should work with the couple on how to accept a sick child with love and provide the best possible treatment. It was hard to stop this discussion, which involved what the students learned as well as their personal values.

One more activity that aimed at increasing relevancy was the fundraising activity that served as an opening and summarizing assignment in the adapted module. In their responses to this task, the students expressed empathy, and referred to their responsibility as citizens.

> After we learned about CF, we realize that the public awareness is not sufficient, so we would like to participate and contribute to increasing awareness (gr. 2).

In the post-task the students were asked to provide a recommendation using the money raised. In their answers, the students not only addressed scientific research but also included better equipment and facilities for patients. They advocated for establishing cross country services for parents, which would make their own lives easier, support groups for future parents of sick babies to prepare them and help them in the first months, and fun activities for sick youth.

> We'll go out [to participate in fundraising] because [cf] is actually a chronic disease… it is important that the patient will have the best possible quality of life, and that can be achieved by physiotherapy and donations of people who think it's important.

We observed not only within group discussions about the different purposes, but we also saw many between group discussions about this issue. Evidence from students' work in the module and the observation data indicated deep engagement on the part of students and thoughtful group discussions that were the result of enhancing relevancy and including controversies in teaching the Simple Inheritance module.

Contributions of Field Trip and Online Interaction

To answer the second and third research questions, we describe and compare the contribution of the two enhancements (online interaction with a patient and the visit to the hospital) to students' interest and understanding of scientific ideas in genetics. The analysis of the open-ended responses to the question "In what way/s has the online interaction with David contributed to your learning of genetics in the Simple Inheritance module?" allowed highlighting the contribution of this addition to students' learning (research question 2). A few topics emerged in the students' responses that elucidate this contribution. Major themes are identified below with quotes excerpted from questionnaires that students completed following the experience:

The ability to ask questions improves learning. "The talk with David, in the forum allowed me to ask him questions that interested me about how he copes with the disease. It helped me learn the topic."

Learning new things. "Although his answer to my question was not very clear, he told us many things we did not think about so never asked about."

Understanding the patient challenges. "Talking with David helped me realize what these people go through every day."

The responses to the same question that addressed the field trip provided stronger evidence for the field trip supporting meaningful learning and in general, were more clearly articulated. The topics that emerged were:

Complementarity. "Learning through the CF module and the field trip complemented each other, because things that were in the module were not in the field trip. Both were interesting and contributed."

Meaningful learning. "I learned some background about the disease, which helped me understand the topic. When I wrote my answers in the module, I wasn't sure, but when I went back to the module after the hospital visit, it was more interesting and I better understood"; "No doubt, the hospital visit helped me learning. The presentation of Diana (the social worker) provided many details on the disease and how kids cope with it in their daily lives. The meeting with the sick girl - it certainly helped me to prepare for the test."

Relevancy. "I have two sick friends with CF. Through learning with the module, the field trip and the staff's presentation, I now understand what happens in the disease and what the patients go through every day. While we can do everything we want, they have to do inhalations, eat enzymes..."; "Being there at the hospital and observing the daily routine certainly clarified the stuff. The presentation and the questions we asked summarized the topic perfectly."

While the students who were engaged in the online interaction addressed mainly affective contributions, the students who visited the hospital referred to deeper learning and understanding, as well as to affective contribution of the visit. Moreover, they better connected the out-of-school experience to learning with the module. An analysis of the contribution of the two additions to student learning, as reflected in the sophistication of their responses is presented in Table 2.2.

Table 2.2 indicates that the hospital visit was better perceived as a contribution to the students' learning than the online interaction. The vast majority of the students who visited the hospital provided detailed arguments for why and how the visit helped them learn genetics. The "interaction students" addressed mainly affective aspects in their responses, and more than one-third acknowledged the contribution of the online interaction to their learning only to a limited extent.

As indicated, another way in which we analyzed student engagement was exploring a question about the students' tendency to participate in fundraising efforts for CF. Students worked in groups to negotiate this issue, and our analysis focuses on social responsibility, acquired knowledge, and affect. Table 2.3 presents the classification to the three justification levels described above.

Prior to learning the Simple Inheritance module, 8 groups out of 18 gave poorly justified answer to the question posed compared to 4 groups that gave such an answer at the end of the module. Only 1 group provided a response characterized by the highest level of complexity at the beginning compared to 5 groups that provided a well-established response at the end. More groups of students who visited the hospital provided the highest level of responses than the students who participated in the online interaction. These responses incorporated statements about what they learned or/and what they felt about taking part in fundraising. This is another evidence for stronger effect of the field trip.

Knowledge Acquisition

As noted above, student knowledge acquisition was assessed by: (a) the knowledge integration test developed by the WISE group at Berkeley, (b) another open-ended item that we added to this test (item 4), and (c) analysis of the responses

Table 2.2 Analysis of student answers regarding the contribution of the online interaction and the field trip to their learning

Answer orientation (positive vs. negative) and justification		Example	Frequency (%) Online interaction (N=22)	Field trip (N=28)
Negative	Unjustified	N/A	4	
Negative	One justification	Talking with David did not help me understand genetics better, it only helped understanding his everyday functioning (online interaction) No, talking to David did not help me understand, while seeing his family tree helped better (online interaction)	19	0
Limited contribution	At least one justification	Not much, since David did not add anything new. He mainly talked about how the disease affects his life. What he did add was that he explained how a carrier or a sick person calculates the odds of a sick offspring, so David will not have kids with a carrier (online interaction) Not so much, because the visit took place after we completed most of the module, but it did help me connect what we already learned (field trip)	14	4
Positive	Generic or one justification	Yes, David helped me understand the disease characteristics Yes, because learning in school with the teacher – the routine, is less effective than real interacting with patients (field trip)	53	69
Positive	A claim and detailed justification	Talking with David in the forum helped because he explained about CF, and answered the question that interested me such as… (online interaction) Yes, the field trip helped learning the subject. In the hospital, they showed a presentation that gives clear explanation of the disease. For example, we learned that in order to increase their life expectancy the patients need high calorie diet, they need to consume enzymes to catalyze the fat, practice out, do physiotherapy and inhalations… we were physically much close to the whole thing, and it made us want to learn more about the disease, its symptoms, cure and so forth (field trip)	10	27

Table 2.3 Justifications for fundraising activity

Justification	Example	Frequency			
		Online interaction		Field trip ($N=10$)	
		Pre ($N=8$)	Post ($N=6$)	Pre	Post
Unjustified generic response	Yes, we agree to participate (in the fundraising program) in our free time	2 (25%)	3 (50%)	6 (60%)	1 (10%)
Justified response supported by sense of responsibility expression of feelings OR acquired knowledge	Yes, we will participate, as we understand how severe the disease is	6 (75%)	2 (33%)	3 (30%)	5 (50%)
Justified response supported by sense of responsibility that addressed acquired knowledge AND affect	Yes, it's important for us to save lives, and now we care much more, because we know about CF symptoms	0	1 (17%)	1 (10%)	4 (40%)

to questions that students answered using "notes" in the Simple Inheritance module. In this section, we present the outcomes from the analysis of these three data sources.

Knowledge Integration Test

Figure 2.2 shows students' responses to items 1–3 in the test. As can be seen from Fig. 2.2, the differences between the online interaction group and the field trip group were not large in magnitude. The differences were not statistically significant.

The additional question developed for the Israeli version of the knowledge integration test (item 4), required students to suggest why many CF patients in Israel have difficulties in identifying members from their larger families who had CF in the past. The complete answer to this question could include a few possible reasons: In past generations, especially in underdeveloped countries where many of the Jews lived, it was common for individuals not to know the accurate reasons for why death occurred at early ages. In particular, with the case of CF, many deaths were attributed to pneumonia and other infections. Being a recessive disease, it was also possible that there was only evidence of carriers (and no evidence of diseased individuals) in the immediate past generations of a family. A third possible reason for this lack of knowledge can be attributed to the Holocaust. Many family lineages were almost extinct and in general, the ability to know the life circumstances of ancestors is limited. The main themes that emerged from the analysis of students' answers to question 4, with respect to the reasons for the scarce knowledge base

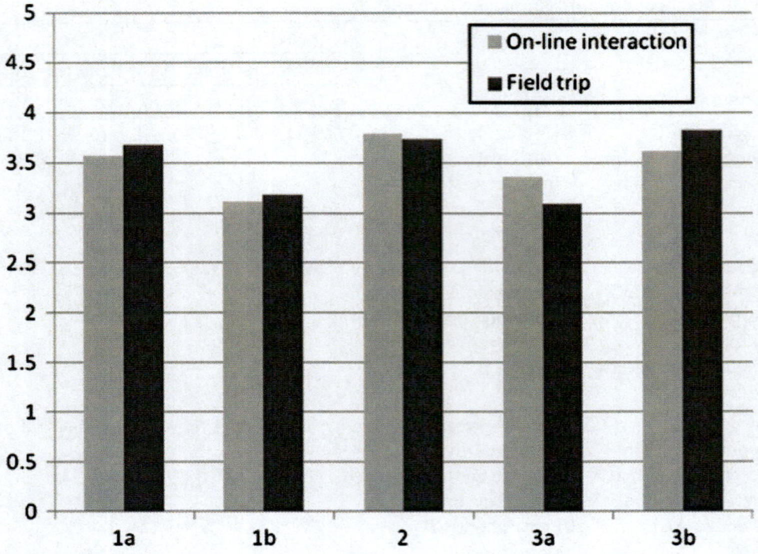

Fig. 2.2 Scores of the test items (max = 5)

about sick relatives of CF patients were: (a) insufficient awareness of the disease and lack of advanced technological means in the past for diagnosis and treatment, (b) low probability of having the disease, and (c) insufficient family background information due to the Holocaust (although this response was only offered by one student). Interestingly, the distribution of students' answers, as presented in Table 2.4, indicates more genetics-based justifications from the online interaction students (55%) in contrast to more justifications related to technology, diagnosis, and awareness brought up by the field trip students (55%). In other words, while the majority of the online interaction students based their answers on the scientific aspect, the majority of the field trip students founded their answers on the social-technological aspect. Additionally, Table 2.4 shows that irrelevant answers were provided by more field-trip students than by online interaction students.

Simple Inheritance Module Notes

One example of the Simple Inheritance module tasks was the family tree task, in which students had to predict how a sick child "got" the disease. The categories we employed (Table 2.1) were: (a) claim (wrong/correct); (b) evidence (correct /incorrect tree); (c) justification (explaining the claim based on the information from the family tree). Table 2.5 presents the distribution of responses of the groups of students who studied with the two enhancements. The maximum points available for each group was 4, and the number of groups was eight in the online interaction version and 10 in the field trip.

Table 2.4 Answers to the Israeli-context item about history of CF patients

Type of response	Online interaction (N=22)	Field trip (N=29)	Examples
Difficulties in diagnosis and poor technology; lack of knowledge about the disease	9 (41%)	16 (55%)	• There was no awareness to the disease and no treatments • People died at early age of various reasons including CF with no distinction • Many Ashkenazy families were exterminated in the Holocaust, so no one really knows
Probability of having sick people is low; more people being carriers than sick	12 (55%)	8 (28%)	• Probably, in past generations in these families there were only carriers • As the probability (to be sick) is not high, because it's a recessive trait, the disease did not express • The disease was not known then, so • people were not diagnosed properly, and their death was attributed to something else
Not relevant	1 (4%)	5 (17%)	

Table 2.5 Distribution of the answers to the family tree task

	Claim		Evidence		Justification		
Rank	0	1	0	1	0	1	2
Online interaction (N=8)	3.375	5.625	4.5	4.5	30.375	1.125	4.5
Field trip (N=10)	2.2	8.8	2.2	8.8	1.1	5.5	4.4

The distribution of the answers shows some advantage to the field trip students with more groups providing accurate claims, suggesting correct family trees as evidence and more groups justifying the claims with the tree-evidence. Outcomes of multiple choice items, embedded within the module that enabled us to assess students' understanding of the concepts of genotype/phenotype and recessive/dominant genes revealed no difference between the two enhancements. Overall, using all the data available to us, there seems to be no significant gap between the performances of the online interaction group and the field trip groups regarding knowledge acquisition.

Student Attitudes

Comparison between students' attitudes toward learning with the module with each of the two additional components was carried out by employing Mann–Whitney's U test. The results are presented in Fig. 2.2; the term "addition" in the figure refers to each of the two additions: the online interaction or the field trip.

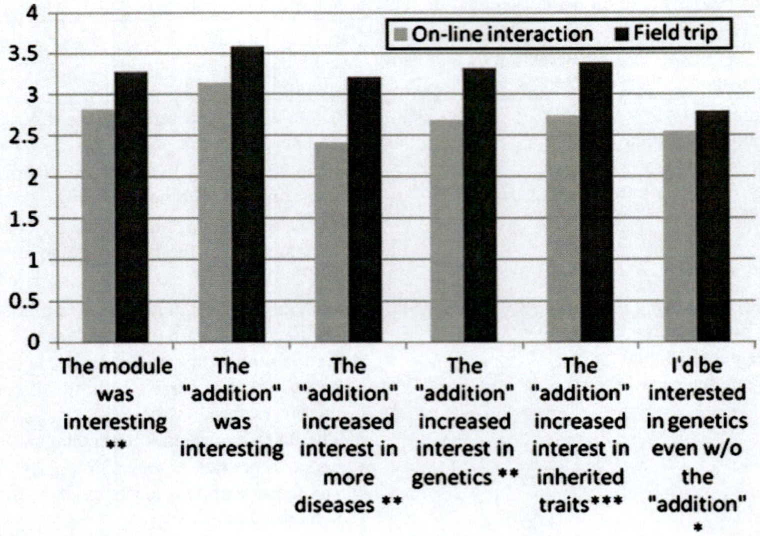

Fig. 2.3 The students' attitudes toward learning with the online interaction and the hospital field trip

Figure 2.3 shows that in most items, field trip students expressed attitudes that were significantly more positive toward the module than the online interaction students.

A brief summary of our findings shows that: (a) the adapted module, even without the additions, created interest and motivation among students to learn about genetics, (b) the field trip addition was more productive than the online interaction addition in enhancing student interest and self-viewed learning, and (c) no differences were found in students' knowledge acquisition as measured by the test and the module tasks when learning with the modules with each of the two additions.

Discussion

The findings described above show that the design of the adapted module, even without the additions of the field trip and the online interaction with a patient, was successful in getting students interested in understanding the science behind the CF disease. The findings indicate that features in the project, such as incorporating the real story of Shefa, involving students in making decisions (even though these were fictitious decisions) about whether they would participate in a fundraising program, or what they would recommend to a family confronted with the possibility of having a baby afflicted with CF were crucial in getting students engaged and promoting their interest in understanding genetics.

Relating science to personally relevant contexts is a well-known instructional strategy for designing learning environments that can make science accessible (see for example Duschl, Schwiengruber & Shouse, 2007; Linn, Davis, & Bell, 2004; Kali et al., 2008). In fact, the design of both the original module and the adaptations introduced into the module for the Israeli audience described in this chapter, were based on this design principle. This design principle is very much in line with some of our previous work, in which we aimed at increasing student engagement in science by developing learning materials, which were based on STS ideas and incorporated a variety of socioscientific issues about genetics, and the Mediterranean coast environment (Dori et al., 2003; Tal & Kedmi, 2006). Nevertheless, we view the contribution of the current study, in enabling a critical analysis of means for enhancing authenticity. By comparing what students thought of their learning with each of the additions to the module that were designed to increase authenticity, we were able to closely investigate what it is that makes successful or less successful means of increasing authenticity. We would like to stress that we do not view this comparison as one that would enable us to say that either field trips or online interactions are superior means of increasing authenticity. This would be an oversimplification of our findings. Rather, we take a design stance (see for example Kali & Linn, 2007) to make sense of our findings. Since we have two designs, the field trip-enhanced module and the online interaction-enhanced module; the first which elicited a higher degree of interest and engagement among students than the latter, we can identify important design elements that support science learning in socioscientific contexts. In the next sections we elaborate on these design elements.

Diversified Interactions

As described above, during the field trip, students had an opportunity to interact not only with a CF patient, but also with a social worker, a nurse, a physiotherapist, and a doctor, and to experience real tests and exercises that CF patients need to go through. The online interaction on the other hand, was limited to interaction with David, the CF patient. We assume that the diversified interactions in the field-trip enhancement were highly important in providing students with a holistic understanding of this socioscientific issue, and thus, brought to increased authenticity. This assumption is based on several findings: (a) answers to the open-ended question provided by the field trip students, which indicated a stronger connection to the genetics contents than those provided by online interaction students, (b) the hospital visit that was better perceived as a contribution to the students' learning than the online interaction (Table 2.2), and (c) the stronger, and more content-related justifications that field trip students provided in the fundraising activity (Table 2.3). An improved design can definitely include such diversification, even when constrained to a web-based module. We suggest that adding relevant clips to the online environment (such as clips of practitioners or practices in the field), with prompts for reflection or discussion, can provide students with additional aspects and a broader picture of the topic they are exploring, and also to capture a bit of the authenticity of a field trip.

Live Communication

During the field trip, students were able to communicate with all the people described above in real life. Such social interactions are advocated in the informal science education literature (see for example Ash, 2004; Bamberger & Tal, 2008; Schauble et al., 2002). Asynchronous discussions have the advantage of enabling students to carefully articulate their thoughts, and to reflect before replying. This is definitely an added value in many curricular settings (Hoadley & Linn, 2000). However, based on the same findings indicated in the *diversified interactions* design elements, it seems that in the particular setting of the current study, and perhaps in other SSI, when one of the goals is to engage students emotionally, the disadvantages of asynchronous discussions, which lack the dynamics, the body language, and the liveliness of a face-to-face discussion, are more dominant. An improved design in a technology-enhanced solution, could take advantage of synchronous meetings with people in the field, preferably with audio and video.

Time on Task

The field trip, which was a half-day event, and was preceded by a preparation in class (DeWitt & Storksdieck, 2008; Orion & Hofstein, 1994), required more time than the online interaction, which took place in about two teaching sessions (of 50 min each). Although this might sound obvious, in an atmosphere in which schools are pressed by high stakes measures, teachers are discouraged from devoting time to topics that are not included in the core curriculum or face too many organizational challenges (Dillon et al., 2006; Tal, 2008). We find it important to note that productive socioscientific activities can be time-consuming. When students spend more time on getting to know the details of a real-world problem, they have the opportunity to perceive the complexities involved, and get a realistic sense of the scope of the problem they are studying.

Another interesting finding of this study is that although there was a significant difference between the way students in the two groups (fieldtrip and online interaction) perceived their learning with the module, there was no difference in their knowledge acquisition. However, even though a connection between student interest and knowledge acquisition was not found in the current study, it does not mean that such a connection does not exist. We believe that when students are more interested and engaged, they will explore the problem that they are studying in greater depth. If they are scaffolded properly, there is a greater possibility that they will develop the mental connections required for understanding complex science, and integrate the pieces of knowledge to a coherent and integrated understanding (Blumenfeld, Marx, Patrick, & Krajcik, 1997; Roseman, Linn, & Koppal, 2008; Singer, Marx, Krajcik, & Clay Chambers, 2000; Solomon & Thomas, 1999). One explanation to the lack of such a connection in the current study, is that in both cases, student interest and engagement were high (they were high even before we added the field trip and online interaction enhancements). Perhaps the difference in interest found between the two enhancements was not large enough to show a difference in their knowledge acquisition. Another possible reason for not finding this difference can be attributed to timing of the field trip. Orion (1993) argues that in order to get the maximum effect on learning, the field trip should be carried out at the beginning stages of the learning unit. However, due to organizational constraints, we were able to carry out the field trip only toward the end of the unit. In any case, we would like to stress that we view the goal of enhancing student interest not only as means for supporting their understanding of complex topics but also as an educational goal per-se, especially when socioscientific issues are involved. The literature shows that learning socioscientific issues contribute to the development of a wide range of higher order thinking skills (not necessarily those we assessed in the current study), promote learning of the nature of science, and encourage good citizenship (Dori et al., 2003; Ratcliffe & Grace, 2003; Sadler & Zeidler, 2004, 2005; Tal & Kedmi, 2006; Zeidler & Sadler, 2008).

The three design components articulated in this study, together with the design principle "Connect to personally relevant contexts" (Krajcik, Slotta, McNeill, & Reiser, 2008), which served as a basis of the design of the Simple Inheritance module, are crucial for designing science instruction in the context of socioscientific issues. That said, we would like to stress that we view the educational field trip itself, as an instructional strategy, which serves as excellent means to support the instruction of socioscientific issues. We would like to encourage educators to make the effort involved in having students augment the learning that occurs in class with outdoor experiences. However, we are also aware of difficulties involved in taking students to educational field trips. Thus, we recommend educators to take advantage of online authoring environments, such as WISE and others, in order to design productive online teaching activities for socioscientific issues that build on the design components identified in the current study.

Appendix 1. The Knowledge Questionnaire

1a. Sarah and Michael are going to have a baby. Both of them are completely healthy, but they know that Sarah's dad (the baby's grandfather) has a genetic disease called cystic fibrosis, which affects the lungs. Should they be worried about their child being born with cystic fibrosis?

(Choose one) ____**Yes** ____**No**

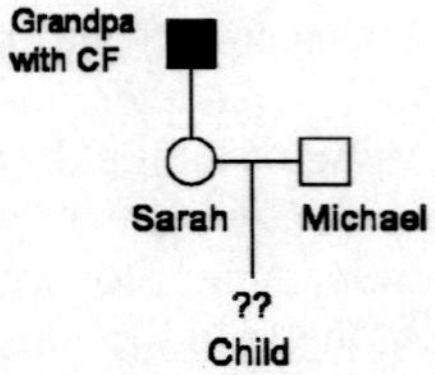

1b. List two pieces of information you need in order to accurately predict the chances that Sarah and Michael will have a child with cystic fibrosis?

2. There are two main phenotypes (physical appearance) for the trait for hairline, which is a genetically inherited characteristic:

Or

Look at the family tree below; is it possible for two parents with widow's peaks to have a child with a straight hairline? **Explain why or why not.**

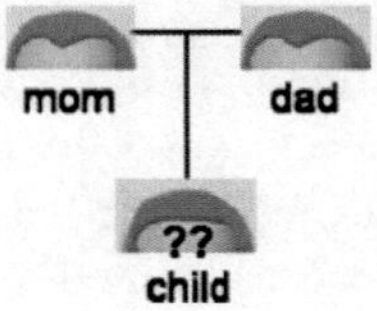

3a. Some humans have a trait (characteristic) for curling their tongues. You observe that a mother and father can curl their tongues, but their child cannot. Which of the traits below is the dominant trait?

(Choose one) ____**Tongue-Curling Ability** ___**No Tongue-Curling Ability**

Please explain how you determined this.

3b. What is the probability that these parents will have a child that will have the tongue-curling ability?

Explain how you got your answer.

In the SI module you got to know a few CF patients. According to the information that X gave, he is the only person in his large family known to have CF. Today, in Israel, in most families of CF patients no one knows about sick relatives in previous generations. Can you suggest a reason for that?

Appendix 2. Attitude Survey

	Do not agree at all	Not agree to some extent	Agree to some extent	Fully agree	Comments
Learning with the SI module was interesting					
The field trip[a] to the hospital was interesting					
The field trip made me learn about other inherited diseases					
Talking with the patient made me interested in how traits are being inherited					
I was interested in genetics even without the visit to the hospital					

[a]In the online interaction version, the words field trip were switched by "the online interaction"

Did the visit to the CF unit at the hospital, meeting with the patient and the staff contributed to your learning of genetics in addition to the SI module?

Please write any feedback or comment about the SI module and your own work

----

References

Aikenhead, G. S. (1994). What is STS science teaching? In J. Solomon & G. Aikenhead (Eds.), *STS education: International perspectives in reform* (pp. 47–59). New York: Teachers College Press.

Aikenhead, G. (2005). *Science education for everyday life: Evidence based practice*. New York: Teachers' College Press.

Ash, D. (2002). Negotiations of thematic conversations about biology. In G. Leinhardt, K. Crowley, & K. Knutson (Eds.), *Learning conversations in museums* (pp. 357–400). Mahwah, NJ: Erlbaum.

Ash, D. (2004). Reflective scientific sense-making dialogue in two languages: The science in the dialogue and the dialogue in the science. *Science Education, 88*, 855–884.

Ash, D., & Wells, G. (2006). Dialogic inquiry in classrooms and museums. In Z. Bekerman, N. C. Burbles, & D. Silberman-Keller (Eds.), *Learning in places: The informal education reader* (pp. 35–54). New York: Peter Lang.

Bamberger, Y., & Tal, T. (2008). Multiple outcomes of class visits to natural history museums: The students' view. *Journal of Science Education and Technology, 17*, 264–274.

Benemann, K. S. (2005). *Promoting students to make connections between inheritance and probability principles within a WISE learning environment*. Unpublished thesis, University of California, Berkeley.

Ben-Zvi Assaraf, O., & Orion, N. (2005). Development of system thinking skills in the context of Earth system education. *Journal of Research in Science Teaching, 42*, 518–560.

Bingle, W. H., & Gaskell, P. J. (1994). Scientific literacy for decision making and social construction of scientific knowledge. *Science Education, 78*, 185–201.

Blumenfeld, P. C., Marx, R. W., Patrick, H., & Krajcik, J. S. (1997). Teaching for understanding. In B. J. Biddle, T. L. Good, & I. F. Goodson (Eds.), *International handbook of teachers and teaching* (pp. 819–878). Dordrecht: Kluwer.

Calabrese Barton, A. (2003). *Teaching science for social justice*. New York: Teachers' College Press.

DeWitt, J., & Storksdieck, M. (2008). A short review of school field trips: Key findings from the past and implications for the future. *Visitor Studies, 11*, 181–197.

Dillon, J., Rickinson, M., Teamey, K., Morris, M., Choi, M.-Y., Sanders, D., et al. (2006). The value of outdoor learning: Evidence from research in the UK and elsewhere. *School Science Review, 87*, 107–111.

Dori, Y. J., & Herscovitz, O. (1999). Question posing capability as an alternative evaluation method: Analysis of an environmental case study. *Journal of Research in Science Teaching, 36*, 411–430.

Dori, Y. J., & Tal, T. (2000). Industry-environment projects: Formal and informal science activities in a community school. *Science Education, 84*, 95–113.

Dori, Y. J., Tal, T., & Tsaushu, M. (2003). Learning and assessing biotechnology topics through case studies with built-in dilemmas. *Science Education, 87*, 767–793.

Driver, R., Newton, P., & Osborne, J. (2000). Establishing the norms of scientific argumentation in classrooms. *Science Education, 84*, 287–312.

Duncan, K. M. (2007). *Factors affecting student learning of genetics from the revised simple inheritance WISE module*. Unpublished thesis, University of California, Berkeley.

Duschl, R. A., Schweingruber, H. A., & Shouse, A. W. (Eds.). (2007). *Taking science to school: Learning and teaching science in grades K-8*. Washington, DC: National Academies Press.

Falk, J. H., & Dierking, L. D. (2000). *Learning from museums: Visitor experiences and the making of meaning*. Walnut Creek, CA: AltaMira Press.

Hoadley, C. M., & Linn, M. C. (2000). Teaching science through on-line, peer discussions: Speak easy in the knowledge integration environment. *International Journal of Science Education, 22*, 839–857.

Hodson, D. (1994). Seeking directions for change: The personalization and politicisation of science education. *Curriculum Studies, 2*, 71–98.

Hodson, D. (1998). *Teaching and learning science: Towards a personalized approach*. Buckingham: Open University Press.
Hodson, D. (2002). Some thoughts on literacy: Motives, meanings and curriculum implications. *Asia-Pacific Forum on Science Learning and Teaching, 3*(1), 1–20.
Jiménez-Aleixandre, M. P., Rodríguez, A. B., & Duschl, R. A. (2000). Doing the lesson or doing science: Argument in high school genetics. *Science Education, 84*, 757–792.
Kali, Y. (2006). Collaborative knowledge-building using the Design Principles Database. *International Journal of Computer Support for Collaborative Learning, 1*, 187–201.
Kali, Y., Fortus, D., & Ronen-Fuhrmann, T. (2008). Synthesizing design knowledge. In Y. Kali, M. C. Linn, & J. E. Roseman (Eds.), *Designing coherent science education: Implications for curriculum, instruction, and policy* (pp. 185–200). New York: Teachers' College Press.
Kali, Y., & Linn, M. C. (2007). Technology-enhanced support strategies for inquiry learning. In M. Spector, M. D. Merrill, J. J. G. V. Merriënboer, & M. P. Driscoll (Eds.), *Handbook of research on educational communications and technology* (3rd ed., pp. 445–490). Mahwah, NJ: Erlbaum.
Kali, Y., Orion, N., & Eylon, B.-S. (2003). Effect of knowledge integration activities on students' perception of the Earth's crust as a cyclic system. *Journal of Research in Science Teaching, 40*, 545–565.
Krajcik, J. S., Slotta, J., McNeill, K. L., & Reiser, B. J. (2008). Designing learning environments to support students' integrated understanding. In Y. Kali, M. C. Linn, & J. E. Roseman (Eds.), *Designing coherent science education: Implications for curriculum, instruction, and policy* (pp. 39–64). New York: Teachers' College Press.
Linn, M. C., Davis, E. A., & Bell, P. (Eds.). (2004). *Internet environments for science education*. Mahwah, NJ: Lawrence Erlbaum Associates.
Linn, M. C., Lee, H.-S., Tinker, R., Husic, F., & Chiu, J. L. (2006). Teaching and assessing knowledge integration in science. *Science Education, 313*, 1049–1050.
Liu, O. L., Lee, H.-S., Hofstetter, C., & Linn, M. C. (2008). Assessing knowledge integration in science: Construct, measures, and evidence. *Educational Assessment, 13*, 33–55.
Margel, H., Eylon, B.-S., & Schetz, Z. (2004). We actually saw atoms with our own eyes. Conceptions and convictions in using the scanning tunneling microscope in junior high school. *Journal of Chemical Education, 81*, 558–566.
McNeill, K. L., & Krajcik, J. (2007). Middle School students use of appropriate and inappropriate evidence in writing scientific explanations. In M. C. Lovett & P. Shah (Eds.), *Thinking with data* (pp. 233–266). New York: Lawrence Erlbaum Associates.
Orion, N. (1993). A model for the development and implementation of field trips as an integral part of the science curriculum. *School Science and Mathematics, 93*, 325–331.
Orion, N., & Hofstein, A. (1994). Factors that influence learning during a scientific field trip in a natural environment. *Journal of Research in Science Teaching, 31*, 1097–1119.
Pea, R., & Collins, A. (2008). Learning how to do science education: Four waves of reform. In Y. Kali, M. C. Linn, & J. E. Roseman (Eds.), *Designing coherent science education: Implications for curriculum, instruction, and policy* (pp. 3–12). New York: Teachers' College Press.
Ratcliffe, M., & Grace, M. (2003). *Science education for citizenship: Teaching socio-scientific issues*. Maidenhead: Open University Press.
Rennie, L. J., & McClafferty, T. P. (1995). Using visits to interactive science and technology centers, museums, aquaria, and zoos to promote learning in science. *Journal of Science Teacher Education, 6*, 175–185.
Rennie, L. J., & McClafferty, T. P. (1996). Science centers and science learning. *Studies in Science Education, 27*, 53–98.
Roschelle, J., Pea, R., Hoadley, C., Gordin, D., & Means, B. (2000). Changing how and what children learn in school with collaborative cognitive technologies. *The Future of Children, 10*, 76–101.
Roseman, J. E., Linn, M. C., & Koppal, M. (2008). Characterizing curriculum coherence. In M. C. Linn, J. E. Roseman, & Y. Kali (Eds.), *Designing coherent science education: Implications for curriculum, instruction, and policy* (pp. 13–38). New York: Teachers' College Press.
Roth, W.-M., & Barton, A. C. (2004). *Rethinking scientific literacy*. New York: Routledge Falmer.

Sadler, T. D. (2004). Informal reasoning regarding socioscientific issues: A critical review of research. *Journal of Research in Science Teaching, 41*, 513–536.

Sadler, T. D., Barab, S. A., & Scott, B. (2007). What do students gain by engaging in socioscientific inquiry? *Research in Science Education, 37*, 371–391.

Sadler, T. D., & Zeidler, D. L. (2004). The morality of socioscientific issues: Construal and resolution of genetic engineering dilemmas. *Science Education, 88*, 4–27.

Sadler, T. D., & Zeidler, D. L. (2005). Patterns of informal reasoning in the context of socioscientific decision making. *Journal of Research in Science Teaching, 42*, 112–138.

Schauble, L., Gleason, M., Lehrer, R., Bartlett, K., Petrosino, A., Allen, A., et al. (2002). Supporting science learning in museums. In G. Leinhardt, K. Crowley, & K. Knutson (Eds.), *Learning conversations in museums* (pp. 425–452). Mahwah, NJ: Erlbaum.

Singer, J., Marx, R. W., Krajcik, J., & Clay Chambers, J. (2000). Constructing extended inquiry projects: Curriculum materials for science education reform. *Educational Psychologist, 35*, 165–178.

Slotta, J. D., & Linn, M. C. (2009). *WISE science*. New York: Teachers' College Press.

Solomon, J., & Thomas, J. (1999). Science education for the public understanding of science. *Studies in Science Education, 33*, 61–89.

Tal, T. (2004). Using a field trip as a guide for conceptual understanding in environmental education: A case study of a pre-service teacher's research. *Chemical Education Research and Practice, 5*, 127–142.

Tal, T. (2008). Learning about agriculture within the framework of education for sustainability. *Environmental Education Research, 14*, 273–290.

Tal, T., & Alkaher, I. (2008). *Environmental projects of Jewish and Arab youth in Israel – The adult leaders' views*. Paper presented at the American Educational Research Association, Washington D.C., DC.

Tal, T., & Alkaher, I. (2010). Collaborative environmental projects in a multicultural society: Working from within separate or mutual landscapes? *Cultural Studies of Science Education, 5*, 325–349.

Tal, T., & Hochberg, N. (2003). Reasoning, problem-solving and reflections: Participating in WISE project in Israel. *Science Education International, 14*, 3–19.

Tal, T., & Kedmi, Y. (2006). Teaching socioscientific issues: Classroom culture and students' performances. *Cultural Studies of Science Education, 1*(4), 615–644.

Tal, T., & Morag, O. (2009). Action research as a means for preparing to teach outdoors in an ecological garden. *Journal of Science Teacher Education, 20*, 245–262.

Tate, E. D. (2008). *The impact of an Asthma curriculum on students' integrated understanding of biology*. Paper presented at the American Educational Research Association, Washington D.C., DC.

Tate, E. D., Clark, D., Gallagher, J., & McLaughlin, D. (2008). Designing science instruction for diverse learners. In Y. Kali, M. C. Linn, & J. E. Roseman (Eds.), *Designing coherent science education: Implications for curriculum, instruction, and policy* (pp. 65–93). New York: Teachers' College Press.

Varma, K., Husic, F., & Linn, M. C. (2008). Targeted support for using technology-enhanced science inquiry modules. *Journal of Science Education and Technology, 17*, 341–356.

Zeidler, D. L., & Sadler, T. D. (2008). The role of moral reasoning in argumentation: Conscience, character, and care. In S. Erduran & M.-P. Jiménez-Aleixandre (Eds.), *Argumentation in science education: Recent developments and future directions* (pp. 201–216). New York: Springer.

Zeidler, D. L., Sadler, T. D., Simmons, M. L., & Howes, E. V. (2005). Beyond STS: A research-based framework for socioscientific issues education. *Science Education, 89*, 357–377.

Zohar, A. (2004). *Higher order thinking in science classrooms: Students' learning and teachers' professional development*. Dordrecht: Kluwer.

Zohar, A., & Nemet, F. (2002). Fostering students' knowledge and argumentation skills through dilemmas in human genetics. *Journal of Research in Science Teaching, 39*, 35–62.

Chapter 3
Metalogue: Using Issues and Participatory Experiences to Enhance Student Learning and Interest

Siu Ling Wong, Tali Tal, and Troy D. Sadler

Comparing Issue-Based Movements

Sadler: To begin this chapter, the authors discuss the socio-scientific issues (SSI) movement in terms of related approaches to the contextualization of science education in issues that matter to students, teachers, and the broader population. They highlight relationships between SSI and the Science-Technology-Society (STS) movement and discuss different ways in which the movements overlap and share consistency: "The essence of all these ideas [SSI, STS, and other approaches that promote progressive visions of scientific literacy] is that the science content should be situated in real, important and often—controversial issues that gain the public's interest" (p. 1). Later in the introduction, the authors extend the links between STS and SSI to include Education for Sustainability and Environmental Education. Historically, many of the issues addressed in Environmental Education (EE) are also issues featured within STS and SSI approaches. For instance, issues related to water pollution and quality fit easily into curricula labeled as EE or SSI. However, I see the purpose of a SSI-oriented curriculum and an EE-oriented curriculum as being significantly different. An SSI approach supports the development of individual learners and emergent communities of learners in terms of decision-making, participation in democratic processes, and reasoning. The focus is on student development and not on the promotion of a particular point of view or orientation. An EE approach may support similar processes but does so toward a desired result, that is, proenvironment attitudes and behaviors.

S.L. Wong (✉)
Faculty of Education, University of Hong Kong, Pokfulam Road, Hong Kong
e-mail: aslwong@hku.hk

T. Tal
Department of Education in Technology and Science, Technion–Israel Institute of Technology, Haifa, Israel

T.D. Sadler
School of Teaching and Learning, University of Florida, Gainesville, FL, USA

T.D. Sadler (ed.), *Socio-scientific Issues in the Classroom: Teaching, Learning and Research*, Contemporary Trends and Issues in Science Education 39,
DOI 10.1007/978-94-007-1159-4_3,

Writing in the recent *Handbook of Research on Science Education*, Hart (2007) quotes from international conferences, which helped to establish EE, in defining the goals of environmental education:

1. To foster clear awareness of, and concern about, economic, social, political, and ecological inter-dependence in urban and rural areas.
2. To provide every person with opportunities to acquire the knowledge, values, attitudes, commitment, and skills needed to protect and improve the environment.
3. To create new patterns of behavior of individuals, groups, and society as a whole, towards the environment (UNESCO, 1977).

In my view, a legitimate outcome of a successful SSI-based learning opportunity may be development of ideas that are not consistent with proenvironmental attitudes and behaviors. For instance, in Chap. 5 of this volume, my coauthors and I present a SSI-based unit on global climate change. The measure of success of this unit was related to how students understood the issue of climate change, were able to think about the scientific data and ideas featured within the climate change debate, and considered the social dimensions of the issue including politics and economics. It was perfectly reasonable for students to participate in the unit and emerge with ideas not supportive of reductions in greenhouse gas emissions. An EE climate change curriculum framework would very likely have a goal of student development of attitudes and behaviors to combat climate change.

Whereas EE has a long history, Education for Sustainability (EfS) is a much more recent development and represents a seemingly rapidly growing movement within education. Based on my limited experience with EfS, it strikes me that it shares many commonalities with EE including a set of normative prescriptions for learner attitudes and behaviors, which seems to be somewhat at odds with a SSI approach. I wonder if others agree with these distinctions or conceptualize the relationships among these different movements in different ways.

Tal: I totally agree that SSI and EE are not identical, but they are different not only because of the end goals that Sadler highlighted. What we meant in the introduction to Chap. 2 was that there are several important commonalities between SSI, STS, and EE in terms of critical thinking, using real life issues and controversies as teaching contexts, promoting citizen's involvement, and preparing students to take part in democratic decision-making. Sadler reminds us the traditional goals of EE that were formulated in major international conferences in the 1970s. However, the EE literature, especially from the 1990s and onward, extensively discusses the tension between the former behaviorist goals of early versions of EE and more affective goals that are related to personal transformation of the learner. The EE literature gives voice to various streams that highlight different emphases of EE. Lucie Sauve (2005), for example, maps 15 currents in EE. One of these currents focuses on cognitive goals such as developing systems thinking, problem solving, and decision-making capabilities. Another focuses on action competence and learning by doing and another is related to developing a sense of appreciation for nature, art, and humanistic thinking. Thus, the main difference between EE as whole and SSI is that EE has many more noncognitive goals. While looking at the more cognitive-oriented currents, I find many similarities between SSI, STS, and EE. Given these overlaps, scholars in the 1990s proposed integrating EE and STS suggesting the acronyms science, technology, environment, society (STES) (Zoller, 1991) and Science-Technology-Society-Environment (STSE) (Hodson, 2003). These scholars pointed to some of the similarities between scientific literacy, environmental literacy, and science education for future citizens:

> [t]he broadening conception of STS to include environmental education (STS becomes STSE), extending the definition of scientific literacy to encompass a measure of political literacy, prioritizing the affective, and making much greater use of informal and community-based learning opportunities (Hodson, 2003, p. 648).

Therefore, the point we wanted to make in our introduction was that teaching socio-scientific issues is congruent with many ideas in EE as well.

Enhancing Authenticity and Interest

Wong: I am particularly interested in the favorable impact of the field trip on students' attitudes toward the module as compared to the online interactions and how educators might enhance online experiences such that they become more effective. Maximizing effects of online experiences may help address issues related to the logistics of organizing field trips to special centers or organizations, organizational challenges, and limited class time. These issues are particularly serious in the East where average class size is 40 or more. The authors' suggestion to enrich the online environment with diversified interaction and enhance authenticity by including clips of practitioners or practices in the field is an excellent starting place. In my chapter (Chap. 14 in this volume), we discuss the SARS case as a means of teaching about NOS and critical analysis of SSI. I found incorporating of video clips of experts, including medical doctors and scientists, talking about their experiences during the SARS outbreak did captivate learners' attention and enhance the authenticity of the SSI under discussion. I also found that resources involving people that the students knew (e.g. interviews that I conducted) generated more interest.

As I reflected on the chapter and my own experiences, I thought of a new model for integrating field trips and student projects that I would like to share here. Below I offer a brief outline of the model and invite feedback:

(a) Draw up a list of science topics in the curriculum and relevant field trip sites for the topics (e.g. topic: simple inheritance; site: CF unit at the children hospital).
(b) Each group of students studies the science topic and develops plans for teaching the key ideas to their fellow students. Groups also plan what should be obtained from the field trip site to enrich their teaching.
(c) The group that works on a particular topic visits the site and collects information and resources (e.g. pamphlets, videos of the site, and interviews of key persons).
(d) Student groups teach their topics to fellow students based on their field trip experiences.

This approach may generate more enthusiasm and interest for students who participate in field settings themselves as well as the other students who get to experience the visits through their peers.

Tal: Wong's idea of having students plan and carry out field trips is very interesting; although, I assume that only a small number of schools would take such an initiative due to logistical challenges. There is a growing body of literature on out-of-school learning that identifies the many challenges teachers face while going on field trips (Dillon et al., 2006; Tal, 2008) despite the accumulated knowledge on their contribution in the cognitive, affective, and social aspects.

A couple of years ago I had the opportunity to learn a simple strategy from Dr. David Zandvliet from Simon Fraser University in British Columbia. In an environmental education course for preservice teachers, David sends the students, in small groups, to investigate neighborhoods in Vancouver. Some of these neighborhoods are affluent while others are of extreme poverty and crime. While accompanying one group that wandered through homeless shelters and streets of crime, I listened to the prospective teachers' ideas on how to teach future students about their experience. After returning to Israel, I adopted David's idea and since then, I send my EE students to explore the neighborhoods of Haifa. Having used this strategy with three classes, I am convinced that having students share and reflect on these experiences has a great value in enhancing interest and learning. I therefore agree that sending out students to explore relevant out-of-school learning environments is a good idea that hopefully can be achieved by certain schools.

Wong: Although you did not find significant difference in students' knowledge acquisition as measured by the test and the module tasks, the qualitative data provided in Table 2.2 convincingly reveal that the field trip group had deeper learning and understanding of the disease (cystic fibrosis) than the general understanding of genetic disease as tested by the test and tasks. I suspect that such deeper understanding might also have indirectly resulted in more justified responses related to affective reasons for their tendency to participate in fundraising efforts for CF as revealed in Table 2.3. This inference is based on the following observations:

(a) Several students from the online interaction group (19%) suggested that David did not help them in understanding genetics but there were no similar responses from the field trip group.
(b) The frequency of positive responses from the field trip group (96%) regarding their learning was considerably higher than that of the online interaction group (63%). Such a difference in response rate supported the qualitative data which suggested that the hospital presentation was very helpful for their learning. Also, the more students knew about the disease, the more they wanted to learn about it.
(c) If we focus on the online interaction group and the field trip group separately in Table 2.3, the picture of the online interaction group was disappointing. The number of groups with unjustified generic response increased from two to three. The number of groups with justified responses supported by sense of responsibility, expression of feelings or acquired knowledge fell from six to two.

Such changes could hardly be compensated by having just one group achieve a response characterized by the highest level of complexity.

(d) A possible reason that students in the online version did not find the interaction with a patient helpful in enhancing their interest as much as the field trip group might be related to the fact that online interaction is not novel for most students. Most students engage in various online interactions (e.g. Facebook, Twitter, Skype, etc.) on very frequent bases, so they may not have perceived as much interest in the online interaction because it is something they regularly do.

It might also be revealing to conduct in-depth interview with students from the groups with decreased complexity in their justifications regarding fundraising activities. This may reveal factors associated with the online interaction that might have negatively impacted the level of justifications for or against the fundraising activity. Avoiding negative factors is arguably as important as introducing positive components in achieving the intended learning outcomes.

Tal: I agree that there is enough data to support a claim that the hospital visit had greater contributions not only in the affective domain but in the cognitive as well. Wong's idea that students who use virtual interactions of various sorts in their daily lives will benefit less from online interaction makes much sense. Taking into account the reduced number of field trips, we need to distinguish between the online interactions we intend to develop from the students' already known interactions. It is true that we did not invest in the design of the online interaction as we did with the module itself. We believed that talking with a real patient would promote learning by itself. As Wong noticed, in the more straightforward test items we did not find a difference between the groups, but in a more refined analysis, there is enough evidence to support the claim that the field trip created better conditions for meaningful learning.

References

Dillon, J., Rickinson, M., Teamey, K., Morris, M., Choi, M.-Y., Sanders, D., et al. (2006). The value of outdoor learning: Evidence from research in the UK and elsewhere. *School Science Review, 87*, 107–111.

Hart, P. (2007). Environmental education. In S. K. Abell & N. G. Lederman (Eds.), *Handbook of research on science education*. Mahwah, NJ: Lawrence Erlbaum Associates.

Hodson, D. (2003). Time for action: Science education for an alternative future. *International Journal of Science Education, 25*(6), 645–670.

Sauvé, L. (2005). Currents in environmental education: Mapping a complex and evolving pedagogical field. *Canadian Journal of Environmental Education, 10*, 11–37.

Tal, T. (2008). Learning about agriculture within the framework of education for sustainability. *Environmental Education Research, 14*, 273–290.

UNESCO. (1977). *First intergovernmental conference on environmental education: Final report.* Paris: UNESCO.

Zoller, U. (1991). Problem solving and problem solving paradox in decision-making oriented environmental education. In S. Keiny & U. Zoller (Eds.), *Conceptual issues in environmental education* (pp. 71–87). New York: Peter Lang.

Chapter 4
Learning Science Content and Socio-scientific Reasoning Through Classroom Explorations of Global Climate Change

Troy D. Sadler, Michelle L. Klosterman, and Mustafa S. Topcu

The goals for our research related to socio-scientific issues (SSI) have always been related to the promotion of scientific literacy (see Chap. 1) and the improvement of science learning experiences. However, the work has not always been centrally situated in classroom environments. For much of our early research, we explored students' moral perspectives (Sadler & Zeidler, 2004), reasoning (Sadler & Zeidler, 2005), understandings of science (Sadler & Fowler, 2006), and argumentation (Sadler & Donnelly, 2006) related to SSI in contexts not necessarily connected to students' experiences in science classrooms or other learning environments. We were interested in building an empirical understanding of how science learners made sense of complicated socio-scientific dilemmas, how they made decisions about these issues, and what factors influenced their thinking practices. We engaged students in reasoning and argumentation collecting data through interviews and instruments, but did not explore classroom practices or the possible effects of intervening in learning environments. In an attempt to advance the SSI research agenda and create stronger connections among theory, research, and practice we began working on projects situated in science classrooms.

To make this shift, we started to work on a large classroom-based research project focused on the implementation and study of a technology-based educational innovation. The centerpiece of the project was a multi-user virtual environment (MUVE) designed with contemporary gaming principles. The innovative teaching and learning technology, known as *Quest Atlantis*, provided opportunities for middle school learners to engage in critical thinking and explore content in several different academic areas (Barab et al., 2007). Within the MUVE, players could immerse themselves in different "worlds" each of which had unique curricular goals. One of these

T.D. Sadler (✉)
School of Teaching and Learning, University of Florida, Gainesville, FL, USA
e-mail: tsadler@coe.ufl.edu

M.L. Klosterman
Department of Education, Wake Forest University, Winston-Salem, NC, USA

M.S. Topcu
Department of Elementary Science Education, Yuzuncu Yil University, Van, Turkey

T.D. Sadler (ed.), *Socio-scientific Issues in the Classroom: Teaching, Learning and Research*, Contemporary Trends and Issues in Science Education 39,
DOI 10.1007/978-94-007-1159-4_4, © Springer Science+Business Media B.V. 2011

worlds placed students in a virtual park. The park rangers, charged with managing the park and balancing the health of the natural environment as well as the needs of several different groups of park patrons, enlisted the student players to investigate and propose courses of action to address the deteriorating health of a river flowing through the park. Various groups of park patrons relied on the river and its fish populations but these groups used the resources in different ways, impacted the health of the river in different ways, and had very different perspectives on preserving the resource. In short, this learning environment placed students in the midst of an unfolding SSI.

Our research efforts associated with *Quest Atlantis* involved case studies of classroom implementation including analyses of science content learning (Barab, Sadler, Heiselt, Hickey, & Zuiker, 2007). We also explored ways in which students conceptualized novel SSI and the extent to which learners transferred understandings and practices developed in the context of one socio-scientific issue to other issues (Sadler, Barab, & Scott, 2007). As a means of framing transfer relative to SSI, we proposed a new construct, *socio-scientific reasoning*. Socio-scientific reasoning (SSR) was designed to capture the practices in which citizens can be expected to engage across multiple SSI. That is, socio-scientific reasoning was developed as a means of understanding student practices relative to the invariant features of SSI. (We will take up this topic in greater detail in a later section of the chapter.)

The *Quest Atlantis* findings certainly informed use of SSI by teachers and learners, but the research generated as many questions as it did answers. One important question for us related to the technology platform used to frame the curriculum. *Quest Atlantis* was a powerful teaching and learning environment, but the vast majority of secondary science classes were not well positioned in terms of using this kind of innovation. Therefore, we became interested in studying the implementation of SSI-based learning experiences in the context of classroom environments and resources that were more typical of today's schools.

Based on these experiences and the questions that they generated for us, we developed a new classroom-based study of the implementation of SSI-based curriculum and instruction. We initiated this new study, which serves as the focus of the current chapter, by first developing partnerships with two local high school science teachers. Based on assessments of the teachers' needs and interests as well as our goals for the overall project (to be discussed in the next section), we collaboratively decided that global climate change would serve as the SSI focal point for the project. As a part of the project, we created new curriculum and assessment tools for supporting teaching and learning in the context of SSI. The project became known as the Curriculum and Assessment Tools for Socio-scientific Inquiry (CATSI) project.

Project Goals

We had four goals for the CATSI project: (1) Design and implement a SSI-based curriculum in partnership with local teachers to meet the specific needs of these teachers and their students. (2) Develop an understanding of the implementation of

SSI-based instruction from an up-close perspective (i.e., from within the classroom as the learning experience unfolded). (3) Investigate how SSI-based instruction supports (or fails to support) student development of scientific content knowledge. (4) Extend previous work on socio-scientific reasoning through the exploration of how students may improve their SSR and experimentation with new SSR assessment strategies.

For the CATSI project, we were interested in creating materials and tools designed to meet the specific needs of our teacher partners. One of the consistent limitations cited by teachers interested in enacting educational innovations, particularly SSI-based instruction, was a lack of useful materials (Cross & Price, 1996; Lumpe, Haney, & Czerniak, 1998; Sadler, Amirshokoohi, Kazempour, & Allspaw, 2006). Teachers often concluded that textbooks and other curricular materials did not meet the needs and/or interests of their students. Teachers also expressed skepticism regarding the extent to which curricula and specific learning activities embedded within these curricula could be enacted in their teaching contexts whether it be because of a lack of appropriate facilities, materials, funds or other context-dependent issues (Bryce & Gray, 2004). For these reasons, we chose to initiate teacher collaborations at the outset of the project in order to involve them in the design process with the goal of creating materials geared specifically for their use.

Our goals included not only the design of materials with collaborating teachers but also the study of classroom enactment of these materials. We took advantage of the working relationships that we had developed with the partnering teachers in order to gain an intimate vantage point from which to observe the implementation of SSI-based instruction. We sought to build understandings of how instructional materials built around socio-scientific themes would be used in classrooms with an eye toward distinctions between the intended curriculum, as shaped by our perspectives as designers and researchers, and the enacted curriculum as it unfolded in actual classrooms. By attending to these issues, we wanted to better understand how local contexts influenced the use of and results associated with SSI-based instruction.

Our third goal for the project related to the use of SSI-based instruction as a means of developing student understandings of science content. Researchers and educators have frequently cited the potential for using SSI as a means of providing meaningful contexts for learning science content knowledge (Cajas, 1999; Kolstø, 2001; Zeidler, Walker, Ackett, & Simmons, 2002). These authors (and others) have argued that SSI offered situations that connect science to the lived experience of learners thereby providing an impetus to understand the underlying science. Findings from studies of SSI have provided some initial support for the proposed link between SSI and the learning of science content (Barber, 2001; Zohar & Nemet, 2002). These studies documented content learning through comparison of postintervention assessments from students participating in SSI intervention classes and comparison classes. A possible critique of these studies was the lack of preintervention data that could have established the equivalency of groups and the change within groups. Other studies employed pre/posttest designs to document learning gains in the context of SSI (Dori, Tal, & Tsaushu, 2003; Yager, Lim, & Yager, 2006). These studies provided evidence of learning, but it was important to note that the assessments used directly aligned with the SSI intervention. In the

CATSI project, we aimed to examine content knowledge gains related to the SSI but also to the scientific generalizations that can be abstracted beyond the specific context of a particular SSI. In order to achieve this end, we adopted a multilevel assessment framework in which we created assessments at variable "distances" from the curriculum (Ruiz-Primo, Shavelson, Hamilton, & Klein, 2002; this approach will be discussed in greater detail in a later section). Using these tools we sought to develop a more nuanced understanding of how SSI-based instruction supported student development of science content knowledge.

The final project goal was to further explore socio-scientific reasoning as a measurable construct for use in teaching and research contexts. As referenced above, we introduced SSR in our work on the *Quest Atlantis* project as a tool for conceptualizing student practices related to the negotiation of SSI (Sadler et al., 2007). We proposed four aspects of SSR that captured practices (some but not all) that are necessary for thoughtful negotiation and resolution of complex SSI. These SSR aspects included (1) recognizing the inherent *complexity* and multifaceted nature of SSI, (2) analyzing issues from multiple *perspectives*, (3) appreciating the need for ongoing *inquiry* relative to SSI, and (4) employing *skepticism* in the review of information presented by parties with vested interests. In the *Quest Atlantis* project, we explored SSR by engaging students in interviews during which they considered and discussed complex socio-scientific scenarios. In this initial work we demonstrated that some of the SSR aspects could be measured (i.e., inquiry and complexity) reliably and provided preliminary evidence that these aspects related to an underlying latent variable. In the CATSI project, we wanted to explore other SSR assessment options (noninterview based) with the aim of improving the reliability of measurement and to determine the extent to which students' SSR practices could be improved in response to SSI-based instruction.

Setting

Before discussing issues associated with design, teaching, and research, we have chosen to discuss the settings in which the CATSI project was conducted. This decision was deliberate because the setting was a key determinant for decisions made in the design, teaching, and research. We felt it important that our work as designers and researchers fit well with the needs and interests of the classroom communities with which we worked. We prioritized classroom context in the design and implementation of the CATSI project and wanted this prioritization to be reflected here, in this description of the project.

We conducted the CATSI project in the science classrooms of two local schools. Although both schools were located fairly close to our university in the southeastern United States, they were parts of distinct communities. The first school, to which we referred to as Fields High School, served high school learners (grades 9–12) in a small, rural town. The second school, to which we referred to as Creek Academy, was a developmental research school affiliated with our university. Creek

Academy served students in kindergarten through senior high school (through grade 12) and was located in a moderate-sized city.

At Fields HS, we worked with Molly (pseudonym), a fourth year teacher who had spent her entire young career at Fields. During the year of the CATSI project, Molly taught environmental science and anatomy and physiology. Based on Molly's recommendation and the needs of her students, we developed the CATSI project for implementation in the environmental science courses (three sections). The daily schedule at Fields HS followed a traditional format in that classes met daily in 50-min blocks. Molly worked in a spacious classroom with separate areas for seated deskwork or whole-class instruction and laboratory investigations. The room was decorated brightly and displayed many examples of student work. Environmental science was a course taken primarily by 11th and 12th grade students looking to satisfy a science graduation requirement. Most of these students had taken an integrated science course in 9th grade and a general biology course in 10th grade. Environmental science students tended not to take advanced, college-track science courses including physics and chemistry. As a generalization, most of these students were considered to have average to remedial academic histories. The environmental science students tended not to exhibit enthusiasm for studying science or future careers in science and most did not plan to pursue postsecondary degrees.

At Creek Academy, we worked with William (pseudonym), a fifth year chemistry teacher. Creek Academy employed a block schedule in which each class met on Mondays for 50-min and in 100-min sessions on two other days each week. William taught multiple sections of honors chemistry and "regular" chemistry. Given curriculum constraints associated with the honors classes, we chose to implement CATSI with the regular chemistry classes (two). William's classroom was also spacious providing enough room for individual desks and a separate laboratory area. Simply stated, this classroom had the look and feel of a chemistry lab. A visitor would find various pieces of laboratory equipment, a fume hood with solutions waiting for mixture or display, and an adjoining, well-stocked storeroom. The most dominant classroom "décor" were the three large periodic tables situated so that they could be seen from any position within the classroom. According to state assessment tests, Creek Academy was a higher performing school than Field HS and a significantly greater percentage of Creek graduates matriculated to universities or community college. The highest achieving students and those who had shown strongest aptitude for an interest in school science took honors chemistry. Students who chose not to take the honors track took the regular chemistry course that William offered during their 11th grade year. These chemistry classes were best described as mixed-ability. The school's lower achieving students took the course along with high achieving students who were not particularly interested in science.

The settings in which we chose to work shared common features in that they both served students that were not the highest achieving science students in their schools. Early-career teachers interested in collaborating with university-based researchers led both classrooms, and the schools were geographically close. However, the classrooms were quite distinct in terms of subject area (environmental science and chemistry), schedule (traditional and modified block), and the communities served (rural town and university-based city). More information pertaining to how CATSI was developed for use in each of these settings will be shared in the following section.

Teacher–Researcher Relationships

We approached Molly and William with proposals to collaborate on a SSI-related project over a year prior to the initiation of CATSI. The initial conversations related to a large-scale professional development project dependent on grant funding, and the funds were ultimately not approved. However, these conversations revealed high levels of interest among all parties to collaborate on a project related to SSI-based instruction. The following year, we submitted and were awarded a small grant to partner with a local teacher to develop curriculum and assessment tools for socio-scientific inquiry, and the CATSI project was initiated. Because of the congruence between Molly's teaching focus and the principal investigator's area of expertise (i.e., biology education), we asked Molly to collaborate with us. At the time, Molly was completing an academic year teaching biology and anatomy and physiology with the expectation that her teaching assignment would remain the same in the following year. We made plans to develop the CATSI project around a unit on gene therapy to be featured in Molly's biology class. Over the summer, the project team began developing curriculum materials and creating assessments specific to the gene therapy focus. A few weeks prior to the start of the new school year, administrators at Fields HS informed Molly that she would be teaching environmental science classes rather than biology. Suddenly the focus on gene therapy was not as relevant to Molly and her students. Molly was willing to work with us such that our initial development efforts could be included in her remaining anatomy and physiology classes, but while she would have made this work for us, it was not what she really wanted for her classes. After reflecting on her plans and student needs, we collaboratively decided to shift the focus of our development efforts from gene therapy to global climate change. The new global climate change unit would fit naturally within Molly's environmental science classes and would address science content that she felt was essential for her students.

In order for students to understand the issue of global climate change, it was important that they understand several science concepts including the particulate nature of gases, combustion, atmospheric composition, and energy transformation. Because of the heavy focus on chemistry content, we decided that it was necessary to bring in a chemistry educator as a part of the project team. In addition to maintaining his classroom duties, William had begun work toward a graduate degree in science education at our university, so he became a natural addition to the team. Following an initial organizational meeting, William decided that he wanted to participate as a full-scale classroom collaborator and not just as an advisor to the project. At this point the project team consisted of a principal investigator who was a former biology teacher and had conducted a fair amount of research related to SSI, two full-time science education graduate students with a wealth of experience teaching science in secondary school and college settings, and two high school science teachers. Another graduate student joined the project team after the curricula had been developed and implemented. He assisted in the analysis of data.

The project team met regularly throughout the fall semester on the design of curriculum. Initially, the focus of meetings related to understanding the classroom contexts in which this work would be situated as well as helping the teachers understand our perspectives on SSI and goals for the research. As a part of this, we developed ideas about what our teacher partners were looking for in the collaboration and what could likely be accomplished in their classrooms. We also examined available curriculum materials related to global climate change. Ultimately, we created an instructional sequence that spanned approximately 15 h of class-time. During this timeframe, we also created assessment materials (to be described later). Whereas the teacher partners were intimately involved in the development of the curriculum materials, they played a less prominent role in the development of assessment instruments, which were specifically designed for research purposes as opposed to classroom assessment purposes.

We implemented the CATSI unit on global climate change during the middle of the following spring semester. This timing was determined by the classroom schedules of Molly and William and the curricular flow of their classes. Implementation occurred in Molly's environmental science classes first. At least one member of the research team (the PI and/or one of the graduate students) was present in Molly's classroom over the 3-week period in which the global climate change unit was implemented. Molly maintained her role as the classroom leader, but the other research team member periodically assisted with instruction particularly when students completed lab experiences or computer-based exercises. Otherwise, the research team member observed and videotaped classes without drastically altering the classroom environment. The research team collaborated with William during his implementation of the curriculum, but we were not able to be present on a daily basis.

Intervention

We designed the CATSI intervention to meet the following instructional objectives.

As a result of participation in the CATSI unit, we expected students to

1. Develop an understanding of what global climate change is and why various parties think that this is a significant issue. (Unlike some curricular efforts that originate from the environmental education community, we did not promote advocacy on any side of the global climate change debate. We did not assume a goal of pushing students to adopt a particular position on global climate change but we did intend for students to understand why different groups adopted strong positions on the issue).

2. Develop understandings of scientific principles and concepts related to global climate change. The scientific content that we focused on related primarily to chemistry and earth science including the particulate nature of gases, climate and temperature, atmospheric composition, and combustion as a chemical reaction.
3. Engage in scientific practices including creating and interpreting scientific models, conducting inquiry-based investigations, and graphing.
4. Develop understandings related to why global climate change is controversial and appreciate social factors that contribute to this controversy including economic, political, and ethical concerns.
5. Develop skills for finding and analyzing web-based resources related to SSI.
6. Formulate a personal position on global climate change that is informed by scientific principles and concepts as well as the students' own perspectives on social factors including economics, politics, and ethics.
7. Improve their socio-scientific reasoning practices in contexts beyond the scope of global climate change.

In order to achieve these objectives, we designed a 3-week unit (approximately 15 classroom contact hours) of instruction. As discussed in the section above, we worked with Molly and William to design materials that would meet their specific needs and the needs of their students. We examined published materials and found exercises within the American Chemical Society's *Chemistry in the Community* (ACS, 2006), *Chem Connections* (Anthony, Brauch, & Longley, 2007), and *Climate Change* published by the Lawrence Hall of Science (Sneider, Golden, & Gaylen, 2004) to be particularly useful. Ideas gleaned from these sources were modified and customized such that they fit our goals and classroom contexts. We also drew on the expertise of the project team to create new materials to ensure that the full range of objectives had been met.

The resulting unit was made up of nine unique, although interconnected, lessons (see Table 4.1). The lessons ranged from approximately 1 h in duration to 4 h. The actual implementation times varied between the two classrooms. The first lesson introduced students to global climate change by demonstrating the presentation of climate change in mass media. Students also explored the personal stories of several individuals with vastly different perspectives on the issue including a boy from a coal-mining region of the USA, an Inuit girl from Alaska, a US bureaucrat, a college student in the Maldives island chain, and several others. The second lesson engaged students in a jigsaw activity in which they explored the positions of various political interest groups and ultimately crafted recommendations for US policy related to global climate change. In the course of the first two lessons, we suggested that the scientific community shares general agreement (not without exceptions) that the earth's climate is changing and that human activity affects these changes. The question of whether societies should do anything in response to these changes and if so, what should be done were presented as open questions for students to explore throughout the unit.

The next four lessons were designed to help students build understandings of the science underlying global climate change. In the first of these lessons, student

Table 4.1 Overview of the CATSI curriculum

Sequence	Brief description	Approximate time (h)[a]	Target objective(s)[b]	Source[c]
1	Introduction to global climate change (GCC): Teacher presents recent media reports on GCC and provides basic overview. Student groups explore a resource that describes why various individuals are concerned by GCC or plans to address GCC.	1–2	1, 7	CATSI[d]
2	Complexifying GCC: Students worked through a jigsaw activity in which they analyzed policy recommendations of five special interest groups with distinct perspectives on GCC.	1–2	1, 4, 7	CATSI (Sadler & Klosterman, 2009)
3	Gases and the atmosphere: Student groups moved through four lab stations at which they carried out activities designed to demonstrate the particulate nature of gases. Teacher made presentation providing an overview of earth's atmosphere.	1–2	2	Sadler et al. (2005)
4	Combustion: Brief presentation on the carbon cycle and production of CO_2 through combustion. Student groups used molecular models and chemical equations to model combustion of various hydrocarbons.	1–2	2, 3	Anthony et al. (2007), ACS (2006)
5	CO_2 concentration: Student groups collected gas samples (including car exhaust) and performed titrations to determine CO_2 concentrations. Teacher prompted groups to think about CO_2 and the greenhouse effect.	2–3	2, 3	Sneider et al. (2004)
6	Greenhouse effect: Student groups created model greenhouses and measured temperature changes over time.	1–2	2, 3	Anthony et al. (2007), Sneider et al. (2004)
7	Consequences of GCC: Classroom discussion of previous activities. Students completed a worksheet that prompted them to consider possible consequences of GCC as well as proposals to address GCC. Teacher presents synthesis of the science concepts covered.	1	1, 2	CATSI

(continued)

Table 4.1 (continued)

Sequence	Brief description	Approximate time (h)[a]	Target objective(s)[b]	Source[c]
8	Graph analysis: Student groups analyzed graphs showing temperature, atmospheric CO_2, and economic trends over varying time spans.	1	3, 4	Anthony et al. (2007)
9	Webquest and position development: Student groups worked through a webquest designed to support their use of the Internet as a source of information for GCC and other SSI. The culminating activity of the webquest prompted students to create a product (e.g., website, poster, presentation) that presented their own position on national policy related to GCC.	3–4	1, 4, 5, 6, 7	CATSI (Klosterman & Sadler, 2008)

[a]Time ranges represent variation across the two teachers' implementations
[b]The numbers correspond to objectives introduced in the text
[c]All of the exercises were modified to fit our classroom contexts and instructional goals
[d]Indicates that the materials were designed by the CATSI team

groups rotated through lab stations at which they performed a series of directed inquiries designed to highlight the particulate nature of matter. At the end of class, the teacher brought all of the students back together and presented a lecture on the earth's atmosphere and synthesized the content students had seen in the lab activities particularly as it related to atmospheric conditions. The next lesson introduced combustion complete with an explosive demonstration. Students used ball and stick molecular models to simulate combustion of various fuel types. They also created and balanced chemical equations corresponding to their physical models. The teachers directed their students to carefully note variations in CO_2 production based on fuel types and amounts. In the follow-up lesson, students collected various gas samples including exhaled breath, the product of an acid–base reaction, ambient air, and car exhaust; and they performed simple titrations to determine CO_2 concentrations. The teachers then led the classes through a discussion of the links among human activity, CO_2 emissions, and greenhouse effects. The next lesson built on this introduction to greenhouse effects by having student groups create model greenhouses using glass jars and heat lamps. Students measured temperature changes over time in models with varying gas contents (ambient air, air saturated with water vapor, and pure CO_2).

Following these four lab-based lessons, both teachers used a class period to help students review and synthesize the content that had been covered and make explicit connections to the overarching issue of climate change. Teachers also prompted students to extend their thinking about the possible consequences, both environmental and societal, of climate change as well as actions proposed to stem climate change. In the next class period, students worked in groups to interpret a series of graphs displaying temperature trends, concentrations of greenhouse gases, and economic data. The graphs displayed trends over variable time frames and have been interpreted by parties with vested interests in the climate change issue in very different ways. Students responded to a series of prompts designed to help them attend to issues that can affect interpretation including graph scale and data sources. The final lesson, which extended over several class periods, was designed to help students build media and web literacy skills relative to global climate change particularly with respect to the use of Internet sources. Students completed a webquest that guided them through use of criteria for selecting and evaluating web-based media. The culminating activity for the webquest challenged students to use Internet resources as well as their experiences and findings throughout the unit to create a product promoting a particular course of action (or inaction) relative to climate change. In terms of products, students had freedom to choose their own format but were encouraged to consider creation of a website, a slide presentation, a poster, or narrative that would effectively communicate their intended message. This culminating activity challenged students to create policy positions like they had observed in the unit's second lesson. The intent was to provide an opportunity for students to synthesize all that they had learned in the unit and articulate their own perspectives on global climate change.

Research

Research Questions

Research conducted as a part of the CATSI project related primarily to the third and fourth project goals (presented in a previous section). The third goal called for an investigation of how SSI-based instruction supports (or fails to support) student development of scientific content knowledge. The fourth goal related to an extension of work on socio-scientific reasoning. More specifically, we were interested in exploring methodological improvements for the assessment of SSR. We also intended to test whether SSR could be improved among students participating in the global climate change intervention. To achieve these goals, we developed a research plan around the following research questions.

RQ1. How does SSI-based instruction support student learning of science content knowledge?
RQ2. How can the assessment of SSR be improved?
RQ3. How does SSI-based instruction support student development of SSR?

Content Knowledge

Within the science education community, there has been a long history of calls for using socially relevant issues (i.e., SSI) as contexts for teaching and learning science (DeBoer, 1991). The work of several researchers has supported the long-held contention that SSI-based instruction can support student development of science content knowledge (Barber, 2001; Dori et al., 2003; Yager et al., 2006; Zohar & Nemet, 2002), but like all studies, these had limitations. Namely, some of these researchers relied on curriculum-based tests of content and others did not employ pre/post designs. These researchers chose to focus on certain factors in the design of their research and in so doing created affordances for the investigation of some issues and limitations for others. For the design of the CATSI study, we chose to prioritize the investigation of content knowledge. In doing so, we implemented a pre/postintervention assessment strategy in order to measure gains over time. We also adopted a multilevel assessment framework in which we created content assessments at variable distances from the curriculum.

Assessing the effects, particularly content learning, of curricular innovations presents serious challenges for researchers (Ruiz-Primo et al., 2002). The designers of innovations frequently create assessments directly aligned with their interventions. These assessments, which can be thought of as close to the interventions, can be useful for classroom teachers and students as well as for formative assessment of the intervention. However, these measures are typically critiqued when used as summative assessments. Intervention-specific assessments are limited in scope, do not allow for the assessment of transfer, and do not permit comparisons among curricula.

More distant measures of content such as national exams are much broader in scope, can be used to make inferences regarding transfer, and allow for comparisons among curricula. However, assessments that are as distanced from an intervention as national exams are very insensitive to change. It is not reasonable to expect a national exam to detect significant learning gains associated with a 3-week intervention.

To address these problems, assessment specialists working in science and mathematics education have proposed a multilevel assessment framework for better understanding the effects of curricular innovations (Hickey & Pellegrino, 2005; Hickey, Zuiker, & Taasoobshirazi, 2006; Ruiz-Primo et al., 2002). This framework calls for assessment at various distances from a particular intervention. The approach calls for researchers to collect data through instruments closely aligned with an intervention, but the value of these data can be significantly improved when they are interpreted in conjunction with more distanced measures. In the CATSI project, we collected data at four unique curricular distances: immediate, close, proximal, and distal. The immediate data (classroom observations of student discourse and practices) and close data (student artifacts produced in the midst of the climate change unit) were useful as we considered the design and progress of the intervention. For research purposes, we relied primarily on the proximal (test results using items directly aligned with the curriculum) and distal data (results from a test created with items sampled from state and national exams), the sources of which will be described in the methods section.

Socio-scientific Reasoning

We introduced socio-scientific reasoning as a tool for researchers and practitioners to more effectively operationalize and assess the practices in which students engage as they negotiate SSI (Sadler et al., 2007). Science educators, including us, have argued that SSI-based education has potential to better prepare students for interacting with social issues, making decisions on complex issues, and developing character (Berkowitz & Simmons, 2003; Kolstø, 2001; Zeidler & Sadler, 2008). However, the community lacks conceptual and assessment resources to investigate the extent to which these claims are supported by evidence (Orpwood, 2007). We developed SSR as an initial attempt to address this gap in the SSI research agenda. The basic idea underlying the SSR construct was that most, if not all, SSI regardless of their specific scientific and social contexts share certain features. We hypothesized that as learners interact with specific SSI contexts, they become more aware of and better prepared to respond to the implications of the invariant features of diverse SSI. In developing a framework for assessing SSR, we sought to provide a mechanism for empirically documenting the extent to which learner practices relative to negotiation of SSI developed over time.

We explored existing science education research in order to identify the invariant features of SSI that could be leveraged to operationalize SSR. We initially identified four SSI features to serve as the basis for distinct aspects of SSR. We present the SSR aspects below with a brief sample of the literature used to identify and substantiate each aspect.

1. Recognizing the inherent *complexity* of SSI (Hogan, 2002; Pedretti, 1999).
2. Examining issues from multiple *perspectives* (Sadler & Zeidler, 2005; Zohar & Nemet, 2002).
3. Appreciating that SSI are subject to ongoing *inquiry* (Bingle & Gaskell, 1994; Yang & Anderson, 2003).
4. Exhibiting *skepticism* when presented potentially biased information (Kolstø, 2001; Zeidler et al., 2002).

Our initial investigation of SSR, within the *Quest Atlantis* project, focused on clarifying the construct, establishing assessment protocols, and documenting a baseline of student practices. In this work, we assessed SSR by providing students with two brief SSI scenarios and asking questions designed to elicit ideas and practices related to the SSR aspects. These interviews were audio-recorded and transcribed. Our analyses were based on the full transcripts. We created four-level ordinal scales for each of the SSR aspects. These scales provided a means of classifying the variability of student responses relative to each of the invariant SSI features highlighted by the SSR construct.

In order to explore the measurement properties of SSR, we conducted correlation analyses for inter-scenario aspect scores. These correlations provided a measure of how consistent student performance was for each SSR aspect. In order for an aspect to be meaningful from a measurement perspective, it would need to be fairly stable across contexts. Correlation coefficients (Spearman's rho) were relatively high for the complexity and inquiry aspects (.76 and 0.73 respectively) but low for perspectives and skepticism (.42 and 0.37). We also computed correlations among aspects within the scenarios to explore the extent to which the aspects were related. High correlations among student practices on different aspects would indicate relatedness and support the idea that SSR represented a unidimensional construct. The correlations between complexity and inquiry were relatively high in both scenarios. Correlation coefficients between all other aspects were positive but fairly low. The complexity and inquiry aspects had performed as we had expected; that is, individual student practice was measured consistently across independent contexts and the results suggested that they may both be related to an underlying latent variable that we described as SSR. The other aspects, perspectives, and skepticism, did not perform as expected in terms of inter-scenario consistency or relatedness to a common latent variable. We concluded that future research would have to approach assessment of the perspectives and skepticism aspects in different ways in order to provide useful data (Sadler et al., 2007).

To advance SSR as a useful tool for the research community, assessment of the perspectives and skepticism aspects would obviously have to be improved. We also wanted to explore other assessment formats. The interview protocol used for our initial work provided quality data but the resources required for conducting, transcribing, and analyzing interviews would likely prohibit its use in large-scale projects. Therefore, we designed the CATSI project to experiment with new data collection methods. We created an internet-based questionnaire that could vary questions based on students' previous responses. This allowed us to effectively customize prompts to which students responded, one of the advantages of interviewing, as students completed the survey. The electronic data were automatically logged to a database eliminating the need for transcription.

A primary motivation in developing SSR was generation of a tool that could be used to document student development over time. However, our initial research did not explore the extent to which SSI-based instruction improved SSR. As a part of the CATSI project we intended to address this issue. Our goal was to test the hypothesis that student learning experiences in one SSI context (i.e., global climate change) enhances SSR as displayed in different SSI contexts.

Methods

Data Collection and Analysis: Content Knowledge

We adopted a pre/postintervention design for the analysis of both content knowledge and SSR. Consistent with the multilevel assessment framework that helped to guide design of the CATSI project, we collected data related to student learning of science content at different levels of varying "distances" from the curriculum. For the research reported as a part of this chapter, we focus on two of these data sources: a proximal test and a distal test. The proximal test was aligned with the CATSI curriculum; that is, it was designed to assess student understanding of the content taught as a part of the CATSI unit. This curriculum-aligned test consisted of five open-ended questions: (1) What is global warming? (2) What is the greenhouse effect? (3) How does the greenhouse effect relate to global warming? (4) What is the controversy associated with global warming? (5) Is global warming a challenging problem? Why or why not? The curriculum-aligned test was administered electronically and student responses were collected through a database. Because of technical problems with the database, we were only able to analyze data from the curriculum-aligned test for Molly's environmental science students. The numbers of students whose data were included in analyses are displayed in Table 4.2.

Table 4.2 Number of classes and students participating in the CATSI project

Teacher (subject)	Classes	Students	Informed consent	Proximal test (pre and post)	Distal exam (pre and post)	SSIQ (pre and post)	Interview
Molly (Environmental science)	3	75	57	49	49	50	11
William (Chemistry)	2	62	51	–[a]	34	–[a]	0
Totals	5	137	108	49	83	50	11

[a]Technical difficulties prohibited access to these data

Our analysis of the open-ended student response data was guided by the constant comparative method (Strauss & Corbin, 1998) which was an inductive approach that called for iterative cycles of evaluation of interpretive hypotheses and comparisons to data. Analysis progressed in six phases beginning with two researchers independently reviewing ten randomly selected answer sets to look for general trends. By the fourth phase of analysis, the researchers had independently examined 30 transcripts and developed a coding scheme with high inter-rater consistency (<90%). The coding scheme was designed to characterize individual variation associated with each question. Categories for each question were ordinal in nature. In the fifth phase of analysis, a single researcher completed coding on the curriculum-aligned data. In the final phase, we looked for changes in student performance on the pre- and postintervention tests by testing for changes in categorical proportions. We applied a McNemar analysis, which is similar to a Chi-square test but is designed for repeated measures data.

To examine student development of content knowledge from a more distanced perspective, we developed a distal test. The distal test was aligned with state content standards that guided development of the CATSI curriculum. Therefore, it was designed to measure student understanding of the scientific formalisms underlying climate change but individual exam items were not aligned with unit curriculum. In designing the CATSI curriculum, four sets of science standards were identified. They related to the following four general areas of science: (1) climate and temperature, (2) greenhouse effects and climate change, (3) chemical principles and processes, and (4) graphing and graph analysis. We sampled items from publicly-released standardized tests used for international (e.g. TIMSS-Trends International Mathematics and Science Study), national (e.g., NAEP-National Assessment of Education Progress), and state (e.g., FCAT-Florida Comprehensive Assessment Test) assessments. We created item pools associated with each of the four groups of standards. After extensive pilot and reliability testing, we produced a 20-item, multiple-choice instrument (five items for each of the four standards groupings) and administered it to students in the CATSI project before and after the SSI-based intervention. We analyzed student responses with a repeated measures ANOVA and computed effect sizes.

Data Collection and Analysis: Socio-scientific Reasoning

One of our project goals was to experiment with a new format for assessing SSR. Our previous work suggested that interviews could be used to assess SSR but the resource-dependent nature of interviews limited their use for large sample sizes. We wanted to try an open-ended questionnaire but worried that questions designed to elicit some SSR aspects would be too leading. To address these concerns, we developed an online survey, the Socio-scientific Issues Questionnaire (SSIQ), using an adaptive questioning strategy that directed students to specific open-ended questions based on previous forced-choice responses. This strategy limited the extent to which questions were leading and yet challenged students to generate responses that could illuminate their SSR.

In its final design, the SSIQ consisted of a narrative description with an accompanying diagram of a localized socio-scientific scenario. The description was followed by a series of forced choice and open-ended questions (see Appendix 3.1 for an example). Two scenarios were developed; both related to water pollution issues with economic implications and at least three clearly identifiable parties interested in the issue. Students read and responded to one of the SSIQ scenarios prior to the intervention and the other after the intervention. The selection of scenarios (preversus postintervention) was randomized.

Based on our previous work (Sadler et al., 2007), we knew that the manner in which the perspectives and skepticism aspects had been operationalized required modification. For development of the SSIQ and subsequent analyses, we reconceptualized the perspectives aspect. In our initial work, the perspectives aspect captured the extent to which students adopted multiple perspectives in the justification of their own decisions. In the CATSI project, we revised the perspectives aspect such that it assessed the extent to which students could discuss the perspectives and interests of multiple parties involved in the scenarios. The new perspectives aspect incorporated themes suggested by the original perspectives category as well as the original skepticism category.

In analyzing data collected from the SSIQ, we were guided by scoring codes developed in the initial research (Sadler et al., 2007) but given the changes in data collection and the new perspectives aspect, we made significant modifications. In addition to Molly's and William's classes, we had administered the SSIQ to 37 high school students from a different school. We used these data to develop scoring rubrics for SSR. We had a priori notions of possible ranges of student practices but these ideas were shaped significantly by inductive analyses of the pilot data. We developed five-point ordinal scales for each SSR aspect (complexity, inquiry, and perspectives) and used the rubrics to score the SSIQ data. Two reviewers independently analyzed 20 sets of responses, randomly selected from among pre- and posttests. Initial interrater consistency ranged from 60% to 80% by SSR aspect; however, most of the discrepancies were quickly resolved and ascribed to simple misinterpretations. Following this initial negotiation phase, inter-rater consistency exceeded 90% for all aspects. Given the relatively high tendency for rater error, two reviewers independently coded all responses and rating discrepancies were mediated by a third reviewer. In order to test the validity of using the SSIQ, we conducted interviews with a subset of the sample ($n=11$) and analyzed the transcripts using the same set of rubrics.

Results and Discussion

Findings: Content Knowledge

The first research question addressed student learning of content in the context of SSI-based instruction. Student performance on the proximal, curriculum-aligned test was assessed by means of the emergent scoring rubric described above. Table 4.3 presents the coding schemes developed for each of the five questions,

Table 4.3 Assessment rubric for scoring the curriculum-aligned test

Category	Description	Exemplars
Question 1: What is global warming?		
Misconceptions (A)	Describes global warming using a common misconception such as ozone depletion.	Global warming is the ozone layer growing thinner and thinner and the sun is shinning through more and more.
Temperature (B)	Explains global warming as an increase in temperature	Global warming is where the temperature all throughout the earth is rising.
Temperature and Effect (C)	Explains global warming as an increase in temperature and elaborates on response with a description of its effects	Global warming is when the earth is heating up at an alarming rate and polar ice caps are melting.
Greenhouse effect (D)	Explains global warming in terms of the greenhouse effect and/ or greenhouse gases	An increase in the earth's average atmospheric temperature that causes changes in climate and results from the greenhouse effect.
Question 2: What is the greenhouse effect?		
Inaccurate or Unrelated (A)	Uses unrelated concepts or underdeveloped ideas to explain GHE	The greenhouse effect is when there is a lot of green life and things have an easier time growing in the surrounding climate.
Warming (B)	Explains GHE in terms of global warming or increases in temperature	When the sun heats up the earth and causes it to warm at an increased rate.
Gases (C)	Includes direct reference to or implies a relationship between greenhouse gases the GHE	The greenhouse effect is when the greenhouse gases such as carbon dioxide, methane, etc. are burned off into the atmosphere these gases hold in the heat, therefore heating up the earth.
Question 3: How does the greenhouse effect relate to global warming?		
Inaccurate/unrelated/underdeveloped (A)	Uses unrelated concepts or underdeveloped ideas to explain the relationship between GW and GHE	It's bad for earth.
Similar (B)	The GHE and global warming are described as parallel concepts and often as both related to temperature	They are related to each other because both of them have something to do with the rising of the earth's temperatures.
Greenhouse effect (C)	Explains that GW is caused by or is a result of the GHE	The greenhouse effect speeds up the effect of global warming.

Question 4: What is the controversy associated with global warming?		
Consequences (A)	Describes the controversy of GW in terms of its possible consequences	Frozen places like Antarctica will start or have started to melt.
Uncertainty (Bi)	Describes the controversy as a decision between whether or not global warming exists and if humans are responsible	The controversy with global warming is that some feel that it is of no concern while others feel that it could very well shape the future. Many have debated the issue and feel that there is no solid proof of the fact that humans are responsible for it there for feeling no need to take action.
Inquiry (Bii)	Describes the controversy as a result of not knowing how to take action	The controversy associated with global warming is that we do not know how to solve this problem that will eventually end the world.
Complex tradeoffs (Biii)	Describes the controversy as a series of tradeoffs between personal comfort and political or economic change	Global warming is caused by things that humans do and can be stopped but the things that cause it have become a part of our daily lives and are hard to give up and stop all together
Multiple sources (C)	Expresses more than one source of the global warming controversy	The controversy associated with global warming would be that it is causing a lot of ice causing a lot of the icecaps to melt and some of the animals that live by the water are loosing their homes because their homes are melting and one big controversy is whether or not global warming is a problem and whether or not we should do something about it.
Question 5: Is global warming a challenging problem? Why or why not?		
Underdeveloped (A)	Presents an underdeveloped idea to justify whether GW is challenging	There is so much controversy over the issue.
Unchallenging (B)	States that global warming is not a challenging problem	It is not a challenging problem because it doesn't make it hard for people to live from day to day. It really doesn't affect people's daily lives.

(continued)

Table 4.3 (continued)

Category	Description	Exemplars
Uncertainty (C)	States that global warming is not a challenging problem	Most people don't know that this is a problem and if they do it's not like they care.
No Known Solution (Di)	Determines the challenge to be a result of no known or possible solution	Because no one knows if it can be stopped or how to stop it completely.
Widespread effects (Dii)	Explains that solving GW is challenging because of its widespread effects and possible consequences	Global warming is a challenging problem because we would have to get every nation to start working together to stop or slow down the process of global warming.
Complex tradeoffs (E)	Describes the challenge as a series of tradeoffs between personal comfort and political or economic change	Because if people do want to deal with it they have to find away to begin changing the ways we live on this earth which will be very hard to do because we are so settled in our ways of living now. That it would be too hard to change all of a sudden.

brief descriptions of each category, and exemplars excerpted from student tests. Although it was not our intent to create a hierarchical scoring scheme, the codes for each question could be ranked in terms of accuracy. Low-end codes captured inaccuracies, underdeveloped ideas, and misconceptions. As the codes progressed to higher levels, they captured more sophisticated and scientifically accurate responses. Despite the ordinal nature of most categories, two categories within questions four and five were conceptually distinct but equivalent in terms of accuracy. These categories were identified separately, but combined for the follow-up analyses. The number of categories per question ranged from three (question two) to six (question five). A more complete description of the coding scheme can be found in Klosterman and Sadler (2010).

In order to determine whether student performances differed on the pre- and postintervention assessments, we conducted a McNemar's test for correlated proportions (see Table 4.4). The analysis indicated that the proportion of categorical responses prior to the intervention was significantly different than the category proportions following the intervention for the first three items on the curriculum-aligned test ($p_1=0.011$, $p_2=0.008$, $p_3=0.008$). For Items 4 and 5, the number of responses within several categories was less than five. Probability rates are not reliable for variables with cell frequencies less than five (Agresti & Finlay, 1999); therefore, we chose not to apply the test in these two cases. McNemar's analysis is an omnibus test and, applied to these data, indicated changes in response patterns following the intervention. In order to determine possible directions of change, we examined visual displays of the data (see Figs. 4.1–4.5). These graphic displays,

Table 4.4 McNemar analysis of curriculum-aligned test results

Question	χ^2	df	P
Q1	16.6429	6	0.0107
Q2	14.8410	6	0.0215
Q3	14.5859	6	0.0237

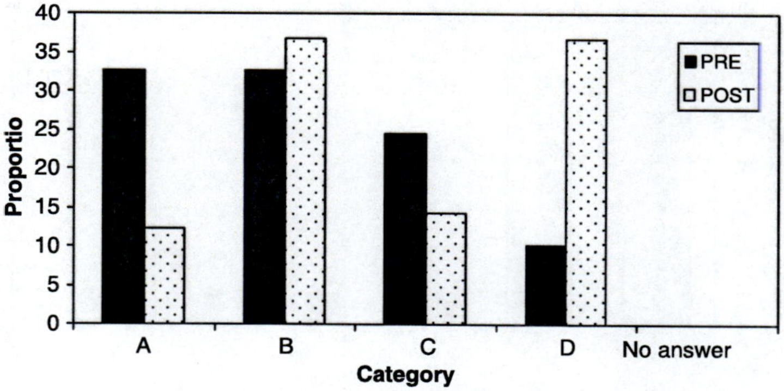

Fig. 4.1 Categorical proportions of pre- and posttest scores for Question 1

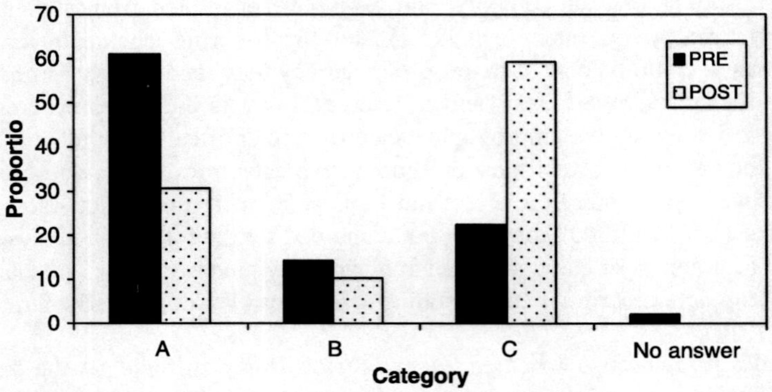

Fig. 4.2 Categorical proportions of pre- and posttest scores for Question 2

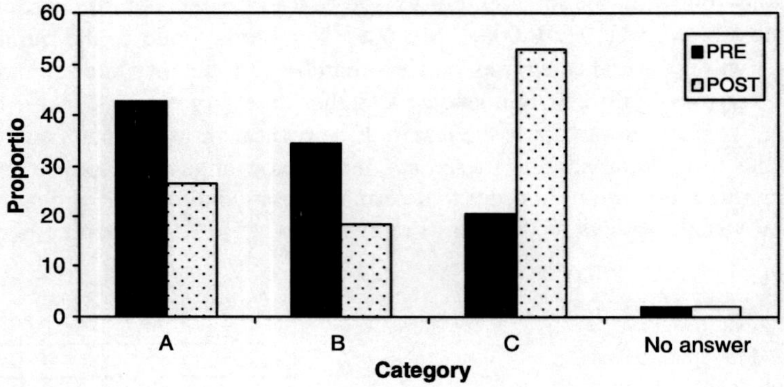

Fig. 4.3 Categorical proportions of pre- and posttest scores for Question 3

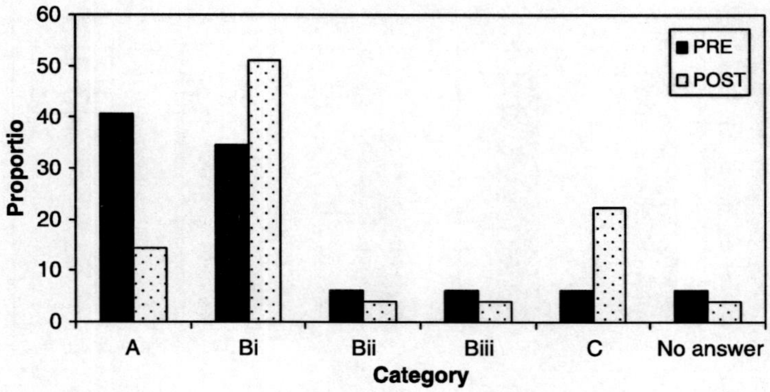

Fig. 4.4 Categorical proportions of pre- and posttest scores for Question 4

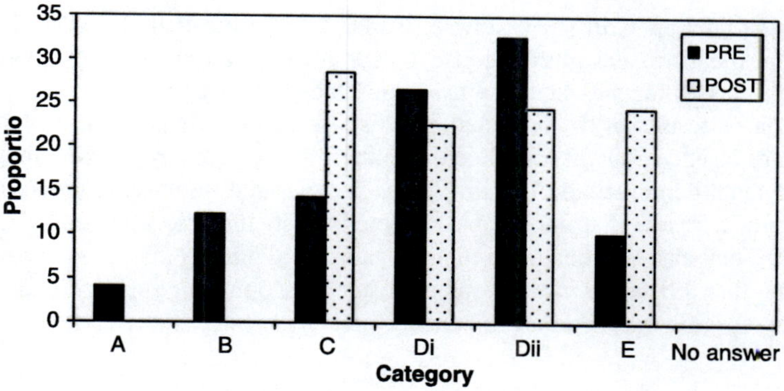

Fig. 4.5 Categorical proportions of pre- and posttest scores for Question 5

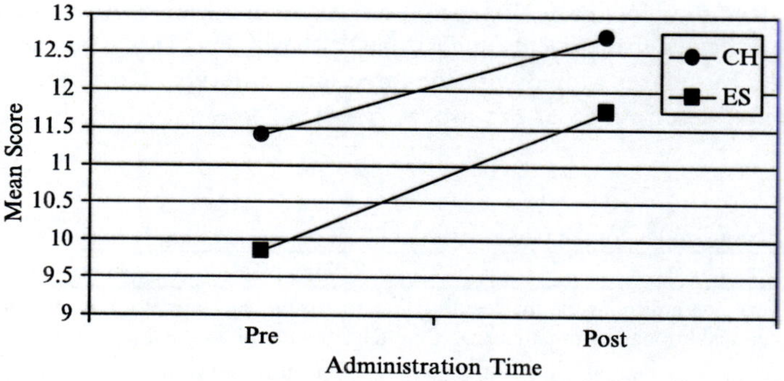

Fig. 4.6 Average scores on standards-aligned exam

which highlighted comparisons between pre- and postintervention performances, revealed shifts toward more accurate responses. The combined results indicated that student performances on the curriculum-aligned content test improved significantly following SSI-based instruction.

Student performance on more distal measures of content was assessed through the standards-aligned test. The average gain (posttest versus pretest) for Molly's environmental science classes was 1.88, and for William's chemistry classes the average gain was 1.29 (total score=20; see Fig. 4.6). Results of a repeated measures ANOVA indicated that posttest scores were statistically significantly different than the pretest scores ($F=15.31$, $p<0.001$). There was no evidence of an interaction effect between the time variable (pre and post) and the course variable suggesting that the intervention produced similar effects in both classroom settings ($F=2.88$, $p=0.094$). Effect sizes were calculated for both sets of classes using the formula for

Cohen's *d* (Dunlop, Cortina, Vaslow, & Burke, 1996) for calculating effect sizes in dependent measures designs. The effect sizes for Molly's classes (d_{ES}=0.49) and William's classes (d_{CH}=0.41) were medium (Cohen, 1988).

Combined results of the proximal and distal tests provided evidence of student learning of science content associated with their experiences in the climate change unit. Students demonstrated learning gains for material directly aligned with the curriculum, a result that should be expected given that the unit extended over 3 weeks. They also demonstrated modest gains on the more distanced assessment indicating that SSI-based instruction can foster development of scientific ideas that transcend specific instructional contexts.

Findings: Socio-scientific Reasoning

A rubric was developed for the assessment of each of the SSR aspects. The three rubrics followed a similar format. Zero level responses indicated that students did not understand the most basic dimension of a particular aspect. Responses scored with a one indicated that students understood the basic aspect but could not provide an example. Responses scored with the three highest levels offered progressively more detailed descriptions of the aspect. Table 4.5 presents the rubric used for

Table 4.5 Scoring rubric for the Inquiry aspect of SSR

Level	Description	Exemplars
Question: If you were responsible for deciding how to resolve the Branville Bay situation, would you need additional information regarding the situation before making your decision?		
0	Suggests that additional inquiry is not necessary.	No. Just ban all boats and fishing in the preserve. That way there won't be any chance for drops in wildlife counts.
1	Suggests that additional inquiry is necessary but does not identify a specific line of inquiry.	Yes. Information like are there any other problems that is going wrong with this situation and can there be more than one solution.
2	Suggests that additional inquiry is necessary and identifies one specific line of inquiry.	Yes. What would happen if we stopped big ships from going back and forth and just allowed smaller boats do all the work?
3	Suggests that additional inquiry is necessary and identifies two specific lines of inquiry.	Yes. Exactly how much are the fish and bird counts declining [and] how much are the ships polluting the water as opposed to the fishermen.
4	Suggests that additional inquiry is necessary and identifies three or more specific lines of inquiry.	Yes. I would need to know the amount of people fishing or going through every day, the amount of fish being caught by people omitted from the fishing laws, and the amount of traffic.

assessing the inquiry aspect and exemplars taken from student products. The complexity and perspectives rubrics followed the same format but attended to the appropriate content.

Because we wanted to use the SSIQ to document pre- and postintervention changes, our goal was to develop two equivalent forms to reduce possible testing effects as a threat to validity. Results from pilot testing indicated that the two forms elicited very similar responses in terms of SSR assessment among students taking both forms. Using data from the CATSI project sample, we conducted *t*-tests between the two scenarios for each of the three SSR aspects. Data used for this analysis was restricted to the 50 students who had responded to both scenarios. Scores on all three aspects between the two scenarios were not statistically significantly different. These results provided empirical justification for our decision to use the scenarios as multiple forms thereby reducing the possibility of testing effects.

In order to check for the validity of our interpretations of student responses to the written SSIQ prompts, we conducted and analyzed interviews with a subset of the sample ($n=11$). These interviews, conducted individually with one of the researchers, took place approximately 1 week after student completion of the postintervention SSIQ. The content of student responses in both assessment contexts (written and interview) were consistent. Analysis of the interview transcripts indicated that students interpreted the SSIQ items in ways that we expected. The interviews also allowed us to check our interpretations of student responses with the students themselves. This form of member checking further supported the validity of our analysis. Finally, we scored the interview responses using the rubrics developed for each of the SSR aspects and compared these ratings with the scores obtained by the same students using the written format. In comparing scores between the written and interview data, only one aspect response (3%) varied by more than one ordinal level. Thirty-nine percent of the responses (equally distributed across the three aspects) varied by one ordinal level, but we detected no consistent patterns in terms of which assessment context tended to afford more advanced patterns. We interpreted these results as indicating a high degree of consistency in terms of the content of student responses and a moderate degree of consistency in terms of performance levels across the assessment contexts.

In the original SSR research, two aspects were highly correlated indicating likely relationships to a common latent variable. To test for these relationships in the current dataset, we conducted correlation analyses among scores on the three aspects. None of the correlation coefficients (Spearman's rho) were statistically significantly different than zero. This result offered no evidence of relationships among the three aspects; therefore, we treated them as separate variables for the subsequent analyses.

To explore the issue of change in SSR associated with the CATSI unit, we conducted paired *t*-tests (pre vs. post) for each of the SSR aspects. Average scores and standard deviations for the pre- and postintervention SSR assessments as well as the *t*-test *p* values are presented in Table 4.6. These data revealed no statistically significantly differences in pre- and posttest performances.

Table 4.6 Average scores (and standard deviations) for pre- and postintervention SSIQ responses

	Socio-scientific reasoning aspects		
	Com	Inq	Per
Pretest	1.4 (1.3)	1.9 (1.3)	2.4 (1.2)
Posttest	1.4 (1.2)	1.7 (1.5)	2.1 (1.1)
p values [a]	0.78	0.30	0.14

[a] *p* values for paired *t*-tests

Implications for…

Teaching and Learning

We begin our discussion of implications with the most obvious inference to be drawn from the research conducted as a part of the CATSI project. Students can learn important science content through SSI-based instruction. The evidence collected as a part of this study support the contentions that students can learn science content directly aligned with the context of an SSI-based unit as well as the more abstracted scientific formalisms represented in standards documents and associated standardized assessments. Given the inevitable limitations of this study, we are not claiming that all students will show similar gains in content knowledge in response to all SSI-based instruction, but we document compelling evidence of science learning for this sample in the context of this issue (global climate change). Classroom-based work like this will likely never produce the kinds of data that could legitimately be considered generalizable for all (or most) students and all (or most) SSI contexts. However, the combined analysis of these results with studies (Dori et al., 2003; Zohar & Nemet, 2002) located in other, diverse settings drawn from different populations and utilizing different issues provide growing support for the efficacy of SSI-based education in terms of promoting the learning of science content.

Most other studies that link SSI-based instruction and science content learning document gains on assessments closely aligned with the context of the SSI under consideration (similar to our proximal level data). These data are certainly important in terms of understanding how SSI-based instruction works, but the high degree of concurrence between the contexts of instruction and assessment leaves open the question of how SSI-based instruction affects learning that transfers beyond the immediate learning environment. In the current political climate, important stakeholders in science education including teachers, school administrators, and policy-makers are particularly interested in how curricular innovations can affect student performance on standardized assessments (Settlage & Meadows, 2002). This study provides some of the first evidence using a pre/post design and a multilevel assessment model documenting student gains on distal level assessments of content that serve as a proxy for standardized assessments. The raw score differences are statistically significant, but the actual change values are relatively small

(1.9 for Molly's students and 1.3 for William's students). We acknowledge the modesty of these changes, but it is important to note the modest nature of the CATSI intervention in terms of its duration. We view the fact that any statistically significant changes in performance on a standards-aligned exam occurred in association with a 3-week unit as an important finding.

In reflecting on what was (and was not) accomplished in this project particularly as related to how the unit was designed and implemented, we believe it critically important for the communities committed to promoting SSI-based instruction to advance new models for introducing SSI into classrooms. Here, we are not arguing for new teaching models (although we certainly would not argue against development of new teaching innovations for SSI-based instruction). Rather, we highlight the need to consider new approaches to support collaborations among teachers, researchers, curriculum designers, and professional development specialists to more efficiently move SSI into classrooms. We, the full CATSI team including university-based researchers and classroom teachers, spent a great deal of time working together in order to create an intervention specifically designed for our target classrooms. This obviously is not a scalable model. As a community, we need to generate better ways to develop and disseminate curriculum and teaching innovations that are responsive to the specificities of individual classrooms.

We did not systematically collect data on student interest in the SSI unit or their motivation to participate in learning activities associated with this unit. However, our classroom observations indicated that the students were not as enthused by and interested in the global climate change issue as we had expected and hoped. One of the arguments made in support of using SSI in classrooms is that SSI provide contexts that connect school science to real-life issues thereby making science more relevant and interesting (Albe, 2008; Harris & Ratcliffe, 2005). Ultimately, the students in the CATSI project were highly engaged, but this engagement seemed not to be related to issue context. The engagement we observed seemed more related to the design of learning experiences than the issue itself. The CATSI unit followed many of the recommendations promoted by the science education community in support of engaging curricular innovations. Students experienced a variety learning experiences in which they were both challenged and supported. They had opportunities to create products, interact with media and technology, and engage in inquiry. We believe that the issue in and of itself was engaging for some students, particularly those who identified themselves as environmental advocates, but many students did not seem any more motivated to learn about global climate change than science topics unrelated to contemporary social issues. We actually do not interpret these observations as suggesting that SSI-based instruction is not or cannot be motivating for secondary students. We believe that these observations highlight the fact that formal science instruction takes place within the larger context of school science. A single, issue-based unit implemented over 3 weeks is unlikely to transform the manner in which students consider and feel about their school-based science learning experiences. This interpretation suggests we exercise care in considering how and the extent to which innovations support the development of interest in and motivation to learn science.

Research

We did not address SSR in discussing implications of this project for teaching. We omitted this discussion because we are uncertain as to what the project's implications for teaching are. We did not document gains in SSR associated with the CATSI unit. This nonresult may indicate that an intervention of this limited time frame cannot affect change in SSR. Changes in SSR may require longer developmental periods. However, an equally plausible interpretation of the results is that the nature of the intervention limited changes in SSR. A 3-week SSI-based unit may have potential to affect changes in SSR, but the CATSI unit itself may not have included necessary elements for these changes to be actualized in student practices. For example, the CATSI unit encouraged students to consider the complexity, inquiry, and perspectives aspects of global climate change, but instruction did not explicitly encourage students to think about how these elements emerge across multiple issues. The unit did not present climate change as a model for other complex SSI or encourage comparisons of similarities among this specific issue and other SSI. We believe further studies designed to explore the effects of specific elements of SSI-based instruction along with possible developmental trajectories of SSR would be fruitful work.

In order for the work called for in the previous paragraph to be done well, we believe the conceptual and assessment tools associated with SSR must be improved. The work described in this chapter represents only the second iteration in the development of SSR as a measurable construct. This second iteration has offered improvements in terms of the assessment context of SSR and the internal consistency of the rubrics used to track levels of SSR; however, much work remains. Some of the issues that demand further attention include possible expansion of the SSR aspects, fuller exploration of possible relationships among SSR aspects, and continued experimentation with forms of assessment. In this iteration, we highlight three aspects of SSR resulting from a rearticulation and merging of two aspects from the original research conducted on SSR. Additional aspects may need to be added to the framework to more validly capture practices associated with the thoughtful negotiation of SSI. In fact, we suspect that the practices associated with the complexity, inquiry, and perspectives aspects vastly under-represent the full range of practices associated with negotiating SSI. The challenge is in creating frameworks that support the valid and reliable assessment of these other aspects.

In our original work on SSR, we produced data suggesting a relationship among aspects of SSR. Based on these results, we postulated an underlying latent variable to which the aspects mapped. In essence, we proposed SSR as a single construct with interrelated subconstructs (i.e., aspects). The results produced in the CATSI project did not support this interpretation. The aspects did not show significant relationships; therefore, we treated the aspects as individual variables. Additional research designed to explore these possible relationships are essential in terms of providing guidelines for how these variables should be handled in future analyses.

We believe that in the second iteration of the study of SSR, we made improvements in the process of assessing SSR. We moved from an interview protocol to an adaptive, online survey the effect of which is to make SSR assessment more feasible and likely for larger groups of learners. However, much remains to be done for the optimization of SSR assessment. Assessing SSR using a broader range of issue contexts is one possible advancement; refinement of the aspect rubrics is another. Here, we offer these two specific suggestions as initial possibilities, but much more could be done to enhance the reliability, validity, and usability of SSR assessments. We leave it to the broader community of researchers interested in SSI and assessment to further define directions for advancing this agenda.

Conclusions

At its core, the CATSI project was a collaboration among teachers and researchers committed to enhancing science learning experiences of specific groups of high school students as well developing more robust understandings of how SSI-based instruction supports progressive goals of science education. We produced evidence regarding how SSI-based instruction can support science content learning at variable distances from the curriculum. The research related to socio-scientific reasoning raised more questions than answers, but we believe that the work significantly advanced the discussion of what ought to be assessed as a part of SSI-based instruction and how it ought be assessed. As university-based researchers, we found the experience of partnering with teachers and maintaining a sustained presence in their classrooms to be productive and informative. It allowed us to better support the needs of teachers as well as better understand their concerns and constraints associated with implementing SSI-based instruction. We gained better perspectives on the politics and processes of local schools. For example, the reassignment of one of our partnering teachers a few weeks before the start of a new academic year leading to the scrapping of our initial development efforts presented a significant setback and a new appreciation for the challenges associated with working in actual classrooms situated in broader local, district, and state contexts. Despite these challenges, we think this kind of situated work is essential for building understanding of how SSI-based instruction can be used productively in classroom settings and what SSI-based instruction can afford in terms of student learning and continued development as active participants in societies increasingly shaped by science and technology.

Appendix 3.1. SSIQ Prompt and Questions

Branville Bay is located on the Gulf of Mexico. The city of Branville has built up along the northern border of the Bay and a wildlife preserve has been established along the southern border. The Branville area was the ancestral home for

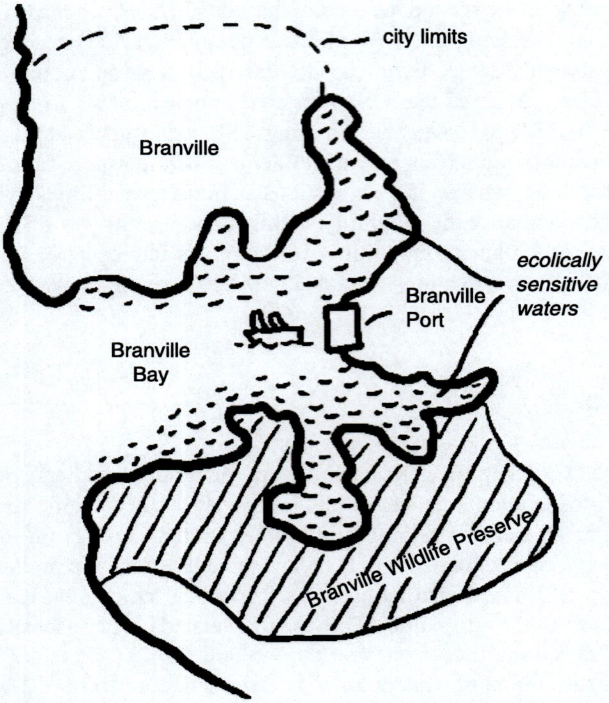

Fig. 4.7 Map of Branville Bay and the surrounding area

several tribes of Native Americans. More recently, Branville has become a major shipping port. Ships from all over the world dock at Branville Port delivering products like oil, clothing, toys, and fruit. These products are then distributed throughout the USA. Businesses in the USA also use the port to send their products around the world (see Fig. 4.7).

Branville Bay is a sensitive ecological area serving as the breeding grounds for many fish, birds, and other wildlife. There are strict laws that govern fishing in the most sensitive areas of the bay. However, these laws do not apply to the Native Americans still living in the area because they have claimed ancestral fishing rights in the area.

Managers of the Branville Wildlife Preserve have started reporting declines in fish counts, bird counts, and water quality measures. These managers have concluded that the heavy ship traffic moving in and out of Branville Port is damaging the Branville Bay ecosystem. Port Authorities claim that their ships stay in deep water channels and do not travel into the most sensitive waters of the bay. They argue that the Native American fishers are the most likely culprits because they use boats and fish in the bay's most sensitive waters.

Local leaders are trying to decide what to do.

Questions:

1. Can the Branville Bay situation be solved easily?

   A. YES
   B. NO

If A, then: Explain why you think the Branville Bay situation should be easy to solve.

If B, then: Explain why you think the Branville Bay situation cannot be solved easily.

2. If you were responsible for deciding how to resolve the Branville Bay situation, would you need additional information regarding the situation before making your decision?

   A. Yes, I would need to have additional information to make a decision.
   B. No, I have sufficient information to make a decision.

If A, then: What kinds of additional information would be necessary for you to make a decision regarding the Branville Bay situation?

If you were responsible for deciding how to resolve the Branville Bay situation, what would you recommend doing as a next step? Please explain why this would be an effective strategy.

If B, then: If you were responsible for deciding how to resolve the Branville Bay situation, what would you recommend doing? Please explain why this would be an effective strategy.

3a. In the previous prompt, you were asked to suggest a course of action for the Branville Bay situation. Describe the strengths of your proposed approach.

3b. Describe the weaknesses of your proposed approach.

4a. A group of concerned Branville citizens gathered to discuss a solution for the Branville Bay situation. The group suggested that Native American fishing permits in the most sensitive waters of the Bay be reduced by half and that ship traffic be reduced by one-third (that is, only two-third of the current number of ships traveling in the bay could continue coming into the Bay).

4b. How do you think Branville Port Authorities would respond to this suggestion? Please explain your response.

4c. How do you think Native Americans in Branville would respond to this suggestion? Please explain your response.

4d. How do you think managers of the Branville Wildlife Preserve would respond to this suggestion? Please explain your response.

5. In response to the previous questions, you commented on how three different groups (Port Authorities, Native Americans, and Wildlife Managers) would respond to a proposed solution. Which of the following statements most accurately reflects your responses?

   A. The Port Authorities, Native Americans, and Wildlife Managers would have similar responses to the proposed suggestion.

B. The Port Authorities, Native Americans, and Wildlife managers would have different responses to the proposed suggestion.

If A, then: Explain why you expect the Port Authorities, Native Americans, and Wildlife Managers to have similar responses to the proposed suggestion.

If B, then: Explain why you expect the Port Authorities, Native Americans, and Wildlife Managers to have different responses to the proposed suggestion.

References

ACS. (2006). *Chemistry in the community* (5th ed.). New York: W. H. Freeman.

Agresti, A., & Finlay, B. (1999). *Statistical methods for the social sciences*. Upper Saddle River, NJ: Prentice Hall.

Albe, V. (2008). When scientific knowledge, daily life experience, epistemological and social considerations intersect: Students' argumentation in group discussion on a socio-scientific issue. *Research in Science Education, 38*, 67–90.

Anthony, S., Brauch, T. W., & Longley, E. J. (2007). *Chem connections: What should we do about global warming?* New York: W. W. Norton.

Barab, S. A., Sadler, T. D., Heiselt, C., Hickey, D. T., & Zuiker, S. (2007). Relating narrative, inquiry, and inscriptions: Supporting consequential play. *Journal of Science Education and Technology, 16*, 59–82.

Barber, M. (2001). *A comparison of NEAB and Salters A-level Chemistry: Students views and achievements*. UK: University of York.

Berkowitz, M. W., & Simmons, P. (2003). Integrating science education and character education. In D. L. Zeidler (Ed.), *The role of moral reasoning on socioscientific issues and discourse in science education* (pp. 117–138). Dordrecht: Kluwer.

Bingle, W. H., & Gaskell, P. J. (1994). Scientific literacy for decision making and the social construction of scientific knowledge. *Science Education, 78*, 185–201.

Bryce, T., & Gray, D. (2004). Tough acts to follow: The challenges to science teachers presented by biotechnological progress. *International Journal of Science Education, 14*, 717–733.

Cajas, F. (1999). Public understanding of science: Using technology to enhance school science in everyday life. *International Journal of Science Education, 21*, 765–773.

Cohen, J. (1988). *Statistical power analysis for the behavioral sciences* (2nd ed.). Hillside: Lawrence Earlbaum Associates.

Cross, R. T., & Price, R. F. (1996). Science teachers' social conscience and the role of controversial issues in the teaching of science. *Journal of Research in Science Teaching, 33*, 319–333.

DeBoer, G. E. (1991). *A history of ideas in science education: Implications for practice*. New York: Teachers College Press.

Dori, Y. J., Tal, R., & Tsaushu, M. (2003). Teaching biotechnology through case studies – Can we improve higher order thinking skills of nonscience majors? *Science Education, 87*, 767–793.

Dunlop, W. P., Cortina, J. M., Vaslow, J. B., & Burke, M. J. (1996). Meta-analysis of experiments with matched groups or repeated measures designs. *Psychological Methods, 1*, 170–177.

Harris, R., & Ratcliffe, M. (2005). Socio-scientific issues and the quality of exploratory talk-what can be learned from schools involved in a 'collapsed day' project? *The Curriculum Journal, 16*, 439–453.

Hickey, D. T., & Pellegrino, J. W. (2005). Theory, level, and function: Three dimensions for understanding transfer and student assessment. In J. P. Mestre (Ed.), *Transfer of learning from a modern multidisciplinary perspective* (pp. 251–293). Greenwich, CT: Information Age Publishers.

Hickey, D. T., Zuiker, S. J., & Taasoobshirazi, G. (2006). Balancing varied assessment functions to attain systemic validity: Three is the magic number. *Studies in Educational Evaluation, 32*(3), 180–201.

Hogan, K. (2002). Small groups' ecological reasoning while making an environmental management decision. *Journal of Research in Science Teaching, 39*, 341–368.

Klosterman, M. L., & Sadler, T. D. (2008). Information literacy for science education: Evaluating web-based materials for socioscientific issues. *Science Scope, 31*(7), 18–21.

Klosterman, M., & Sadler, T. D. (2010). Multi-level assessment of content knowledge gains in the context of socioscientific issues based instruction. *International Journal of Science Education, 32*, 1017–1043.

Kolstø, S. D. (2001). 'To trust or not to trust,…' -pupils' ways of judging information encountered in a socio-scientific issue. *International Journal of Science Education, 23*, 877–901.

Lumpe, A. T., Haney, J. J., & Czerniak, C. M. (1998). Science teacher beliefs and intentions to implement science-technology-society (STS) in the classroom. *Journal of Science Teacher Education, 9*, 1–24.

Orpwood, G. (2007). Assessing scientific literacy: Threats and opportunities. In C. Linder, L. Ostman, & P.-O. Wickman (Eds.), *Promoting scientific literacy: Science education research in transaction: Proceedings of the Linnaeus Tercentenary Symposium* (pp. 120–129). Uppsala: Uppsala University.

Pedretti, E. (1999). Decision making and STS education: Exploring scientific knowledge and social responsibility in schools and science centers through an issues-based approach. *School Science and Mathematics, 99*, 174–181.

Ruiz-Primo, M. A., Shavelson, R. J., Hamilton, L., & Klein, S. (2002). On the evaluation of systemic science education reform: Searching for instructional sensitivity. *Journal of Research in Science Teaching, 39*, 369–393.

Sadler, T. D., Amirshokoohi, A., Kazempour, M., & Allspaw, K. (2006). Socioscience and ethics in science classrooms: Teacher perspectives and strategies. *Journal of Research in Science Teaching, 43*, 353–376.

Sadler, T. D., Barab, S. A., & Scott, B. (2007). What do students gain by engaging in socioscientific inquiry? *Research in Science Education, 37*, 371–391.

Sadler, T. D., & Donnelly, L. A. (2006). Socioscientific argumentation: The effects of content knowledge and morality. *International Journal of Science Education, 28*, 1463–1488.

Sadler, T. D., Eckart, T. M., Lewis, J. E., & Whitley, K. M. (2005). It's a gas! An exploration of the physical nature of gases. *Science Scope, 29*(3), 12–14.

Sadler, T. D., & Fowler, S. (2006). A threshold model of content knowledge transfer for socioscientific argumentation. *Science Education, 90*, 986–1004.

Sadler, T. D., & Klosterman, M. L. (2009). Exploring the socio-political dimensions of global warming. *Science Activities, 45*(4), 9–12.

Sadler, T. D., & Zeidler, D. L. (2004). The morality of socioscientific issues: Construal and resolution of genetic engineering dilemmas. *Science Education, 88*, 4–27.

Sadler, T. D., & Zeidler, D. L. (2005). Patterns of informal reasoning in the context of socioscientific decision making. *Journal of Research in Science Teaching, 42*, 112–138.

Settlage, J., & Meadows, L. (2002). Standards-based reform and its unintended consequences: Implications for science education within America's urban schools. *Journal of Research in Science Teaching, 39*, 114–127.

Sneider, C., Golden, R., & Gaylen, F. (2004). *Climate change*. Berkley: Lawrence Hall of Science.

Strauss, A., & Corbin, J. (1998). *Basics of qualitative research: Techniques and procedures for developing grounded theory*. Thousand Oaks: Sage Publications.

Yager, S. O., Lim, G., & Yager, R. (2006). The advantages of an STS approach over a typical textbook dominated approach in middle school science. *School Science and Mathematics, 106*, 248–260.

Yang, F.-Y., & Anderson, O. R. (2003). Senior high school students' preference and reasoning modes about nuclear energy use. *International Journal of Science Education, 25*, 221–244.

Zeidler, D. L., & Sadler, T. D. (2008). The role of moral reasoning in argumentaion: Conscience, character, and care. In S. Erduran & M.-P. Jiménez-Aleixandre (Eds.), *Argumentation in science education: Recent developments and future directions* (pp. 201–216). New York: Springer.

Zeidler, D. L., Walker, K. A., Ackett, W. A., & Simmons, M. L. (2002). Tangled up in views: Beliefs in the nature of science and responses to socioscientific dilemmas. *Science Education, 86*, 343–367.

Zohar, A., & Nemet, F. (2002). Fostering students' knowledge and argumentation skills through dilemmas in human genetics. *Journal of Research in Science Teaching, 39*, 35–62.

Chapter 5
Metalogue: Issues in the Conceptualization of Research Constructs and Design for SSI Related Work

Troy D. Sadler, Vaille M. Dawson, Michelle L. Klosterman, Jennifer L. Eastwood, and Dana L. Zeidler

Socio-scientific Reasoning

Sadler: As I reflect on the project featured in this chapter while simultaneously considering the future of my own work in the area of SSI and the SSI research agenda more generally, I am drawn to the socio-scientific reasoning (SSR) aspect of the project. In some ways, this part of the project was not very successful. First, the students demonstrated no gains in SSR. And second, the subcomponents of the larger SSR construct (i.e., complexity, inquiry, and perspectives) did not show evidence of association to an underlying latent variable. In other words, these data suggest that the SSR aspects ought to be treated as independent variables as opposed to related subconstructs. However, despite these results, I think the work around SSR may be the most important contribution of the project. As we mention in the chapter and elsewhere (Sadler, Barab, & Scott, 2007; Sadler & Zeidler, 2009), I think that there is a real need for tools to help us as researchers and educators better operationalize what it is that we are trying to do with SSI. I do not think it is enough for us to continue to argue that we need to enhance scientific literacy. I do not think that anyone contributing to this book would disagree that promoting scientific literacy is important, but given the political climate in which schools are currently situated, I think that our community (i.e. those of us who advocate the

T.D. Sadler (✉) and J.L. Eastwood
School of Teaching and Learning, University of Florida, Gainesville, FL, USA
e-mail: tsadler@coe.ufl.edu

V.M. Dawson
Science and Mathematics Education Centre, Curtin University, Perth, WA, Australia

M.L. Klosterman
Department of Education, Wake Forest University, Winston-Salem, NC, USA

D.L. Zeidler
College of Education, University of South Florida, Tampa, FL, USA

T.D. Sadler (ed.), *Socio-scientific Issues in the Classroom: Teaching, Learning and Research*, Contemporary Trends and Issues in Science Education 39,
DOI 10.1007/978-94-007-1159-4_5, © Springer Science+Business Media B.V. 2011

contextualization of science education through SSI) has a responsibility to move beyond the rhetoric of scientific literacy as a rationale for SSI-based education.

Do others agree with the assertion that the community needs more sophisticated conceptual and assessment tools for supporting and justifying SSI-based education? Is SSR a useful construct in this regard? Does the discussion around this theme within the chapter provide useful insights? What steps should be taken to advance this work?

Zeidler: Let me first say that this study provides many key insights into important areas of SSI instruction and reasoning. The study advances the claim that even a limited experience with SSI can produce conceptual learning outcomes in students. More directly related to the question raised as to whether the science education community needs more sophisticated conceptual and assessment tools related to SSI-based instruction, my response would be "yes," though not necessarily for the premise of "supporting and justifying" the practice of SSI pedagogy. It is my observation that the SSI framework has been amply supported and justified both on analytic and empirical grounds in previous research. However, I do agree that we need more sophisticated conceptual tools to better understand the nuances of factors associated with SSI reasoning and more robust assessment tools that examine both contextual and cultural differences in reasoning about SSI. I think this study moves us a good step in the right direction.

Ironically, as I sit here writing this in May 2010, I cannot help but to think of the human and environmental crisis unfolding in the Gulf of Mexico with the BP oil spill (although the word "spill" seems like a quaint euphemism where we only need to dab up the offending toxic pollutants). This is at once a Real-World Problem and a Real-World catastrophe in every sense. It is also, obviously, a socioscientific issue. In this reality, it probably matters not whether to treat Socioscientific Reasoning (SSR) as a unitary construct or as independent variables (although for theory development it certainly does); rather, it seems clear to me that at this very moment, we can find the need for students to be able to orchestrate the components of SSR—understanding complexity of a SSI, examining issues from multiple perspectives, realizing the need for ongoing inquiry, and evoking skepticism when presented with potentially conflicting and/or biased information, of critical importance.

Sadler: I want to challenge Dana's contention that "the SSI framework has been amply supported and justified both on analytic and empirical grounds." In many respects, I agree with the statement: many studies (including those featured in this volume) provide empirical support for the use of SSI as contexts for science education. The SSI movement seems to be growing in that there is evidence of increased classroom use of SSI as well as more frequent SSI contributions to the science education literature base. However, in discourse at international and (some) national levels, SSI remain marginalized. With the parenthetical insertion in the previous sentence, I am hedging a bit because I know that some national educational systems are incorporating SSI in substantive and meaningful ways, but in our national context, the USA, SSI are not a prominent element of the national discourse around science education. My analysis of current policy and standards document in the USA leads me to think that SSI receive, at best, lip-service in framework-type statements. When those frameworks become translated into standards and benchmarks that become reified in standardized assessments which drive system-wide (local and state school systems) decisions regarding curriculum and pedagogy, the focus on all that is significant with respect to SSI falls away. It is in this context that I contend that our community has much to do in terms of justifying the SSI framework as an important aspect of science education.

A reasonable question in response to this argument in light of the previous comments regarding socio-scientific reasoning is how might a new construct (i.e., SSR) help to justify use of SSI in science education. I think that if we want to advance the SSI movement in terms of making SSI a more prominent aspect of science education, then we need conceptual tools to help translate the lofty goals that often feature SSI-related themes in policy frameworks to the standards, assessments, and curricula that ultimately get enacted. Currently, vague links are made between teaching science in the context of SSI and learner development of scientific literacy and reasoning. But scientific literacy and reasoning can mean any number of things, so assessments and curricula focus on scientific formalisms (i.e., facts and principles) that have been clearly defined and are uncontroversial. In proposing socio-scientific reasoning, we sought to be more precise in identifying a specific suite of practices that could be featured in SSI-based learning experiences and assessed. If we want to move science education systems from an exclusive focus on scientific formalisms, then we have to provide options that fit within the political constraints of those systems. Scientific literacy and reasoning (presented generally) do not fit within the current political constraints because they are ambiguous and practically impossible to assess, at least when they are presented as ambiguously as they are in policy documents. By specifically defining socio-scientific reasoning in terms of measurable subconstructs, we were attempting to create a construct that would fit within the political constraints of modern school systems and better position SSI within those systems. This is why we viewed the finding that the SSR subconstructs were not correlated as being problematic. If the subconstructs are not related then it challenges our definition and ultimate use of the socio-scientific reasoning construct. However, despite the results, I am not completely convinced that the aspects are not related. It seems unlikely to me that how individuals think about the complexity, inquiry, and skeptical aspects of a particular controversial issue are not related. It seems more likely to me that our approaches to measuring these aspects are not sufficiently valid and reliable which, of course, demands additional study.

Eastwood: I certainly agree that the goals of SSI do need to be defined beyond "scientific literacy," and that better instruments for measuring these outcomes are needed. I agree that the construct of socioscientific reasoning has great potential for this and can facilitate the development of effective assessment tools. I see the three-part construct as extremely consistent with King and Kitchener's reflective judgment model. It does make sense to think that the subconstructs of perspectives, inquiry, and complexity would be related, but I can understand why they may come out as separate constructs. For example, students might be inclined to discuss different perspectives on an issue because they were instructed to do this in class—they know to look for multiple perspectives. They could easily still be confused about how scientific knowledge is developed. I wonder if subconstructs would be more related in adults/college students, since clear patterns emerged with these groups in King and Kitchener, Perry, and Baxter Magolda's work. I also wonder if classroom scaffolding favoring certain aspects affects the outcome. It seems important to consider how the aspects are addressed explicitly and contextually in classroom discussions.

Dawson: It has been interesting to read this chapter again and in the same week as the Gulf of Mexico oil spill. I recently received the *International Journal of Science Education* issue containing the Klosterman and Sadler (2010) paper which outlines the benefit of SSI such as climate change and its role in enhancing conceptual development. I want to comment first on SSI in curriculum documents and second, ways of improving the teaching and assessment of SSI. Twice, I have heard quotes that the teaching of SSI is "woolly science". The first is apparently from the new Minister for Education in the UK suggesting that this "type of science is woolly". The second was on the front page of our local (Perth, Australia) paper when our new national science curriculum was released and our local Nobel Prize laureate in Medicine was asked to comment on the curriculum. He also described parts of the science curriculum as being woolly. The section he referred to was "science as a human endeavour". Interestingly, he is actually mentioned by name as an example of a scientist in that section. The point is that many of the people who decide what is taught in our schools are not science educators and have a narrow view of what school science is.

It seems there are two audiences that the outcomes of research in SSI need to reach: (1) power brokers and curriculum writers in central offices and (2) teachers in schools. I was involved in writing a new biology curriculum which included many aspects of SSR such as multiple perspectives, skepticism, evaluating risk, etc. By the time the curriculum was published, these SSR-related elements had all been removed mainly because they were considered too difficult to assess or too far removed from "real science". However, when I speak at teachers' conference, I receive warm receptions from many teachers who are keen to make their lessons more interesting, relevant, and contextual.

Sadler: After reading Vaille's comments, I checked an online dictionary to make sure that I knew what "woolly" meant; it is not a term that I hear very often. The most pertinent entry presented the following definition: "lacking in clearness or sharpness" (Merriam-Webster, 2010). I certainly disagree with pronouncements that SSI-based education is "woolly" in the pejorative sense evident in the quotes that Vaille mentioned. However, I do think that our approaches to defining learning outcomes and assessments in the context of SSI have lacked clearness and sharpness. The socio-scientific reasoning construct may help bring these issues into better focus. Even if the construct does not end up yielding fruitful results, I hope these discussions help the community attend to these issues and better address the perceived shortcomings of SSI-based education.

Zeidler: Not to add any more ad hominem arguments to those woolly-headed reactionaries who cannot conceive the notion that science education may exist in a social context, SSI, I believe, does have ample support and justification. The question may be, who is listening? Now by this I do not mean to imply that the SSI paradigm is now "normal science," and we merely are left with "mopping up operations" to tidy up a few loose ends. And I would agree that it is incumbent upon us—those that are advocates for this progressive scheme—to add clarity, refinement, and where necessary, dismantle and reintroduce more robust ideas about how to engage children in the activity of science, facilitate public understanding of science through the everyday use of SSR, and provide better indicators of the effectiveness of this approach. Questions obviously do remain as to whether the SSI approach is compatible with standardized assessment (as Troy alludes to) or whether models of authentic assessment may gain a foothold in the political hegemony of education (as Vaille seems to suggest). However, I do think there exists some promising protocols to help document progressive classroom environments where SSI would flourish. For example, a modified version of the NCOSP Science Classroom Observation Guide (2008), which seems sensitive to observational records of classroom inquiry dynamics and growth consistent with contemporary science education goals, is promising because it allows for the identification of "indicators" of practices that can be observed in effective classrooms, while providing a differentiation of evolving practices for teachers to become more informed over time in providing student support in the learning of science. The overarching categories (that are broken down into numerous subcategories of classroom instruction) include: Classroom Culture is Conducive to Learning, Science Content is Intellectually Engaging, Instruction Fosters and Monitors Student Understanding, and Students Organize, Relate, and Apply Their Scientific Knowledge. I am not suggesting that this particular protocol is the answer, but I am suggesting that conceptualizations of authentic assessment may, on the one hand, be realized; on the other hand, such assessments may be fundamentally at odds with the type of large-scale assessments (e.g., PISA) that are so prominent within our current system.

Research Design

Klosterman: Over the last year, we have received extensive feedback on the CATSI project. One commonly expressed concern is the lack of a control group in our research design. We acknowledge that control groups are ideal in most educational research. However, as we expressed in our paper (Klosterman & Sadler, 2010), we were not working in an experimentally ideal situation, but "in the situated world of real schools" (p. 1040). Given our local context, limited available time, and desire to work closely with a limited number of teachers, finding another classroom that was similar in terms of size and population was only one issue. In consideration of our limitations, finding another classroom that was addressing the same standards, in the same time frame, and one that did not align with or highlight any SSI was practically improbable.

We contend that the significance of our study lies in the fact that we now have empirical evidence to support our hypothesis that a SSI-based curriculum can impact student content learning. We did not investigate if a SSI-based curriculum can improve content learning MORE than a curriculum void of a SSI focus, in which case we acknowledge a control would be required. To our knowledge, previous SSI research has not looked at student learning gains (both proximally and distally) as a result of a classroom-implemented SSI-based curriculum. This type of research is critical. To influence the policy-makers and other educational gatekeepers, we need to continue to push the SSI agenda forward with concrete evidence of its impact on student learning.

Eastwood: As long as you are not trying to compare improvement in content learning to a traditional approach, I would not say that a control group is "lacking." Since there are so many variables to consider in finding reasonable comparison groups, it makes sense for a study that takes a more in-depth approach to assessment to focus on the students receiving the intervention. The approach using different "distances" of assessment is very useful to identify how and what students are learning, especially since particular assessments tend to favor one group or the other depending on teaching strategies and assessments used in the classroom.

I would say not having a control is justified with more in-depth case studies, those that take into consideration the situational aspects of a complex learning environment. To me, it was effective to have more detail than previous studies on how students gained content knowledge in the SSI intervention. When the description of the intervention and results are given in detail, the reader can make reasonable inferences about how big or small these content gains are in relation to other teaching approaches.

I do not believe the strength of the findings is limited if you are arguing that SSI promotes content learning (you showed this in a very convincing way). This is appropriate for a study using a novel approach in assessing content gains with SSI. I do think repeating the intervention with other groups could increase influence with stakeholders, strengthening the argument of other authors (especially given the various distances of assessment) that content learning is not compromised in SSI.

Dawson: Using a control or not is a very tricky point. You [the chapter authors] did argue convincingly as to why a control group was not possible in the research. Like all good qualitative research reports, you provided a great deal of contextual information and allowed the reader to decide the verisimilitude of the research. It is worth noting that different reviewers for the IJSE paper may well have rejected the paper. I approve of research that acknowledges the 'messiness' of real classroom research and it is time we, the science education community, were more honest about the nature of the work we do. The more necessarily complicated, classroom-based research is presented at conference and in journals by reputable researchers like yourselves, the more it will be accepted.

Benefits of Research

Dawson: One of the desired outcomes of quality research can be gains in understanding by the participants. I would like to ask the authors what they think the participating teachers and students gained as a result of participating in this research and how they know this. I would also like to ask what they found to be the most rewarding aspect. Finally, if they did this study again, what would they do differently?

Sadler: The student participants developed new understandings of global climate change and the scientific concepts underlying this issue. The data presented in this chapter provide support of this claim in that we document statistically significant gains in student performance on a standards aligned test as well as qualitatively distinct shifts in understandings on curriculum-aligned assessment prompts. This is the easy answer to Vaille's first question, but I do not think that it really gets to the point that she is raising. Vaille is asking about the benefits of teacher and student involvement in the research process. For this question, we have less compelling evidence because generating this kind of evidence was not one of our primary goals. We had not thought a lot about this issue beyond an expectation that the partnering students would learn some science and become better prepared for dealing with complex SSI and that teachers with whom we worked would become more comfortable using SSI in substantive ways in their classrooms. To help address this question, I asked William, one of our partner teachers, to share his thoughts on what he and his students might have gained through their participation in the project. (After the school year during which CATSI was implemented, Molly moved to another region of the country and we did not stay in contact.) William reported that his students expressed genuine enthusiasm at being a part of a research project. Many of the students felt empowered because they were contributing to something "bigger" than their typical classroom experiences. They asked questions about how their tests and information would be used and what we might learn from the results. In comparing the seriousness with which the students approached assessments associated with the project and their normal approach to classroom assessments, William felt confident that students exerted a level of effort and sincerity not usually observed.

In terms of the influence of project participation on William himself, he indicated that the experience made him more interested in educational research. At the time of the project, William was in the midst of completing a specialist degree and was considering continuing on to earn a doctoral degree. By the end of the project, William had become very interested in research and the potential roles he could play in conducting science education research. Since the CATSI project, William has continued his graduate studies and he is currently a full-time Ph.D. student. He has developed a research agenda associated with science learning in the context of authentic research opportunities.

Vaille also asked about what we, as the researchers, found most rewarding in the project. For me, the opportunity to collaborate with the teachers as extensively as we did and to be in their classes as they worked through the curriculum that we jointly developed was a great experience. For several years, I have been working with other researchers to develop empirically based understandings about how students negotiate SSI and how to situate SSI in classrooms. I visited classes and interacted with students and teachers, but this project allowed me to collaborate with teachers at a new level. In many ways, it was an opportunity to put much of what we had learned into action. I found it to be challenging but also fun and rewarding.

The last question that Vaille posed challenges us to reflect on what we might do differently if we did the project again. I have two things that I would do differently; although, only one of the two would have actually been possible given the constraints we experienced. The first suggestion, which we could have accomplished but did not think to do so at the time, would have been to collect data that could have informed questions related to student interest and motivation. SSI literature consistently claims that students become more interested in science when they can explore it through contextualized learning opportunities like SSI. I would have liked to have developed strategies for collecting student level data related to this issue. Surveys or focus group interviews focused on these topics would have been relatively easy to conduct and may have yielded valuable insights. The second change that I would like to have implemented but could not given the constraints of the specific contexts within which we worked is administering a follow-up assessment several months after the conclusion of the CATSI unit. Students showed content gains on pre/post-tests given immediately before and after the unit. A follow-up test administered 3 months after the unit would have enabled inquiry into the long-term effects of the experience. As we mentioned in the chapter, timing of the implementation and the need to work around the school and class calendars made this kind of follow-up testing impossible.

Dawson: Just one brief comment about follow-up tests: A concern I have is that students may have assessment fatigue if asked to write too much about one topic even if they are happy to participate in the research. Some students may write less on the premise that they have already told you the answer. Of course knowing whether gains in learning are sustained is important. Maybe we have to be creative about how we find this out. The other important point is whether students who have a greater understanding of climate change issues actually change their behavior in any way.

Klosterman: As Troy mentioned, we admittedly lack evidence to support any claims about how students or teachers benefited from our research in terms of interest or motivation. Nor do we have interview or survey data about the impact of our study outside of content learning. Student content learning and socio-scientific reasoning were the foci of our study and therefore drove our research design. However, this study reminded me of the power that comes from collaborating directly with teachers on projects that immediately impact their classrooms and student learning. For me, this was the most rewarding aspect. As a result of this study and through direct collaboration with teachers, we developed a tangible product that was immediately usable by teachers and was loaded with science content and highlighted the social, economic, and political aspects of global climate change. At the most basic level, the teachers benefited by working as a part of a team to develop this product that was theirs to keep.

In my opinion, having conversations about the theoretical underpinnings of our work and future research possibilities is stimulating conversation. But I believe that our work is truly limited if it does not clearly translate to classroom practice. In this study, we worked with teachers from the outset. We worked together to choose a SSI that was relevant to the teachers' classrooms and spent a considerable amount of time designing lessons that aligned to the course, state, and national standards and personal teaching styles. In fact, at times it felt like we were participating in a classroom service project rather than a research project because our efforts were so focused on the tangible curriculum. But to me, that makes sense. Ultimately, the goal of research is to improve student learning. The closer we can get to students and to the teachers that work with them, the more likely we are to impact student learning.

Although I felt like this project was certainly a step in the right direction, I feel like we still could have pushed the envelope farther in terms of the practical utility of our research findings. The results of this study obviously contribute to the science education research field and its understanding of how SSI-based instruction can impact student learning. However, what did the teachers gain from these findings? We mentioned that the teachers were admittedly less involved in the design of assessment instruments than in curriculum development. We did not make a deliberate decision NOT to involve the teachers; the timing and amount of effort required to do so made it impractical. However, I am left to wonder what the teachers would have thought of the students' responses to the proximal (curriculum-aligned) assessment and their ability to make connections to the broader scientific concepts on the distal (standards-aligned) assessment. Would teachers have used those results? And if so, how?

Dawson: The benefit of educational research to the participants is something that I ponder often in my classroom-based research. When the focus of the research is related to morals, ethics, values, and multiplicity of views, then it becomes even more pertinent. We would like students to consider other stakeholders' points of view, show empathy so it is important that we do the same.

In regards to motivation and enjoyment, certainly observing classes where students are debating SSI, the excitement is palpable (perhaps difficult to collect evidence about though). In addition, this motivation may well be one of the reasons that students' conceptual understanding improves when SSI are used even if not as much time is spent on learning content.

References

Klosterman, M., & Sadler, T. D. (2010). Multi-level assessment of content knowledge gains in the context of socioscientific issues based instruction. *International Journal of Science Education, 32*, 1017–1043.

Merriam-Webster. (2010). In *Merriam-Webster's online dictionary*. Retrieved June 4, 2010, from http://www.merriam-webster.com/dictionary/woolly.

Sadler, T. D., Barab, S. A., & Scott, B. (2007). What do students gain by engaging in socioscientific inquiry? *Research in Science Education, 37*, 371–391.

Sadler, T. D., & Zeidler, D. L. (2009). Scientific literacy, PISA, and socioscientific discourse: Assessment for progressive aims of science education. *Journal of Research in Science Teaching, 46*, 909–921.

Chapter 6
Effects of an Interdisciplinary Program on Students' Reasoning with Socioscientific Issues and Perceptions of Their Learning Experiences

Jennifer L. Eastwood, Whitney M. Schlegel, and Kristin L. Cook

Preparing students to take informed positions on complex problems through critical evaluation is a primary goal of university education (Association of American Colleges and Universities, 2007; Baxter Magolda, 1999) and an important aspect of scientific literacy (Sadler & Zeidler, 2009; Roberts, 2007). In approaching contemporary problems, the ability to understand and position oneself in interdisciplinary issues is essential (Mansilla & Duraising, 2007; Klein, 1990). This is especially true where the study of biology meets contemporary global problems. For example, to understand the nature and impacts of disease, it is essential to examine psychological and socioeconomic aspects as well as biology and pathology. These goals of university education are highly consistent with the framework of socioscientific issues (SSI), which seeks to integrate science concepts and their social significance, facilitate reasoning with complex problems, and promote content learning (Sadler, 2009; Zeidler, Sadler, Applebaum, & Callahan, 2009).

This study focuses on a 4-year university program designed to integrate biology with social aspects of the human, scaffolding students to develop their reasoning related to complex issues and advocate for their own committed positions. The primary goal for this research was to compare reasoning and perceptions of students who participated in this program and those who chose a traditional biology major. We intended to determine whether students who experienced a sustained approach to teaching through SSI differed in their reasoning or incorporation of different perspectives into their thinking about science issues with social significance. We also aimed to discover common themes in students' perceptions of their experiences with SSI and overall outcomes of their majors.

A second goal of the study was to illustrate how the pedagogy was enacted to meet program goals. Many factors contribute to the complex learning environment of an SSI unit or course. In the SSI classroom context, variables contributing to

J.L. Eastwood (✉)
School of Teaching and Learning, University of Florida, Gainesville, FL, USA
e-mail: jleastwood@coe.ulf.edu

W.M. Schlegel and K.L. Cook
Indiana University, Bloomington, IN, USA

T.D. Sadler (ed.), *Socio-scientific Issues in the Classroom: Teaching, Learning and Research*, Contemporary Trends and Issues in Science Education 39,
DOI 10.1007/978-94-007-1159-4_6, © Springer Science+Business Media B.V. 2011

student outcomes cannot be easily isolated to reveal their direct contributions to student outcomes. Particular aspects of instruction can influence students' knowledge and perceptions to different degrees and complicate findings. Describing the curriculum and detailing how instruction is carried out helps to explain the nuances or anomalies in results. Where other research reports discuss goals or general strategies of SSI instruction, we sought to provide a more in-depth description of the context to create a more complete picture for interpreting results. We also envisioned the study as an opportunity to address a current gap in the literature, showing how SSI is enacted in a college level environment.

A third goal was to apply an SSI framework to an existing program developed using resources from outside the science education literature. The program under study was developed using literature on development in the college years and interdisciplinary education. It sought to challenge students to explore different perspectives and develop positions to which they are committed. Professors used an interdisciplinary, case-based approach to help students learn to reason and take positions on controversial issues with both scientific and social implications. Although the goals and pedagogical strategies employed were consistent with SSI, the term "socioscientific issues" was unfamiliar to the program faculty. Through our collaboration, a science education researcher and the program director, a biology professor and well-established scholar in biology teaching and learning, hoped to connect congruent concepts from our different perspectives.

A fourth goal was to provide an opportunity to open the SSI discussion to include reasoning with socioscientific issues for students of science and health professions. Most of the SSI literature has focused on science education for citizenship. Although it is important to help all students understand the science behind issues, reason effectively when called upon to vote, and advocate for their positions, different approaches may be needed for students entering fields like healthcare and environmental science. Their decisions will affect patients' outcomes and policy on important issues, so ability to reason across disciplines and understand different perspectives is essential for these students. For science majors, content-heavy required courses and busy programs of study may leave little room for SSI instruction, so it is important to illustrate how SSI may be effectively integrated into college science curricula.

Establishment of Teacher–Researcher Collaboration

This section is written from the point of view of the first author, a researcher in science education. This collaboration was unique in that the teachers, including the program director and several faculty members, were already independently conducting and presenting teaching and learning research. Dr. Schlegel (second author on this chapter) is a professor of biology with a record of scholarship in case-based learning, team-based learning, and interdisciplinary education. Before beginning this study, Dr. Schlegel and I had established a relationship over

4 years of my graduate degree, having carried out studies to investigate student interactions, content learning, and the role of reflective activities in her team-based and case-based physiology course. In our meetings over these years, we discussed the development and enactment of the interdisciplinary program (SSI). She and other program faculty were involved in studies of student engagement and outcomes in relation to pedagogical strategies including team-based learning and interdisciplinary team teaching, and specific learning tools including a longitudinal reflective portfolio.

Having understood the goals of the program as well as the theoretical background for the curriculum and pedagogical strategies, I chose to situate my dissertation research in this program, which at the time was in its fourth year and holding its first senior level course. Rather than developing an intervention, I hoped to document a novel approach to college biology education, connecting pedagogical practices to the science education literature and detailing both what took place in the classroom and student outcomes and perceptions. As a researcher with no official affiliation with the program, I could provide a different perspective from those centrally involved, comparing intended practice with observed practice. My position as researcher could also increase credibility of feedback from students, considering that they may be inclined to respond differently to program faculty out of concern for their grades or their professors' feelings. I was also better positioned to compare outcomes and perceptions of students in the program to a comparable group of students who chose a traditional biology major.

Despite these aspects of my role as researcher, I could not entirely position myself as an "outsider." I had become familiar with the program, observing planning meetings, and discussing the theoretical underpinnings and pedagogical strategies in depth with program developers and instructors. Also, students in the SSI group recognized me as a "friend of the program" and knew I worked with their professors, so they may have been cautious about their comments in interviews. My "close" position led to some limitations but also created affordances that would have been unavailable otherwise. I was able to mingle with both students and professors in the SSI group and develop a sense of the student and faculty cultures. Students allowed me to sit in their groups and listen to their informal conversations. They recognized me and seemed to feel comfortable with me. I gained a similar perspective with the senior level biology students, for whom I served as laboratory instructor. As a graduate instructor, I had a "helper" role for these students and was likely perceived as someone they could trust. They informally discussed their successes and frustrations with me, and this provided an insider perspective on being a biology major.

Perhaps the greatest benefit of a close relationship with the SSI program was the opportunity to become familiar with the point of view from which the program was designed. A key feature of the program was a developmental approach, recognizing trends in epistemological and ethical development that influence learning. Professors designed courses with the understanding that college students exhibit consistent patterns in their understanding of knowledge and approaches to learning as they navigate experiences in a college environment. Professors used these patterns to understand how their students think, effectively scaffold student learning, and design learning objectives. Understanding this underpinning of the program helped me to compare the actual learning environment to that intended by faculty members. In addition, my close relationship with the program allowed me to understand the focus on advancing development as being highly consistent with the aims of SSI to enhance students' evidence-based reasoning and consideration of multiple perspectives. This helped me to more clearly articulate my theoretical framework for the study.

Research Questions

This study investigates the effects of this interdisciplinary undergraduate program. Although the program incorporates many theoretical constructs and pedagogical strategies, we focus on the SSI context embodied in the program. As SSI students typically chose the program as an alternative to the traditional biology major and were similar in achievement levels and future career paths, our comparison group included biology majors (referred to as the "BIO group"). Research questions include:

1. Do SSI majors reason with socioscientific issues differently from BIO majors?
2. How do SSI and BIO majors' perceptions of their experiences with socioscientific issues differ?
3. How do SSI and BIO majors' general perceptions of their majors, including personal outcomes and the learning environment differ?

Theoretical Framework

We approached the study with a theoretical lens incorporating situated learning and theories of development in the college years. Situated learning posits that knowledge is connected to the context in which it is learned (Brown, Collins, & Duguid, 1989). As a tool is understood through its use, students make sense of a new concept in the context of its application and discipline. Contextualized learning promotes a knowledge structure that allows concepts to be accessed for relevant problems and not remain "inert" (Bransford, Sherwood, Vye, & Reiser, 1986). Effective learning environments offer students opportunities to work with and apply concepts in contexts authentic to their use. They also remain authentic to common practices in the fields, such as collaboration. For example, physicians, nurses, and other health-care professionals work in teams, and scientists collaborate among and between research teams. Through this theoretical lens, pedagogy that emphasizes interpersonal interaction and is contextualized in realistic problems should promote development of concepts, skills, and disciplinary knowledge. In the SSI program, case studies situate learning in realistic problems, promoting development of knowledge less likely to remain "inert." The collaborative nature of the program promotes cognitive development through interaction and explanation, and situates learning in a realistic context, since biological inquiry is generally a collaborative endeavor.

Developmental frameworks for college students and adults, including the Reflective Judgment Model (King & Kitchener, 1994), Perry's Scheme of Intellectual and Ethical Development (1979), and the Epistemological Reflection Model (Baxter Magolda, 1992) also contribute to the theoretical foundations for this study. Although the researchers behind these models used different approaches and arrived at different conclusions, similarities among their findings reinforce and validate the existence of particular trends. Early stages are characterized by conceptions of knowledge as absolute and derived from authority, understanding of reality as directly observable, and difficulty recognizing complexity or different perspectives. Middle stages are characterized by perception of complexity, uncertainty, and multiple perspectives, although reasoning may be inconsistent and decisions or commitment may be hindered by complexity. In the highest levels of development, knowledge is seen as complex, uncertain, and a product of inquiry. Individuals apply consistent criteria to form evidence-based decisions and recognize and incorporate multiple perspectives in their reasoning.

Here, we do not seek to apply developmental stages to participants or document changes in these stages, but to approach the study with the understanding that students' ability to perceive complexity or uncertainty in situations, recognize inquiry-based rather than authority-based sources of knowledge, base personal positions on evidence, and consider multiple perspectives develops over time and with experience. The classroom environment, including opportunities for reasoning with ill-structured problems with instructors and peers, influences students' development to higher levels (Baxter Magolda, 1999; Zeidler et al., 2009).

Context of Study

The study took place in a large, research-oriented university. Participants were recruited from the SSI and BIO majors. These majors differed where, in addition to required and elective courses, all SSI majors took yearly core courses and seminars and maintained a 4-year portfolio. BIO majors took cohesive series of required and elective courses, but were not involved in longitudinal projects or yearly core courses. Outside of core courses, curricula for students in both programs were very similar and students took many of the same courses. In the SSI major, SSI case-based reasoning was deliberately structured into all core courses, with a developmental focus on moving students from exploring different perspectives to position-taking and advocating for evidence-based positions. For the SSI group, we will provide an in-depth description of the class context and activities used to specifically teach SSI. We will then describe the comparison (BIO) group in terms of curriculum and general teaching methods.

SSI Group

The published mission of the SSI program was "to integrate the biological and social sciences with the humanities and the arts in the study of human beings and the human condition." Students enrolled in the program planned to enter life science graduate programs or professional programs, such as medicine, nursing, dentistry, physical therapy, law, or journalism, or pursue careers in teaching, the life science industry, or public policy. The program included selected courses from four concentration areas: human environment and ecology, human origins and survival, human health and disease, and human reproduction and sexuality, as well as a series of interdisciplinary core courses taken each year.

The foundation of the program was established in yearly core courses. These interdisciplinary courses connected primary biological concepts with related social and ethical issues and explicitly addressed epistemological concepts in biology including uncertainty, tentativeness, and the centrality of evidence to knowledge in biology. The interdisciplinary nature of the program responded to recent trends, including converging fields of disciplinary knowledge, professional requirements, and the need to solve problems that are both social and intellectual in nature (Klein, 1990). Core courses were team taught by an expert from the life sciences and an expert from a social science or humanities discipline, and the specific course topics depended on the expertise of the instructors.

Key themes running through these courses included scientific literacy through position-taking on socioscientific issues, collaboration, contextualized learning, and reflection. Scientific literacy includes "informed decision-making, the ability to analyze, synthesize and evaluate information, dealing sensibly with moral reasoning and ethical issues, and understanding connections inherent in socioscientific issues" (Zeidler, 2001). In the SSI core classes, scientific literacy was conceptualized to connect the inquiry process with social context, historical context, and ethical context. These efforts toward scientific literacy echo the goals of SSI. Understanding of issues was developed through discussion, peer evaluation, and service learning.

Student collaboration was a central component of the SSI program. Students worked in teams for entire semesters on case studies, course projects, and even exams. This aspect of the program is based on scholarly work on team-based learning (Fink, 2002; Schlegel & Pace, 2004).

Contextualizing learning in scientific problems through case studies and service learning projects was also an essential aspect of the program. Learning in the context of case studies promotes ethical development through consideration of how power and authority influence scientific endeavors (Zeidler & Keefer, 2003). As established for problem-based and case-based learning, problems addressed in core courses are complex, target multiple skills, and resemble the activities members of a culture actually participate in. Within the case context, the instructors provide models of expert problem solving (Willaims, 1992). The team teaching approach allows students to see integration of different perspectives to solve problems. Reflection and documentation of content learning, collaborative processes, and personal experiences are scaffolded through class discussions and a progressive electronic portfolio which individuals update throughout the entire program. Technology-based reflective scaffolds have been found effective to help students articulate ideas and develop and reflect on explanations (Land & Zembal-Saul, 2003).

The following section will provide context for the teaching of SSI in the two core courses observed: sophomore and senior level courses.

Sophomore Level Core Course

The syllabus describes the primary objectives of the course as follows:

> This course introduces the social and ethical dimensions of human biological experience and the construction of scientific knowledge through in-depth consideration of human death and disease … we will use a collaborative, case-based approach to explore the operationalization of scientific concepts, the logic of scientific inquiry, and the effective communication of evidence, interpretation, and claims.

Learning goals included constructing models integrating different fields of science, gathering, evaluating, and applying scientific data to understand patterns of disease and death, evaluating different perspectives on death and disease and arguing a chosen position using scientific evidence, and developing a portfolio in areas of inquiry of personal interest.

The structure of the course was team-based. Teams of five to six students collaborated in and out of class, and the majority of class time involved teams working together on particular tasks, generally case studies, followed by whole-class debriefing. Since the course depended on active student involvement, participation and peer evaluation made up 20% of the course grade. Other activities included short lectures by instructors, guest lectures, short individual assessment activities, team presentations, and exams. In class, teams were seated around individual tables to encourage team interaction. Each table was equipped with at least one laptop to research topics and create documents. The atmosphere of the classroom was energetic and informal, with the majority of interaction within teams.

Like most of the SSI core courses, the class was team-taught by a life scientist and an expert from the social sciences or humanities. In the sophomore course, the instructors included a neuroscientist and a sociologist with expertise in epidemiology. Generally, one professor presented new information and when cases were discussed, each professor modeled his or her disciplinary perspective and the importance of integrating those perspectives. They were honest about the limitations of their knowledge and participated in information searches when difficult questions arose.

The course was structured into three modules: death and dying, infectious disease, and HIV and AIDS. These modules focused on understanding operational definitions of death and disease for different contexts, physiological and microbiological understanding of these topics, understanding of these topics at multiple scales, and use of scientific information in arguing positions. We will describe activities from the third module in depth to demonstrate how the course was designed to encourage effective reasoning with SSI, teach science content in context, help students evaluate differing perspectives, and develop and argue their own positions.

The third module spanned approximately 4 weeks. The stated goal of the module was, "We use in-depth analysis of HIV/AIDS to investigate how complex pathogens, politics, and ideologies contribute to infectious disease epidemics locally and globally." Specific learning goals included biology content, such as differentiating between viruses and retroviruses, analysis of public epidemiological data, and argumentation regarding political and ideological controversies around the disease. The largest part of the module included discussion of a controversy ignored by most scientists and the media over whether the HIV virus actually causes AIDS, which was presented in two papers from the journal, *Science* (Blattner, 1988; Duesberg, 1988). Before class, students read the position papers and completed a reflective writing assignment asking them to answer why they do or do not believe HIV causes AIDS, what issues were important from the readings, and five areas they needed to learn more about to resolve the problem. Team members were instructed to share their responses, pool knowledge, and research some answers to questions they had identified. They were also asked to develop a consensus list of key issues from the readings. To better understand Duesberg's argument against HIV as the causative agent in AIDS, each team was assigned one or two of Duesberg's ideas to investigate and explain in depth for the next class. Students used additional sources to develop short presentations on such topics as accepted postulates of virology, normal characteristics of retroviruses, and normal presentation of disease after viral infection.

Students' final assignment for "The Duesberg Phenomenon" was to present an argument supporting or refuting Duesberg's argument. One professor explained the assignment, emphasizing that students should focus on making strong arguments. Students' handout read,

> Your goal in your investigation is to evaluate Duesberg's position and that of his critics with reference to current scientific knowledge. As you evaluate their positions, consider what we have learned about evaluating scientific evidence: for which side is the evidence strongest? What additional information is needed to fully evaluate the competing positions?

Students were told they were expected to consult relevant sources other than those provided to the class, finding information that was current and reliable. Each team was expected to define both Duesberg's and Blattner's positions clearly and use evidence to make an argument for their positions. Students made 5–7 min presentations, answered questions posed by their peers and professors, and turned in a summary of their argument.

In the second part of the module, students considered government responses to the AIDS epidemic, using the contexts of Brazil and South Africa. First, the class was introduced to Brazil's position rejecting money from the United States, which was set aside for AIDS programs on the condition that the country make a declaration condemning prostitution. The students were asked to consider, "Is it appropriate for the US government to place ideological constraints on funding for global health initiatives?" Teams were split in half by "yes" and "no" positions, and were instructed to develop "logical evidence-based arguments" for assigned positions. Each side was given about 30 min to research their positions, and then they made their arguments to their team members. As a class they presented preliminary points for "yes" and "no" positions. The teams then developed consensus positions and arguments supporting their positions incorporating background information on the history of US funding for AIDS programs and of AIDS in Brazil. Their handout stated, "The most effective arguments will take into account the points made by the opposition."

For their final projects, teams were assigned a paper comparing and contrasting how AIDS is experienced in Brazil and South Africa, and presenting a "multiscalar model of AIDS." Teams were expected to integrate aspects of the biology of AIDS with epidemiological data, prevention strategies, treatments, and factors influencing treatment of individual patients in each country. Teams were required to provide a visual representation of the model with a complete description.

Senior Level Core Course

This core course titled "Complex Problems of Humanity" was considered the capstone experience for the SSI program and was geared toward student advocacy. Unlike the other three core courses, it was not team-taught, but lead by the director of the program. The course was primarily collaborative and project-based and students took responsibility for the direction of the course. It involved service learning components, working with other organizations to participate in the National Global Warming Teach-In, and work with the local Parks and Recreation Department to assess and research local water quality. This course focused not only on understanding socioscientific issues and arguing positions based on evidence, but challenged students to organize and act based on these positions. The course description in the syllabus read as follows:

> In this course students will focus on significant problems at the interface of science and society, such as global warming, water contamination and scarcity, fossil fuel exploitation, insufficient global healthcare, and inefficient use of dwindling land resources. Students will advocate for change so as to persuade policy makers and community leaders to support change using innovative approaches that reflect the foundations of science.

Learning goals included engaging scholars in an endeavor to educate the community about the complexity of problems and the need for different perspectives in finding solutions, understanding different dimensions of problems from local to global, understanding advocacy through individual reflection and team work, and learning how to confront challenges as "an engaged citizen with an evidence-based approach to advocacy."

Three full modules were completed in the course. The first module was intended to educate the public and focused on global warming. The second module, intended to engage students in the community, included a service learning project with the local Parks and Recreation Department and a state river water quality program. In this module, students researched the definition of a water shed, reflected on documentary films, and completed case studies on ecological and political concerns involving water, and connected this knowledge to local water issues through collecting water samples and conducting original research projects. The final project included group poster presentations of this research, which were open to the university and community. The third module included a personal audiorecorded reflection on advocacy. To illustrate how goals of SSI were enacted in this course I will discuss the first module in-depth.

The 4-week global warming module began with general education on the subject through independent study and discussion of various readings including Gore's (2006) *An Inconvenient Truth*. The majority of class time was spent on planning and holding an on-campus teach-in incorporating the goals of the National Teach-In (http://www.nationalteachin.org). Students were expected to review the goals of the National Teach-In and review the scientific data and human and ecological implications of global warming independently. In class, students were introduced to two graduate students from biology and political science who organized the first teach-in at the university the previous year. These graduate students described the outcomes of the previous teach-in as well as specific challenges and suggestions for the next one.

After having opportunities to brainstorm ideas for the teach-in with input from the graduate students, students were asked to condense and report their goals for the teach-in. These included clarifying myths about global warming, getting good attendance, motivating participants, educating participants on how they could help, providing information on local resources for energy conservation, becoming more fluent with ideas and vocabulary related to global warming, and making impacts like reducing carbon footprints and improving health. Students shared ideas for local groups and professors who might prepare presentations or exhibits for the teach-in. They were also given opportunities to ask questions of the class. They shared what they knew about contacting politicians, how human health can be related to global warming, and what kinds of visual aids might be useful for conveying the environmental impact of global warming. For the remainder of the module, students worked in groups to reserve a room in the student union, book speakers and representatives of community resources to meet goals previously specified, manage funding resources and ask businesses for donations, and decide on schedule and room layout issues.

The teach-in lasted a full day and was well attended by students, professors, and passers-by. Several representatives of local conservation resources were available for consultation, professors gave presentations on relevant research, and brochures and student-developed educational resources were available. At the end of the module, students were asked to reflect on the experience, including conceptual learning as well as personal and team-building experiences. They prepared a list of effective strategies and suggestions for future teach-ins. Overall in this module, students studied the scientific and social implications of global warming, developed informed positions about what could and should be done about the issue, and advocated for their cause through a collaborative effort combining local and national resources. This module not only helped them reason in a socioscientific context, but to put their knowledge and personal positions into action. This required reaching consensus among many different individual perspectives to create and carry out specific goals.

Biology Comparison Group

BIO majors were recruited as a reasonable comparison group to SSI majors, considering they take many of the same courses and pursue similar career and professional paths. This group, though diverse, was considered because they did not participate in the series of core courses, SSI community events, or portfolios designed to promote reflection on interdisciplinary, biology-related issues. Since there were no correlates to SSI core courses and individual programs of study varied greatly, we will describe the basic curriculum and goals of the major.

The BIO major required courses in introductory biology, molecular biology, and evolution. Many of the courses had associated labs. Based on limited observation of biology courses (molecular biology), the first author's participation as a lab instructor (histology), and student interviews, we understood that biology courses were typically lecture-based. Professors delivered slideshow presentations and students took notes. Students were given opportunities to ask questions, and many large courses scheduled discussion sessions with graduate instructors, where students worked on worksheets or brought questions on course material. Lab courses offered students opportunities to learn techniques and verify concepts taught in class, and in some instances were inquiry-based where students independently investigated their own research questions. Although we could not verify that all biology courses did not have SSI components, no biology course descriptions reviewed included SSI, and none of the 16 students interviewed reported having been involved in in-depth discussion of SSI in their biology courses.

Methods

We used a triangulation mixed methods convergence design, in which both qualitative and quantitative data are collected and results are converged. To interpret results, outcomes of each component are compared and contrasted to validate or further explain a phenomenon (Creswell & Plano Clark, 2007).

Participants

Participants included students at the mid-point and end of their college careers. Both groups were chosen because the SSI senior class was very small (19 students), and both sophomore and senior level core classes were in session at the time of data collection. Sophomore level participants included 30 SSI students and a matching sample of 30 BIO students, and senior level included 15 SSI and 20 BIO students. SSI students were recruited from core classes, and BIO majors were recruited from biology classes of comparable levels. The criteria for participation in the study were major and level of progression in the major, as determined by the levels of recruitment courses. These convenience samples, though not ideal, made data collection manageable. Recruitment courses for the BIO group included two sections of sophomore level molecular biology (approximately 200 students each) and four lab sections of senior level Human Tissue Biology (approximately 120 students total). SSI students were recruited first, and recruitment of BIO students continued until nearly equal sample sizes were reached.

Overall, SSI and BIO participants reported similar professional goals and grade point averages (GPAs) (with BIO students approximately 0.2 points higher on a 4 point scale; see Table 6.1). SSI and BIO students were nearly equal in numbers

Table 6.1 Demographic information for SSI and BIO participants

	SSI soph	BIO soph	SSI senior	BIO senior	SSI total	BIO total
	$n=30$	$n=30$	$n=15$	$n=20$	$n=45$	$n=50$
Career path						
Medicine/PA	13	13	5	7	18	19
Nursing	3	2	1	0	4	2
Other grad health profession	4	9	4	8	8	17
Work/other	2	0	2	0	4	0
Graduate school/research	3	4	1	2	4	6
MBA	2	0	0	0	2	0
Public health/social work	0	0	2	1	2	1
Undecided	2	1	0	0	2	1
Law	1	1	0	1	1	2
Minor						
Psychology	9	5	6	5	15	10
Other social science	4	4	3	6	7	10
Human sexuality	2	0	1	0	3	0
Humanities	7	13	1	4	8	17
Business/management	5	0	0	2	5	2
Public health	0	1	1	1	1	2
Biology/exercise Science/nutrition	3	2	3	0	6	2
Chemistry	1	9	4	13	5	22
Information technology	0	1	0	1	0	2
None	5	6	2	1	7	7
Average GPA	3.20	3.41	3.32	3.41	3.26	3.41
Lab experience (n/n reported)	3/27	7/30	3/13	6/18	6/40	13/48
Teaching experience (n/n reported)	4/27	4/30	6/13	8/18	10/40	12/48

of students planning to go to medical school and graduate school or research. Few, but comparable numbers of students from each program planned to study law, public health or social work, or nursing. A greater number of BIO students planned to enter other graduate level health professions, like dentistry, optometry, and physical or occupational therapy. Responses to why students chose their major indicate that the biology major was a close fit to the requirements for these professional programs. SSI students were slightly more likely to plan to enter the workforce or to obtain a business degree.

Minor choices reveal some differences in focus of study between SSI and BIO participants. Minors are reported by adding together all minors listed by students in each group. Some students had multiple minors and some had no minor. Double majors were few, but were included as minors because they illustrate additional expertise. BIO students were more likely to have multiple minors than SSI students (43% vs. 17% respectively in sophomore level classes and 55% vs. 40% in senior level classes). This could be due to the interdisciplinary nature of the SSI program. Since focus areas allow students to explore areas outside of biology, a minor may not have been viewed as necessary to illustrate expertise. SSI participants were more likely to minor in psychology or business. These students may be more interested in behavioral aspects of humans. More SSI participants minored in biology, exercise science, or nutrition, by taking enough additional courses to meet the qualifications. We noted that the biology minor was only an option for SSI students. BIO students were much more likely to minor in chemistry. Informal discussions with students suggest this is because the requirements for the biology major include a great deal of chemistry. BIO students were also much more likely to minor in areas of the humanities, such as foreign language, literature, creative writing, music, or visual arts. The reason for this is unclear, although a more focused curriculum may offer more flexibility to pursue an additional focus area. Groups were similar in minors in social science and public health.

SSI and BIO participants both reported that one-fourth of the groups had experience teaching, including undergraduate teaching assistantships, teaching programs for children, and health-related training programs. BIO students were about twice as likely to have participated in a research group (BIO: 27%, SSI 15%). Although the faculty in the SSI program worked to encourage undergraduate research, surveys suggested biology majors may have participated more to be competitive for professional programs.

Data Collection

General Procedures

Participants completed questionnaires in an online format in a computer lab setting. A subset of sixteen participants (four from each group by year and major) were interviewed on a voluntary basis. SSI participants were chosen to best represent the population of the small program by sex and ethnicity, then to provide a variety of grade point averages and course backgrounds. BIO participants were chosen to match SSI participants as closely as possible by these criteria. No male students from the sophomore level BIO group were available for an interview, so only female students were interviewed. The interviews followed a semistructured protocol, but interviews varied depending on the interests or concerns of the participants. In addition to researcher notes, interviews were audiotaped and transcribed. Course observations and professor interviews served as secondary data sources. To provide context for the core courses, the first author attended more than half of the class sessions for both sophomore and senior level SSI core courses. Field notes were taken describing the activities and atmosphere of classes attended, and all professors for these courses were interviewed to further establish goals for SSI and perceptions of student progress. One interview with both sophomore level professors was audiotaped and transcribed. The director of the program who taught the senior level class participated in the writing of this report. In addition, copies of syllabi, assignments, and handouts used in these classes were collected.

Decision Making Questionnaire

Participants took a modified version of the Decision Making Questionnaire (DMQ; Bell & Lederman, 2003) near the end of the spring semester. The questionnaire development was based on various resources and validated through review by an expert panel of four science educators and two scientists. Science and technology issues were chosen to represent real, controversial issues in which citizens may need to consider and interpret a great deal of evidence to make decisions.

Although this questionnaire was originally used several years before the current study, we selected it because it was designed to measure socioscientific reasoning in adults. Also, the scenarios and the science behind them were likely to be familiar to an audience of science students. The original questionnaire included scenarios of science and technology-related controversial issues, including fetal tissue implantation, climate change, the relationship between diet and cancer, and smoking. To reduce the time required of participants, and because of the highly emotional nature of the topic, we chose to omit the scenario on fetal tissue implantation. Scenarios were followed by questions that asked students to take positions on the issues and explain the factors influencing their decisions.

Interviews

Within the semistructured interviews, students from both groups were asked to talk about personal outcomes of their majors and what experiences, inside or outside of normal major requirements were most significant in their development. They were asked if they had taken courses in which they discussed or reflected upon science issues with social impacts. They were also asked to discuss their perceptions of the learning environment in their major courses including teaching strategies they found useful, levels of community, and interaction with professors. They were directed to speak primarily of courses within their major, but they were free to discuss experiences in any of their courses.

In addition, probing questions used by Bell and Lederman (2003) in follow-up interviews to the DMQ were used to validate questionnaire responses and evaluate reasoning strategies. We included or adapted these questions to further probe reasoning in each of the three scenarios, and asked clarifying questions about student responses on the DMQ. Follow-up questions specifically asked how participants made decisions in response to opposing arguments, which were still debated at the time of the interview.

Data Analysis

Comparison of Decisions

The modified DMQ was analyzed blindly to groups. "Yes," "no," and "undecided" decisions for each question were totaled for the four groups and compared by percentages of students choosing each decision. Differences in decisions between SSI and BIO students were tested for significance using Fisher's Exact tests for sophomore level, senior level, and total groups. Because undecided decisions were few, they were omitted from the analysis.

Comparison of Factors in Decision-Making

Based on the entire set of DMQ responses, categories of factors considered in decision-making for each question were established through several rounds of inductive analysis and revision. As similar themes in these factors emerged, category codes were developed, refined, and used to re-code questionnaire responses. After the first author refined and reduced the codes, a second researcher confirmed these codes or suggested adjustments to the coding scheme, based on her analysis of an approximate 20% sample of questionnaires. Finally, the first author reviewed all questionnaire responses and adjusted coding to accommodate minor adjustments made in negotiation with the other researcher.

Many of the questions in the DMQ were conceptually related and consequently received similar responses in factors considered. Based on similarity of responses, questions were grouped into emergent clusters that differed slightly from the original grouping of questions by scenario. For each cluster, a list of reasoning categories was developed and refined from the codes designated for the respective questions. For each questionnaire, reasoning categories represented at least once in each cluster were determined. Coded questionnaires were then identified by year and major, and the number of students citing each reasoning category was compared for groups by calculating percentages.

Scoring Rubrics

To assess reasoning with SSI, a scale was adapted from Zohar's and Nemet's (2002) argument analysis, and Tal and Hochberg's (2003) Reasoning Complexity Rubric originally developed from Hogan, Nastasi, and Pressley (2000). The simplified scale included number and explanation of justifications (see Tables 6.2 and 6.3). Like both cited analyses, the rubric rated responses on number of justifications supporting decisions as well as whether students explained an underlying reason or mechanism for their justifications (Tal & Hochberg, 2003; Zohar & Nemet, 2002). No points were awarded when no reason was cited or the reason was nonsensical in the context of the question, one point was awarded for one unelaborated or unexplained justification, two points were given for two or more unelaborated justifications, three points were given for one elaborated justification, four points were given for multiple justifications with one elaboration, and five points were given for multiple elaborated justifications.

Table 6.2 Rubric for reasoning and perspectives applied to DMQ

Reasoning score (R)	Perspectives score (P)	Example: ban smoking?
0—No justification/ nonsensical in context of question	0—No evidence of multiple perspectives	No … tobacco companies are right in saying that smoking is a free choice of the consumer. However, it is not the free choice of the nonsmoker receiving passive cigarette smoke. So, though cigarette smoking should not be illegal, there should, however, be legislation passed that confines cigarette smoking only to smokers
1—One justification of decision: mechanism unelaborated	1—Recognizes other perspectives exist, but does not elaborate them	R: 4; Two reasons support decision: free choice and reasonable alternative (explained)
2—Two or more justifications of decision: mechanisms unelaborated	2—Elaborates on different perspectives, but offers no logical conceptual resolution	P: 3; Resolution incorporates perspectives of smokers, tobacco companies, and nonsmokers
3—One justification of decision: mechanism explained with examples	3—Considers different perspectives in depth and reaches a clear, complex resolution	
4—Two or more justifications of decision: one mechanism explained		
5—Two or more justifications of decision: multiple mechanisms explained		

Table 6.3 Examples of scoring for reasoning scale

Score	Participant response	Explanation for scoring
1	No, that would have a negative effect on the economy	One unelaborated justification
2	Yes, I do not like cigarette smoke in general, and children seem to be starting smoking earlier and earlier—it is very sad	Two separate unelaborated justifications (personal dislike, children smoking)
3	I would be willing to pay increased taxes to provide funding for research on alternative energy resources because in the end by being more efficient and less dependent on foreign oil I will save money	One justification is explained
4	Yes, I believe that more money should be given to this research and IMPLEMENTATION. It is already understood how solar and wind work, but they must be implemented! We must be a leader in the fight on global warming, demonstrating that this issue is at the forefront of our concerns	Two justifications given (focus on implementation and example for world). Implementation is elaborated, but why the USA needs to lead is not
5	No [do not make smoking illegal]. With as much information about the risks of smoking available today, people should be responsible enough to educate themselves and make their own decisions about smoking. I do however think that smoking in public areas should be illegal because then you are exposing others to danger	Two justifications are given for the position (keep smoking legal, but not in public areas). Each justification is explained

Since multiple perspectives were commonly cited in questionnaires, though not explicitly elicited by the DMQ, a perspectives score was adapted from the "synthesis" component of the Reasoning Complexity Rubric (Tal & Hochberg, 2000). Inclusion of multiple perspectives was generally not in-depth due to the short nature of responses, so a simplified 3-point scale was used. No points were given if the participant discussed only one perspective, one point was awarded when another perspective was recognized, two points were given when another perspective was elaborated, but not resolved with the perspective guiding the decision, and three points were given when multiple perspectives were elaborated and incorporated into a resolution consistent with the decision (see Tables 6.2 and 6.4).

The scoring rubric was reviewed for validity by two other science education researchers. The first author analyzed the whole DMQ data set and the second coder independently analyzed an approximate 20% sample. Analysis was conducted blind to groups. Inter-rater reliabilities based on a 20% sample were 78% for reasoning and 85% for perspectives. Discrepancies were resolved to reach 100% agreement and the remainder of the sample was then revised for consistency. Average reasoning scores were computed for each student, and an independent-samples *t*-test was conducted to compare SSI and BIO group totals. Data for average perspectives scores were skewed toward the lower end of the scale, so we conducted a Mann–Whitney test for nonparametric data for this scale.

Table 6.4 Examples of scoring for perspectives scale

Score	Participant response	Explanation for scoring
0	I would be ok with this [tax increases to fund alternative energy] because these techniques are much better for our environment	Only one perspective is evident
1	Yes, people who choose not to smoke should not be put in danger by those who choose to smoke. There should be a special place designated for smokers because they chose to live an unhealthy lifestyle	Recognizes the position of "those who choose to smoke," but does not fairly consider this point of view
2	It should because people that smoke are putting themselves at risk for many types of cancer and diseases. This in turn increases the cost of healthcare that will have to be provided because of their smoking. However, it also is someone's decision or not to smoke, and you can't really pass legislation against it because they have the liberty to do it, even though it is extremely harmful	Expresses health value of legislation, but also recognizes the individual's control over health. However, no resolution is reached
3	On a personal level I believe they should [set limits on emissions], though I do see reasons for not doing so. The United States competes in the global market with nearly every country on earth, and if one country uses methods that are cheaper, though cause more pollutants and greenhouse gasses, they are more likely to perform better economically due to their ability to produce the product at less of a cost for the present (though long term these often will cost us more). If a legally-binding limit can be reached, I do think the United States however, as one of the more powerful nations, has somewhat of an obligation to do its best in complying with whatever may be better for the future of mankind	Explains and offers examples for reasons for supporting and non-supporting positions. A resolution is reached where the influence of the USA is seen as an adequate reason to commit despite the drawbacks

Interview Analysis

Qualitative questionnaire data and interview transcripts were triangulated to enhance validity of interpretations. Analysis of DMQ follow-up questions from Bell and Lederman (2003) was guided by themes developed from King and Kitchener's Reflective Judgment Model (1994), with insights from Sadler, Barab, and Scott (2007), including view of knowledge, recognition of complexity, consideration of perspectives, and use of evidence. Student responses to the three follow-up questions were assessed using the criteria described in Table 6.5. SSI and BIO participants were then compared across groups. Responses to general questions about experiences with SSI were compared within groups for emergent themes, then compared across groups.

Table 6.5 Assessment themes for stages of reflective judgment

Theme	Pre-reflective	Quasi-reflective	Reflective
View of knowledge	Absolute, authority driven	Uncertain, contextual	Tentative and inquiry-based
Complexity	Not perceived	Perceived, frustrated with ambiguity	Understood, criteria applied for evaluation
Other perspectives	Unrecognized	Contextual nature complicates evaluation	Considered across contexts
Evidence	Not considered; truth is directly observable	Used idiosyncratically in reasoning	Evaluated by criteria, applied in context

Text from the student perceptions segments of the interviews were coded independently for major ideas and compared within groups for emergent themes. These themes, supported with quotes and examples, were compared and contrasted between SSI and BIO groups.

Results

Comparison of Decisions

On 11 situational position-taking questions on the modified DMQ, SSI and BIO groups were very similar in their responses (see Table 6.6). Decisions were compared between groups for sophomore and senior level classes and total SSI and BIO students. According to Fisher's Exact probability tests of "yes" and "no" responses for SSI and BIO groups at each level, the only significant difference was for question 8, which asked whether or not participants exercised regularly, in the senior level class (SSI 60% yes, BIO 90% yes, $p = .027$). Although this question tested whether students based their behavior on scientific knowledge, it did not ask them to take a position on a controversial issue. SSI students were less likely to exercise regularly and they commonly reported time constraints as the reason for this behavior.

Comparison of Factors Influencing Reasoning

Distinct categories were developed for factors that students indicated as influencing their reasoning in each of the three clusters of questions. These factors could be positively influencing their decision, negatively influencing it, or simply raised for consideration. In this comparison, the quality of reasoning is not considered, but

Table 6.6 Percentages of SSI and BIO students by decision for questions on the modified DMQ

	Yes				No			
Question numbers	SSI Soph/Sr	SSI total	BIO Soph/Sr	BIO total	SSI Soph/Sr	SSI total	BIO Soph/Sr	BIO total
Climate change and policy								
1	63/87	71	87/85	86	30/13	24	10/15	12
2	33/47	38	47/40	44	63/53	60	53/60	56
3	70/60	67	83/80	82	20/33	24	17/20	18
4	63/73	67	77/60	70	33/27	31	23/30	26
Health research/food choice								
6	73/80	76	80/85	82	23/7	18	17/15	16
7	50/33	44	57/50	54	43/60	49	40/40	40
8	80/60	73	70/90	78	17/40	24	20/5	14
Regulation of food or tobacco								
9	70/47	62	60/60	60	30/53	38	40/40	40
10	40/20	33	30/50	38	57/80	64	67/50	60
11	83/100	90	87/85	86	17/0	11	13/15	14
12	90/93	91	83/90	86	7/7	7	17/10	14
Total		65		70		32		28

simply the types of factors mentioned. Questions from the DMQ were clustered by similarity in responses. Ten categories were developed for the *climate change and policy* cluster, 9 categories were developed for the *diet and health research/food choice* cluster, and 11 categories were developed for the *regulation of food or tobacco* cluster.

For the *climate change and policy* cluster, factor categories included any reference to the environment, evidence related to global warming, the need to hold individuals or nations accountable for their role in climate change, the political influence the USA has in the world, influence of decisions on public perception of the USA, personal values or responsibility, political views, practical outcomes of decisions, suggestions of alternative options to those suggested in the DMQ, or economic aspects of decisions (Table 6.7). For the *diet and health research/food choice cluster*, factor categories included food and exercise choices as part of general lifestyle, preferences unrelated to research, long-term effects like disease prevention, limitations on time and resources, influence of personal experiences (such as witnessing family member's disease), knowledge of research, unawareness of research, distrust of research, and ambivalence toward health knowledge or research (Table 6.8). For the *regulation of food or tobacco cluster*, factor categories included health, scientific evidence, need or desire to remove the problem through regulation, health as personal responsibility, moral or social concerns, personal preferences, smoking/unhealthy foods being appropriate in some situations, the need to be consistent (as comparing cigarettes with alcohol), economic concerns, practical concerns, and alternatives to regulation (Table 6.9).

Table 6.7 Factors influencing reasoning in Climate change and policy question cluster of DMQ

Factors	%SSI-Soph (n = 30)	%BIO-Soph (n = 30)	SSI-BIO-Soph (%)	%SSI-Sen (n = 15)	%BIO-Sen (n = 20)	SSI-BIO-Sen (%)	%Total SSI (n = 45)	%Total BIO (n = 50)	SSI-BIO Total (%)
Environment	66.7	83.3	–16.6	33.3	50.0	–16.7	55.6	70.0	–14.4
Evidence	43.3	33.3	10.0	46.7	35.0	11.7	44.4	34.0	10.4
Accountability	43.3	23.3	20.0	26.7	40.0	–13.3	37.8	30.0	7.8
Political influence	13.3	33.3	–20.0	46.7	35.0	11.7	24.4	34.0	–9.6
Public perception	23.3	6.7	16.6	26.7	25.0	1.7	24.4	14.0	10.4
Personal values	70.0	56.7	13.3	66.7	75.0	–8.3	68.9	64.0	4.9
Political views	16.7	16.7	0.0	26.7	15.0	11.7	20.0	16.0	4.0
Practical outcomes	70.0	63.3	6.7	66.7	70.0	–3.3	68.9	66.0	2.9
Other options	63.3	46.7	16.6	60.0	50.0	10.0	62.2	48.0	14.2
Economic	80.0	90.0	–10.0	80.0	80.0	0.0	80.0	86.0	–6.0

Table 6.8 Factors influencing reasoning in Diet and health research/food choice question cluster of DMQ

Factors	%SSI-Soph (n = 30)	% BIO-Soph (n = 30)	SSI-BIO-Soph (%)	%SSI-Sen (n = 15)	% BIO-Sen (n = 20)	SSI-BIO-Sen (%)	%Total SSI (n = 45)	%Total BIO (n = 50)	SSI-BIO Total (%)
Lifestyle choice	90.0	83.3	6.7	93.3	100.0	–6.7	91.1	90.0	1.1
Personal preferences	30.0	16.7	13.3	46.7	35.0	11.7	35.6	24.0	11.6
Limitations	13.3	10.0	3.3	26.7	20.0	6.7	17.8	14.0	3.8
Long-term effect	16.7	30.0	-13.3	33.3	25.0	8.3	22.2	28.0	–5.8
Personal experience	20.0	16.7	3.3	20.0	5.0	15.0	20.0	12.0	8.0
Research	33.3	40.0	-6.7	26.7	20.0	6.7	31.1	32.0	–0.9
Unaware of research	10.0	13.3	-3.3	0.0	5.0	–5.0	6.7	10.0	-3.3
Distrust research	16.7	10.0	6.7	6.7	15.0	–8.3	13.4	12.0	1.4
Ambivalence	6.7	3.3	3.3	13.3	0.0	13.3	8.9	2.0	6.9

Table 6.9 Factors influencing reasoning in Regulation of food or tobacco question cluster of DMQ

Factors	%SSI-Soph (n = 30)	% BIO-Soph (n = 30)	SSI-BIO-Soph (%)	%SSI-Sen (n = 15)	% BIO-Sen (n = 20)	SSI-BIO-Sen (%)	%Total SSI (n = 45)	%Total BIO (n = 50)	SSI-BIO Total
Health	40.0	36.7	3.3	26.7	35.0	–8.3	35.6	36.0	–0.4
Evidence	20.0	6.7	13.3	20.0	20.0	0.0	20.0	12.0	8.0
Remove problem	53.3	63.3	–10.0	73.3	35.0	38.3	60.0	52.0	8.0
Personal responsibility	56.7	70.0	–13.3	73.3	40.0	33.3	62.2	58.0	4.2
Moral/social	90.0	76.7	13.3	86.7	100.0	–13.3	88.9	86.0	2.9
Personal preferences	13.3	6.7	6.7	20.0	10.0	10.0	15.5	8.0	7.5
Appropriate sometimes	13.3	0.0	13.3	13.3	25.0	–11.7	13.3	10.0	3.3
Consistency	26.7	6.7	20.0	6.7	10.0	-3.3	20.0	8.0	12.0
Economic	13.3	23.3	-10.0	20.0	10.0	10.0	15.5	18.0	–2.5
Practical concerns	23.3	33.3	-10.0	60.0	55.0	5.0	35.5	42.0	–6.5
Other options	40.0	50.0	-10.0	66.7	35.0	31.7	48.9	44.0	4.9

In general, SSI and BIO participants cited similar factors as influencing their reasoning. SSI and BIO groups did not vary by more than 15% on the frequency of citing individual factors within question clusters and they differed by less than 10% on the majority of factors (24/30). This is consistent with Bell's and Lederman's (2003) study of professors with different NOS views, where few differences were found in frequencies of decisions between the two groups. It is unsurprising that positions differ little between similar groups of scientifically literate preprofessionals. However, on average SSI students cited more factors as influencing their decisions, considering the number of citations for the SSI group was higher for most categories of factors (29/30).

Differences of 10% or higher between total SSI and BIO groups were found for four reason categories. In the *global warming* question cluster, BIO participants were more likely to support legally binding limits on carbon emissions (more pronounced at sophomore level). One reason for this may relate to the fact that SSI participants were more likely to suggest alternatives to legislation, like incentives (62% vs. 48%) in this scenario. Perhaps having extensive experience with argumentation helped SSI students to think creatively about alternatives to the suggested response. In this cluster, SSI participants were also more likely to refer to public perception of the US (24% vs. 10%). This difference may relate to their focus on the social aspects of problems and consideration of different perspectives. Many participants discussed how improving global perceptions of the USA could facilitate cooperation in solving problems like global warming. Finally, SSI participants were more likely to refer to evidence (44% vs. 34%) in their reasons for making decisions in this cluster. This difference could relate to the explicit focus on evidence-based arguments in their core courses. Surprisingly, SSI participants were less likely to include environmental factors in their decisions (56% vs. 70% BIO). Perhaps BIO students were more attuned to environmental concerns, whereas SSI participants were more likely to focus on social concerns, or to more seriously consider both socioeconomic and environmental aspects of the problem. Overall, for the global warming cluster, SSI students included more reasons as influencing their decision, so perhaps SSI students viewed the problem as more complex then the BIO groups.

In the *diet and health research/food choice* cluster, SSI participants were more likely to report that they make food and exercise choices according to personal preferences or tastes (35% vs. 24%). This appears consistent with SSI students' lower participation in regular exercise. The reasons for this difference are unclear. Although it seems SSI participants were aware of the benefits of exercise and healthy diets, they were less likely to apply this knowledge to their own lives. It is possible if students felt more comfortable with the researcher, they may have been willing to be more honest about their habits.

In the *regulation of food and tobacco* cluster, SSI students were more likely to discuss consistency with other laws as influencing their reasoning (20% vs. 8%). Although the difference for total groups was smaller, senior level SSI students were much more likely than BIO participants to cite availability of other options, like banning smoking in public places (67% SSI vs. 35% BIO). This may explain why senior level SSI participants were more likely to answer "no," (80% SSI vs. 50% BIO) when asked whether they thought cigarette smoking should be made illegal. This group also more often cited the importance of personal responsibility as a reason why smoking or foods should not be banned (73% vs. 40%). They were also more likely to discuss legislation of unhealthy foods or tobacco as a way to remove the problem, whether they supported it or not (73% vs. 35%). This result is unclear, but may relate to these students' tendency to incorporate different ideas or positions in their responses.

Comparison of Reasoning

Using an independent samples *t*-test, mean reasoning scores were found to be significantly higher for SSI students (M=3.46, SD = .63) than for BIO students (M=3.19, SD = .68), $t(93) = -1.98$, $p = .05$; (see Table 6.10). These results suggest that an SSI-focused interdisciplinary major provides some benefits in reasoning with complex problems. Although most decisions and reasons behind decisions were similar for both groups, we expected that reasoning processes would be more sophisticated in SSI students. Theoretically, SSI students' considerable experience with socioscientific issues helped them to develop reasoning processes that could be applied to other issues. Both groups showed higher reasoning scores for senior level than sophomore level, which would be expected regardless of instruction, since higher levels of reflective judgment, which relates to reasoning in SSI, would be expected with development and experience in a college environment (King & Kitchener, 1994).

For perspectives scores, although the SSI mean (M=1.24, SD = .71) was slightly higher than the BIO mean (M=1.04, SD = .57), the difference was not statistically significant (See Table 6.11). Still, it is worth noting that although the number of DMQ items in which participant responses included any reference to multiple perspectives was nearly equal between groups (SSI mean=3.9; BIO mean=3.8), of these responses, SSI students had a greater proportion of scores of 3 (55% vs. 47%),

Table 6.10 Reasoning scores for SSI and BIO groups

	Sophomore level		Senior level		Total		
	Bio (*n*=30)	SSI (*n*=30)	Bio (*n*=20)	SSI (*n*=15)	Bio (*n*=50)	SSI (*n*=45)	Sig.
Reasoning mean	3.10	3.40	3.33	3.58	3.19	3.46	0.050*
Reasoning st. dev.	0.66	0.61	0.70	0.67	0.68	0.63	

*$p \leq .05$

Table 6.11 Perspectives scores for SSI and BIO groups

	Sophomore level		Senior level		Total		
	Bio (*n*=30)	SSI (*n*=30)	Bio (*n*=20)	SSI (*n*=15)	Bio (*n*=50)	SSI (*n*=45)	Sig.
Perspectives mean	0.93	1.10	1.06	1.26	1.04	1.24	0.154
Perspectives st. dev.	0.54	0.67	0.59	0.73	0.57	0.71	

and lower proportion of scores of 1 (34% vs. 43%). SSI students were somewhat more likely to consider other perspectives in depth and reach a logical conclusion. In responses given a perspectives score of one, an alternative perspective was referenced, but not explained or elaborated. In responses scored 3 (more frequent for SSI), alternate perspectives were considered in depth and the participant was able to reach a resolution incorporating those different perspectives. This result may be related to the focus in the SSI program on fully considering different perspectives before committing to a position. Incorporation of other perspectives was considered an important part of a good argument, and students were assessed on how well this was done. In the sophomore level core course, one handout said explicitly, "The most effective arguments will take into account the points made by the opposition." Consistent with this result, SSI students also noted in interviews that they were more likely to consider other perspectives than before entering the major. They reflected on the importance of being aware of other ideas and hearing from all sides.

The consistently higher scores of the SSI groups suggest that participation in the SSI program may foster more sophisticated reasoning with SSI. SSI students were more likely to use multiple justifications to support their positions and to better explain those justifications. This likely relates to the focus on reasoning with socioscientific issues in core courses. Students were routinely challenged to make evidence-based arguments, using as much relevant and credible data as possible. Their arguments were assessed in position papers and critiqued by their peers and professors in debates.

It should be noted that average scores for both groups were fairly high (SSI: 3.46 and BIO: 3.19 on a 5-point scale). On average, students from both groups were likely to include and elaborate on at least one justification of their positions. Few studies of socioscientific reasoning have been conducted with college students, who would be expected to exhibit higher levels of intellectual development and reflective judgment than middle school or high school students. Perhaps higher developmental levels, especially considering the high achievement levels of our sample, may explain the fairly small, though significant differences in reasoning. In addition, in assessing reasoning with SSI, we could not control for many of the factors that could influence results. Although our comparison group of BIO students did not experience core courses and reported few SSI experiences in their science courses, they varied greatly in their experience reasoning with controversial issues. This could have diminished differences in reasoning between the two groups. Also, perhaps a greater difference between groups could have been found with a more sensitive instrument and coding scheme.

Interviews

In interviews, the majority of both SSI and BIO students' responses to scenarios were consistent with King and Kitchener's (1994) reflective stages. Student responses from both groups revealed their views of knowledge as uncertain, tentative, and inquiry-based, except two SSI and two BIO students whose responses suggested there was a single "best" answer. Responses showed all students to perceive complexity of problems with the exception of one BIO student who was hindered by it. All students considered alternate perspectives across contexts, except one BIO student who had difficulty resolving them. The majority of students in both groups sought resolutions by recognizing similarities in different perspectives, for example, using their biological knowledge to illustrate how both carbon emissions and deforestation may contribute to climate change. All students except one BIO student discussed applying criteria to logically evaluate evidence. For example, most students said that to persuade them make dietary changes, claims must have accumulated a great deal of long-standing evidence.

The finding of few qualitative differences in key aspects of reflective judgment between SSI and BIO interviewees may be partly explained by the fact that all students in this study were high achieving pre-professionals who had reached high levels of development. However, this analysis was limited in that three probing questions could not capture the complexities of students' reasoning. These brief scenarios did not warrant a more thorough evaluation of reflective judgment as used with the rubric for the Prototypic Reflective Judgment Interview (King & Kitchener, 1994; Zeidler et al., 2009).

In spite of this limitation, when probing reasoning processes on scenarios, we found that SSI students tended to provide examples from their core courses when they saw relevance. For example, two of the SSI students interviewed compared a hypothetical argument that a causal mechanism between smoking and cancer has not been found to the classroom debate over whether HIV causes AIDS. SSI students also mentioned many examples of cases from their core courses when asked about their experiences with SSI or aspects of their major that they valued. Perhaps extended experience with SSI, as provided in the human biology major, provides a repertoire of cases students can access in relevant situations (Bransford et al., 1986).

Perceptions of SSI in Majors

Views of SSI Students

All SSI students interviewed mentioned that they highly valued incorporation of social aspects of biology-related problems in SSI (all names are pseudonyms). Gary described the major as follows:

> the premise of our written assignments were to not only include the biology or the physiology or the science aspect of what is going on in these problems [controversial issues], but also to discuss the social implications, whether it's a good or a bad thing, what views are held, if we accept the view, what would that change about society, things like that. And so it's almost like the core of the major is actually to focus on these things. Not simply what is the science, but how does it affect people and why is that important?

Sarah contrasted SSI courses with traditional biology courses:

> Because a regular biology class would be like, oh, this is what a retrovirus does, like AIDS. Well, now in this class, we're learning all about AIDS, what goes into it, how it affects the body more in-depth, how it affects the social aspects, where in another class, we're not really going to talk about, like how people get treatment ... so I feel like what we get is more specific issues, but then within those issues it covers the whole range of the issue.

All SSI students except one senior level and one sophomore level student noted that the ability to consider multiple perspectives is important in science and health fields. Laura cited a recent newspaper commentary arguing that academics had become too specialized:

> [The author] said, you don't have anyone in the world who understands the water issue from every perspective. You have economists who understand it from that perspective, you have politicians who understand it from that perspective, you have people who are thirsty who understand it from that perspective. He said, but you don't have anyone who can put it all together.

Laura argued that to solve eminent social and ecological problems like water conservation, decision makers need to be familiar with different disciplinary perspectives. Kelly also related the SSI major to science careers, explaining that it has helped her decide to pursue a degree in public health as well as medical training.

All SSI students also said that they valued having been exposed to new ideas or controversies in their major. Several students mentioned that SSI core courses illuminated controversy in issues they had never seen as problematic. For example, Laura said, "I didn't know that there was going to be a water crisis until we studied it in [the senior level core course]." Shawna said, "I guess it kind of made me realize how controversial some things are that I've already had pretty opinionated stances on." She said she had become more willing to reevaluate her positions in light of new information and perspectives.

Views of BIO Students

When asked how well prepared they believed they were to make decisions on science issues with social implications, all BIO students said that they had little experience with SSI in traditional science classes they had taken. Polly said,

> In terms of just the biology classes I don't think that they have prepared me very well to have a stand because teachers usually, if they do bring up controversial topics...right now I'm thinking about stem cell research, um so I'm really interested in that, but the bulk of knowledge I do have about that is outside of the classroom. And I took cell biology where they like touched on it, but I mean not into the amount of detail that one would need to have a stand.

In addition, all BIO participants felt that their biology courses focused primarily on mechanisms. They felt it was left to the student to connect social aspects or alternative perspectives. Ellen said,

> I think they've kind of created those opportunities with the genetics and the molecular biologies to, um, understand the mechanisms behind this, but I think it's left upon the

individual to like search for those answers [to SSI-related problems] in other disciplines or in the same disciplines in other fields…

Tracy said that such issues had been briefly addressed in her biology classes, but she had not had opportunities to explore them in depth. The professor gave "just the overall result, what we learned from it and move on." When asked if she had had opportunities to integrate perspectives on socioscientific issues in her courses, Carrie said, "I would probably say no. I mean, they pretty much all focus in their own area. They never really connect to each other, explicitly. I've never really thought about it that much either."

Although they had few experiences discussing SSI in their biology courses, all BIO participants said they had taken classes that explored social aspects of scientific issues to some degree. These courses included anthropology and human genetics, medical sociology, psychology, and religion and evolution. Approaches to SSI in these courses varied. Some students (two of four who discussed this) felt those courses focused on social aspects, expecting an adequate level of science background knowledge or providing quick summaries of the biology content. Carrie said of her medical sociology course,

I think that most of the students in that class were like premed or predental or whatnot, so I think that they probably assumed that you have some biological background, but, we never really got into anything really science, it was more like issues that, like people who were going into the sciences, particularly medical careers and how those related to sociological aspects of how we live.

Relatively few students (two) said that their courses integrated biological concepts and social implications.

Despite having little SSI integrated into biology courses, half of BIO participants at both levels reported feeling more able to make well-formed arguments with SSI. These students integrated their knowledge from science and social sciences courses to consider particular issues in depth. Kevin explained that the emphasis on supporting ideas with evidence in the biology major helped him to reason more effectively with SSI. "Honestly I think my major helped me argue this stuff a lot. Which is good, it gave me a way to [use] my evidence and support my theories with it, in all honesty, supported from a logical view, whether than being too emotional about one of my ideas." Tracy attributed her ability to argue for a position in SSI to her anthropology courses. She said,

I thought at the beginning when you said that a lot of students are tripped up by the first questions [from DMQ] was interesting because I know for a fact that if I didn't have a lot of my anthropological classes, I would get tripped up on those. But since I've applied them, and have to explain them over and over again, it's like I have a deeper understanding, whether they are correct or not, of being able to sit with someone and explain the effect to them. And then, I love the smoking example, because it's definitely something I've thought about and I just, I was very excited to write about what I thought.

Overall, BIO participants expressed concern about SSI and found ways to apply their biological knowledge to such problems.

Student Perceptions of Majors

Perceptions of Personal Outcomes

SSI Group

When asked what they valued about their experience in their majors, SSI participants consistently discussed outcomes of SSI aspects of their programs. In interviews with SSI students, the parts that addressed SSI and the parts that addressed valued outcomes generally overlapped or ideas were repeated. SSI students valued having the opportunity to explore controversial interdisciplinary issues in depth. They felt they had improved their abilities to discuss and take positions on these issues, and they noted that this competency was essential for future careers in science or health care. They also felt that the major offered them opportunities to understand controversies they were not familiar with and situations they never before saw as problematic. When asked about the major student outcomes instructors worked toward, the neuroscientist instructor of the sophomore level core course said,

> The major outcomes that we're looking for mirror the outcomes of the program as a whole, that we're looking for the students to be able to think more critically, more flexibly, about the problems that we talk about in the course, and what makes our course specific is the context that we frame those problems within.

Clearly, SSI participants at both levels grasped this central program goal.

All SSI participants said they felt more able to consider multiple perspectives as a result of participating in the program. Shawna explained, "I think it's been good to see both sides of things. I guess I've kind of gotten a little less opinionated in the sense that I'm a little bit more open to other reasoning that I never saw as logical before but I guess I understand them better now." All SSI students except one sophomore level student reported feeling more able to discuss controversial issues or take a position on such issues. Kelly said,

> I'd say the overall goal is just making us question information we've been given and evaluate our own things that are important to us, and to try to really get out what you believe in. It's helped me grow, having confidence in doubting something. Like I guess before I kind of never even thought to doubt stuff, like one professor told me is true, like what I learn in school is true, and I've never been encouraged to doubt something.

Sarah said, "I just think I have a broader outlook on things now when they're presented to me, instead of just like, oh this is what happens like this, when I can be like, oh well, this [is an] issue that surrounds it and you have to think about what if this happened, and contrast it." Through discussion of SSI issues in core courses, students learned to question information presented, and to explore different aspects of the issue.

BIO Group

BIO participants also valued many aspects of their major, although none of the students voluntarily brought up SSI-related outcomes. More than half of BIO participants said they developed diligence or responsibility in their major. Kevin said, “It’s honestly, the main thing it’s taught me to do is how to establish goals and meet them myself.” Most BIO students (75%) appreciated opportunities to learn about new ideas and discover new passions for science. Ella appreciated being exposed to “new and upcoming ideas.” Ellen said she gained a better understanding of science and its “utility for the world.” Polly said the biology major “has made me love science even more. Learning about it made me want to pursue my goals, I guess.”

Half of the senior level BIO participants said they learned about the challenges of scientific research or graduate school. Ellen said,

> This major has helped me to see that things take time, and um, it has allowed me to see that you’re not always going to have a direct result, sometimes, I don’t like this answer, but more often you’re gonna fail than you succeed, and that’s in experiments as well, so and the importance of diligence, so that’s what this has helped me to see.

She noted that her experience in a research lab as well as in her classes helped her see that scientists must learn to accept failure and frustration as part of their daily experience. After graduation, she intended to gain more experience in a research lab to prepare her for the psychological rigors of medical school. She said, “You need to be able to take it mentally—deal with the stress.”

Although participants did not feel that their biology courses offered opportunities to learn about or discuss socioscientific issues in depth, many felt these courses helped them understand biological mechanisms more thoroughly, which enhanced their understanding of socioscientific issues discussed in other classes. One sophomore level BIO participant explained that she was disappointed in the lack of opportunities to discuss SSI, so she planned to change her major to anthropology while maintaining a premedical curriculum. Tracy said,

> And that’s the problem I fell into was that the classes that I was taking, understandably, were the basics of everything, that was my problem, it was like, when am I going to get to the things that I can truly apply in my every day world? That’s not to say, as I learn right now about the many steps of transcription and translation, it’s beautiful and I like to learn about it, but it would be cool to watch the news and how they talk about global warming and the ice caps melting and exactly be able to say, you know, why is that, why is it that CO2 levels are contributing to the atmospheric thickening, where as I feel like I almost get that information from the media, from watching *An Inconvenient Truth*.

Perceptions of the Learning Environment

SSI Group

All SSI students interviewed valued the team-teaching approach in the program. They appreciated having experts from different disciplines model the integration of perspectives to approach an issue. They also valued the accessibility of knowledgeable teachers to help them apply complicated content knowledge. Dana said, “It’s interesting how they have usually two teachers teaching together … they had a neuroscientist with a sociologist teaching together to just give you a lot of different perspectives on disease … It’s very integrated and they try to give you as much different backgrounds as possible." Laura felt that integration of perspectives varied among her core courses. She felt some professors were too attached to one disciplinary perspective when dealing with issues, and this was an ineffective model for problem solving with socioscientific issues.

The majority of SSI students (three-fourths at both levels) said that they valued collaborative work in their major. Laura said,

> So they [core classes] are very team-based. And the rationale behind it has to do with the fact that when you get out into the real world, nine times out of ten, most jobs you're going to be working in a team … In some ways it's very effective and in other ways it's really annoying. But I guess that's more how real life is anyway.

Sarah said, "Now I know how to depend on other people, and I know how to provide to the team as well, or provide to just anybody who's depending on me." Although they valued teamwork in the major, many of the SSI students interviewed said they found the team work challenging. They cited different expectations and uneven levels of engagement among team members as sources of frustration.

BIO Group

Although they noted that biology classes tended to be lecture-driven, all BIO participants said that they valued opportunities for independent research or participation in a lab. They said those experiences helped them to see how science is really done. Chloe noted that participation in a research lab helped her see how as a scientists, you use "your own underlying process," as opposed to a strict classroom definition of the scientific method, where "they teach you rigid steps." Ellen felt that her experience in a research lab opened her eyes to new opportunities for discovery in biology she would not have known from classes alone. She said, "Others may feel there's one track to go through biology—they've only seen one side of biology." Polly also felt her experience in her major was dramatically enhanced by participation in a research lab. When asked how her experience would have been different if she had "just had the courses," Polly said, "Oh, I think it would be really different. Yeah, I wouldn't know the material as well probably. I wouldn't enjoy it as much because I wouldn't understand really as well as I do, yeah, so that would be unfortunate I'd say."

Other students felt that experiences in both research labs and course laboratories reinforced their conceptual learning in their biology classes. Kevin said, "I know certain people who do learn better from the books, but I believe more so people learn better from the labs … You have to understand it if you do an experiment rather than just regurgitate information you've heard about it." Kevin also felt that the few inquiry-based course labs he took helped him understand how scientists pose questions.

Some BIO students also noted that research lab experience and course labs helped them understand the breadth and big-picture goals of science. Natalie said, "I think that the labs have helped me to have a glimpse of what kind of fields of biology are out there. Like [histology class], I love it because I want to go into that field, pathology, or histology." Tracy noted, "It's interesting [working in a lab] because you see what all this work goes into. You see why you have to have the basic understanding."

As well as laboratory experience, all BIO students said they valued opportunities for discussion or debate. BIO participants described discussion sessions associated with courses as opportunities to review concepts and work through problem sets in a nonthreatening environment with a graduate or advanced undergraduate student. Natalie said, "[I learn best in a] discussion type thing, when the professors, they're not down some stairs all the way at the bottom of the lecture hall." Carrie said, "I think [discussion sessions are] definitely helpful, because you can see stuff hands on that's just not being, like being lectured to you in class, like you can do that kind of stuff. I mean, discussion helps a less threatening environment." Ellen mentioned that she also valued opportunities to discuss controversial issues in her biology and social science courses. She said, "I welcome debates, but I think, I kinda like when your perspectives are challenged because it allows you to prove what you know, why you know it, and why it should matter."

When discussing the level of community in their majors, all except one senior level BIO students said community is not facilitated, but develops, especially in upper level classes. They explained that the large size of the major makes developing a sense of community difficult. Polly said, "I wouldn't say it's like collaborating is discouraged, but it's just not facilitated too much." Kevin noted that the common practice of grading on a curve in biology classes seemed to discourage some students from working with other students. However, he said that many students worked together in spite of this. BIO participants also noted that a stronger sense of community developed in upper-level students as class size decreased. Two discussed a particular social event for senior biology majors that helped them feel more connected.

The majority of BIO students said that interacting with faculty was available when students sought it (half of senior level and all sophomore level). When asked, "Do you have an opportunity to get to know the professors?" Carrie said, "I would say that you have to make an effort to do that, it's not easy to do that. I especially think it's hard when you're doing well in the class, you don't really have any mood to go into office hours, so, you know." Overall, BIO students explained that going to office hours was the primary way to get to know their professors, but they felt that professors clearly wanted to help students.

Summary of Major Findings

Our first research question was "Do SSI majors reason with SSI differently from BIO majors?" We found that SSI students showed higher levels of reasoning with SSI and that they showed tendencies toward incorporating different perspectives into their decision-making (although the differences in perspective-taking between the SSI and BIO groups were not statistically significantly different).

In response to research question 2, "How do SSI and BIO majors' perceptions of their experiences with SSI differ?" interview responses showed that SSI students "bought in" to the importance of recognizing different perspectives in decision-making and were more perceptive to controversy and complexity of science-related situations as a result of participating in the SSI program. In addition, SSI students appeared to use case studies from core courses to reason with new similar problems. These findings together suggest that classroom experience with SSI with a focus on integrating different perspectives, impacts students in their awareness of and approach to real-life problems as well as their ability to reason effectively with such problems.

In response to Question 3, "How do SSI and BIO majors' general perceptions of their majors differ?" interviews suggested that students from both groups valued their majors and were satisfied with personal outcomes. SSI students consistently viewed ability to integrate different perspectives and understand the social aspects of scientific problems as significant program outcomes. SSI students also valued aspects of the learning environment designed to support ethical, cognitive, and epistemological development, including an inclusive, collaborative environment and supportive relationships with faculty. SSI students felt a greater sense of community as compared to BIO students. These findings support the idea that careful structuring of the learning environment provides essential components of learning through SSI.

Implications and Emergent Questions

Incorporation of SSI in College Science

The SSI program represents one thoughtfully developed way to integrate disciplinary concepts and engage students in reasoning with SSI in a college setting. Core courses may be considered a consistent venue for reasoning with SSI throughout the program. The neuroscientist professor of the sophomore level core course explained that the core courses were designed specifically for this purpose, while biology courses should supply more in-depth biology content. Addressing biology content, he said, "That's what you're going to get if you want to learn cell biology, that's why [this university] has a great resources, people who know a lot about cell biology, you go over there and you take those courses, if it makes sense for your curriculum." According to the faculty members involved in this study, core courses are not considered biology courses.

Still, professors in the SSI program recognized a resistance from students who felt their core courses should supply more biology content. The sophomore level neuroscientist professor argued that biology content should be taught, but framed differently from a traditional biology class. He said,

> The idea is that the content that we want them to walk away with is content that will allow them to apply their newly gained knowledge of bacteria to a variety of situations. Which I think is different from the way they approach it in the biology department, and having sat through some of those courses, I know that one of the things is you memorize this, this is this specific bacteria and this is that specific virus and you memorize what they look like. And we don't so much care about what they look like, we care about how is this disease similar to that one, and why would you treat something with antibiotics and one infection with antibiotics and another one you wouldn't. Why is that? Those are the kinds of questions that we want the students to be able to understand because they see things on the news about this infection that's occurring and a lot of lay people say, oh take antibiotics. That doesn't work!

This illuminates an important question for future research: How should content and socioscientific reasoning be balanced in college science majors? Considering that science courses cover a great deal of content, some may argue that SSI should be explored through other venues, like independent projects or other courses like the SSI core courses. However, if SSI are not incorporated into science content courses, only students motivated to pursue such experiences may be exposed to them. This also raises the question of whether socioscientific reasoning should be a mandatory or optional part of undergraduate science training.

Structuring of the Learning Environment

It is important to note that the learning environment played an important role in students' perceptions of their majors and personal outcomes. The SSI and BIO groups discussed different types of class activities as being important to their learning. Both groups valued opportunities for discussion and application of concepts. BIO participants were more likely to cite class discussion sessions and labs and participation in independent research projects as critical experiences for their learning. SSI participants more commonly cited case studies as important opportunities to apply concepts. Throughout the interviews, SSI participants cited many examples of case studies from core courses, and often related them to other issues. This may have implications for transfer of knowledge and skills to new problems.

Another important aspect of the learning environment involved the sense of community. SSI participants valued a sense of family among students and professors. Part of this they attributed to a small program and part they attributed to the team-based nature of core courses. They felt that working in teams prepared them for collaborative work environments and helped them develop confidence in their interpersonal and public speaking skills. BIO participants did not feel that their major was structured to support community, primarily because of its size. Individual interaction with professors was not structured, but could be fruitful when students pursued that interaction.

This raises the question of how SSI learning environments need to be structured. Sadler (2009) argues that participation in communities of practice

(Lave & Wenger, 1991) is central to SSI. Students must come to understand the culture of the community, including its rules and practices, and they adopt and project particular identities within those communities. Our study illustrates fruitful communities of practice established in core courses; however, these courses were rich in resources, including interdisciplinary faculty, low professor to student ratios, and semester-long courses allocated for SSI-based instruction. This leads to the following questions. Can learning environments that purposefully nurture communities of practice in SSI-contexts be structured for large classes, or is a small class size important to this aspect? If effective communities of practice could be established through an SSI framework in large majors like biology, could this help students feel more supported and increase persistence in science majors?

Connecting Complementary Literatures

In this study, we applied an SSI framework to a program developed independently from the literature on SSI. We found the goals, pedagogy, and outcomes extremely consistent with the SSI literature. Considering that scholars from different backgrounds work toward similar goals in college program development, it is important to connect the science education literature on SSI to bodies of literature that other interdisciplinary interventions in college settings may work with. One existing example is a study by Zeidler et al. (2009) that assessed outcomes of an SSI intervention using the Reflective Judgment Model, a work familiar to many college educators. Future work in SSI should make research published in different bodies of literature accessible to teachers and researchers from different disciplinary backgrounds. In-depth reviews from research across disciplines would help researchers recognize and understand nuances of terminology addressing similar concepts. Work from different disciplinary perspectives can provide additional evidence and strengthen rationales for SSI. In addition, combining research and curriculum descriptions could enhance future SSI interventions for students of different age groups.

SSI with Science Majors

This study suggests that incorporation of SSI can influence college students' reasoning with such issues and how they value understanding different perspectives and incorporating different disciplines into their decision-making. One future direction is to investigate how SSI learning environments affect students' thinking about science careers. SSI students who were interviewed in this study were convinced the ability to reason integrating both scientific and social aspects of problems was an essential competency for decision-makers in science fields. Future studies may investigate whether students who experience SSI are more likely to be aware of the complexity of problems faced by science professionals and if they recognize a need to incorporate different perspectives. For example, how might exposure to SSI influence a premedical student to think about addressing different aspects of patients that influence their health, such as availability of healthy food, limitations on their ability to travel to clinics or pharmacies, or psychological states that may influence behavior? This may also have implications for medical education. The Institute of Medicine (2004) has recently stated, "physicians must be equipped with the knowledge and skills from the behavioral and social sciences needed to recognize, understand, and effectively respond to patients as individuals, not just to their symptoms" (p. 16). Clearly, postsecondary education is in need of effective strategies for integrating the life sciences and social sciences to help professionals manage complex problems.

Conclusion

Overall, this study contributes to our understanding of how an SSI-focused, 4-year program can help students reason and make informed decisions on real and complex problems. With the assumption supported that traditional university biology teaching provides few opportunities for critical evaluation of interdisciplinary problems currently called for (AACU, 2007), it provides an example of effective SSI teaching, with modeling of reasoning from social science and biological disciplinary perspectives. The study provides a springboard for new studies investigating SSI in interdisciplinary college contexts.

References

Association of American Colleges and Universities. (2007). *College learning for the new global century*. Washington, DC: AACU.

Blattner, W., Gallo, R. C., & Termin, H. N. (1988). HIV causes AIDS. *Science, 241*, 515–516.

Baxter Magolda, M. (1992). *Knowing and reasoning in college: Gender-related patterns in students' intellectual development*. San Francisco, CA: Jossey-Bass.

Baxter Magolda, M. (1999). *Creating contexts for learning and self-authorship constructive-developmental pedagogy*. Nashville, TN: Vanderbilt University Press.

Bell, R., & Lederman, N. (2003). Understandings of the nature of science and decision making on science and technology based issues. *Science Education, 87*, 352–377.

Bransford, J., Sherwood, R., Vye, N., & Reiser, J. (1986). Teaching thinking and problem solving: Suggestions from research. *American Psychologist, 41*, 1078–1089.

Brown, J. S., Collins, A., & Duguid, P. (1989). Situated cognition and the culture of learning. *Educational Researcher, 18*, 32–42.

Creswell, J. (2009). *Research design: Qualitative, quantitative, and mixed methods approaches*. Los Angeles, CA: SAGE Publications.

Creswell, J. W., & Plano Clark, V. L. (2007). *Mixed methods research*. Thousand Oaks, CA: Sage Publications.

Duesberg, P. (1988). HIV is not the cause of AIDS. *Science, 241*, 514–517.

Fink, L. (2002). Beyond small groups: Harnessing the extraordinary power of learning teams. In L. K. Michaelsen, A. B. Knight, & L. D. Fink (Eds.), *Team-based learning: A transformative use of small groups* (pp. 3–26). Westport, CT: Praeger Press.

Gore, A. (2006). *An inconvenient truth: The planetary emergency of global warming and what we can do about it*. New York: Rodale.

Hogan, K., Nastasi, B. K., & Pressley, M. (2000). Discourse patterns and collaborative scientific reasoning in peer and teacher-guided discussions. *Cognition and Instructional, 17*, 379–432.

Institute of Medicine of the National Academies. (2004). *Improving medical education: Enhancing the behavioral and social science content of medical school curricula*. Washington, D.C.: National Academic Press.

King, P. M., & Kitchener, K. S. (Eds.). (1994). *Developing reflective judgment: Understanding and promoting intellectual growth and critical thinking in adolescents and adults* (pp. 65–84). San Francisco, CA: Jossey-Boss.

Klein, J. T. (1990). *Interdisciplinarity: History, theory, and practice*. Detroit, MI: Wayne State University Press.

Land, S., & Zembal-Saul, C. (2003). Scaffolding reflection and articulation of scientific explanations in a data-rich, project-based learning environment: An investigation of progress portfolio. *Educational Technology Research and Development, 51*, 65–84.

Lave, J., & Wenger, E. (1991). *Situated learning: Legitimate peripheral participation*. Cambridge: Cambridge University Press.

Mansilla, V. B., & Duraising, E. (2007). Targeted assessment of students' interdisciplinary work: An empirically grounded framework. *The Journal of Higher Education, 78*, 215–237.

Perry, W. G. (1979). *Forms of intellectual and ethical development in the college years: A scheme*. New York: Holt, Rinehart & Winston.

Roberts, D. A. (2007). Scientific literacy/science literacy. In S. K. Abell & N. G. Lederman (Eds.), *Handbook of research on science education* (pp. 729–780). Mahwah, NJ: Lawrence Erlbaum Associates.

Sadler, T. D. (2004). Informal reasoning regarding socio-scientific issues: A critical review of research. *Journal of Research in Science Teaching, 41*, 513–536.

Sadler, T. D. (2009). Situated learning in science education: Socio-scientific issues as contexts for practice. *Studies in Science Education, 45*, 1–42.

Sadler, T., Barab, S., & Scott, B. (2007). What do students gain by engaging in socioscientific inquiry? *Research in Science Education, 37*, 371–391.

Sadler, T. D., & Zeidler, D. L. (2009). Scientific literacy, PISA, and socioscientific discourse: Assessment for progressive aims of science education. *Journal of Research in Science Teaching, 46*, 909–921.

Schlegel, W. M., & Pace, D. (2004). Using collaborative learning to decode the disciplines in physiology and history. In J. Mittendorf & D. Pace (Eds.), *Decoding the disciplines: How do we more effectively bring students into the thinking we do in our disciplines?* San Francisco, CA: Jossey-Bass Publishers.

Tal, R., & Hochberg, N. (2003). Assessing high order thinking of students participating in the 'WISE' project in Israel. *Studies in Educational Evaluation, 29*, 69–89.

Willaims, S. (1992). Putting case-based instruction into context: Examples from legal and medical education. *Journal of the Learning Sciences, 2*, 367–427.

Zeidler, D. (2001). Participating in program development: Standard F. In D. Siebert & W. McIntosh (Eds.), *College pathways to the science education standards* (pp. 18–22). Arlington, VA: National ScienceTeachers Press.

Zeidler, D., & Keefer, M. (2003). The role of moral reasoning and the status of socioscientific issues in science education: Philosophical, psychological, and pedagogical considerations. In D. L. Zeidler (Ed.), *The role of moral reasoning on socioscientific issues and discourse in science education* (pp. 7–38). Norwell, MA: Kluwer.

Zeidler, D., Sadler, T., Applebaum, S., & Callahan, B. (2009). Advancing reflective judgment through socioscientific issues. *Journal of Research in Science Teaching, 46*, 74–101.

Zohar, A., & Nemet, F. (2002). Fostering students' knowledge and argumentation skills through dilemmas in human genetics. *Journal of Research in Science Teaching, 39*, 35–62.

Chapter 7
Metalogue: SSI in Undergraduate Science Education

Jennifer L. Eastwood, Troy D. Sadler, and María Pilar Jiménez-Aleixandre

Impacts of SSI-Based Education

Sadler: This is a very interesting chapter in that it is features several unique elements as compared to most other chapters. Most other chapters deal with researcher-designed (or at least influenced) SSI-based interventions that are relatively limited in scope on the order of weeks. The chapter by Dana Zeidler and colleagues addresses a considerably longer curriculum (1 year), but Eastwood, Schlegel, and Cook explore SSI infused across an entire 4-year program. These authors also offer the only study of SSI situated in a college context. The work associated with designing and implementing this program is obviously extensive and the multifaceted research design and execution is equally ambitious. I commend the authors on both aspects of this work and believe that the broader community interested in SSI education can learn a great deal through this presentation.

One of the study results that grabbed my attention was the comparison of reasoning between the two groups. My interpretation of the "take home message" was that the two groups did not seem to engage in reasoning that was qualitatively different (i.e., the groups did not take up significantly different positions or rely on significantly different rationales) but that the SSI group demonstrated higher quality reasoning. This was the same kind of result that we saw in a study comparing SSI reasoning between undergraduate science majors and undergraduates studying nonscience disciplines (Sadler & Zeidler, 2005). Both groups engaged in the same kinds of reasoning patterns but the science majors offered higher quality reasoning in the discussion of SSI. The chapter documents statistically significant differences in the SSI and BIO group reasoning scores on the order of about .4 standard deviations. I would be interested to hear more from Jenny in terms of how practically

J.L. Eastwood (✉) and T.D. Sadler
School of Teaching and Learning, University of Florida, Gainesville, FL, USA
e-mail: jleastwood@coe.ufl.edu

M.P. Jiménez-Aleixandre
University of Santiago de Compostela, Santiago de Compostela, Galicia, Spain

T.D. Sadler (ed.), *Socio-scientific Issues in the Classroom: Teaching, Learning and Research*, Contemporary Trends and Issues in Science Education 39,
DOI 10.1007/978-94-007-1159-4_7,

significant these differences are. Using conventional definitions of effect sizes, I can interpret this gain as moderate, but that clearly would not tell the whole story. Jenny was in the classes and saw the curriculum implemented; she also conducted interviews and poured through the data as the primary researcher. This puts her in a unique position with respect to the interpretation of these results. Jenny, given your vantage point, what do you see as the practical significance of these results?

Eastwood: The difference in reasoning between the two groups is not huge, but it is meaningful in light of my experiences interacting with students and being in the classroom. There are many reasons why finding a large difference between groups would have been surprising. All of the participants in the study were science majors, so they were all good students and presumably all of them cared about science and envisioned futures related to science. The students had diverse backgrounds, and as I learned in interviews, the SSI core courses were not the only courses on campus that involved SSI. Some BIO students had extensively read and discussed SSI on their own initiative and some SSI students tried to slide through their major with as little reflection as possible. Additionally, the questionnaire and rubric was limited in its ability to definitively assess students' reasoning with SSI. Participants were simply asked to take a position on questions related to scenarios and explain their reasoning. Given conditions more authentic to real-life decision-making, like more time for reflection or opportunities to discuss the issues with others, the reasoning outcomes could have been quite different. Accurately assessing reasoning with larger sample sizes is a challenge.

In light of these limitations, a statistically significant result was encouraging. Still, my impressions of the practical significance of the results are shaped more by the student and professor interviews and classroom observations. The aspect of the program that seemed most significant for gains in reasoning was the consistency of the SSI-based learning environment and the way instructors explicitly guided students on a trajectory toward better reasoning. In class, professors constantly reminded students to back up their positions with evidence and they modeled critical evaluation of evidence and conflicting viewpoints through many different examples. Students were even asked to reflect on how they learned to take committed positions on issues and advocate for particular causes.

These emphases in the SSI program came through very clearly to the students and they seemed to internalize the value placed on informed decision-making. SSI students consistently explained that the program changed the way they approached controversial science issues. They discussed being more open to carefully considering different perspectives on issues and some acknowledged that they would not even have been aware of certain controversies if they had not been in the major. They also tended to relate issues they had discussed in class to examples or questions in the interviews and questionnaires. These results taken together suggest that the program helped students learn to seek out different perspectives and use their experiences with SSI as reference points for new issues they encounter. They seemed to recognize their growth toward a more mature way of thinking about SSI.

Students interviewed from the BIO group were also very interested in SSI, but consistently reported that courses in their major were not geared toward preparing them to make informed decisions on such issues. A common perspective (including one of the SSI professors) was that biology courses were there for learning in-depth science content and reasoning with SSI could be sought elsewhere. BIO students may have cared about SSI, but they were not consistently engaged in SSI activities, at least not in their biology courses. They did not volunteer insights on seeking out multidisciplinary perspectives and different points of view. They had varying levels of confidence in their abilities to reason with SSI but did not feel that their college major prepared them to reason with SSI.

Although I feel encouraged by the SSI students' enthusiasm and professed intellectual growth, I realize it is important to differentiate between students' self-reflections on their reasoning and "outside" assessments of their reasoning. Based on my experience with the SSI program, I would gladly advocate for SSI-based programs that build instruction around development of informed and ethical decision-making. However, even if goals of scientific literacy and responsible decision-making are embraced, I think more evidence of gains in reasoning will be needed to justify the kind of restructuring that would need to occur in colleges and universities.

SSI and Interdisciplinary Education

Eastwood: Interdisciplinary education is a popular theme now, especially in college education circles, but little empirical research currently exists on the topic. SSI is by nature interdisciplinary science education that is developing a distinct discourse. However, from my perspective, the term, "socioscientific issues" does not seem to go beyond the science education community, even where interdisciplinary science teaching and learning is discussed. Although the human biology program discussed in this chapter was clearly doing SSI very effectively, the term "SSI" and the related literature was unfamiliar to the faculty involved. Another example is the NSF-funded project, Science Education for New Civic Engagements and Responsibilities (SENCER), which has sought to "improve undergraduate STEM (science, technology, engineering and mathematics) education by connecting learning to critical civic questions" (http://www.sencer.net/About/projectoverview.cfm). The organization provides faculty development opportunities and excellent resources like issues-based course models and encourages educational research. Although this project is also very consistent with the goals of SSI, I found no references to SSI in documents available from the SENCER website. Clearly both the SSI community and members of other disciplines who are carrying out and conducting research on interdisciplinary, issues-based science education can benefit from each others' work. My question is how the SSI community, which values integration of different perspectives to address problems, can become more connected with others who are trying to accomplish many of the same goals for college students. What do you see as hindering these connections?

Sadler: The issue raised here points to the insularity of academic disciplines and is certainly not just a problem for SSI and interdisciplinary education. The lack of communication across disciplines is a problem for many areas; although, it is somewhat ironic that in this case, in particular, more interdisciplinary collaboration is not seen. The "academy" and the established norms and expectations for researchers are structured in ways that work against cross-disciplinary communication. This is not to say that cross-disciplinary approaches cannot or do not exist. I see the recent emergence of the learning sciences as a very nice example of how scholars can create opportunities to draw on the expertise of multiple, previously isolated disciplines or subdisciplines. The learning science movement has created space for scholars to build on research in the cognitive sciences, information and communications technology, science and mathematics education, and instructional design in productive ways. In the case of SSI and interdisciplinary education, I think it will take the efforts of a few dedicated scholars well-grounded in both communities to show how drawing from the two strengthens efforts in either community and ultimately moves the broader field of education toward goals associated with improving science education and promoting scientific literacy.

Jiménez-Aleixandre: I have been involved in interdisciplinary programs when I was a high school teacher, but carrying them out at the university level is exceedingly difficult. Interdisciplinary programs pose many challenges, some related to their implementation, and others to research about them. The chapter authors explicitly identify some of the challenges:

> Many factors contribute to the complex learning environment of an SSI unit or course. In the SSI classroom context, variables contributing to student outcomes cannot be easily isolated to reveal their direct contributions to student outcomes. Particular aspects of instruction can influence students' knowledge and perceptions to different degrees and complicate findings.

This is a very important point. Even in courses focusing on just one subject, learning environments are complex, and interactions among students, teacher, curriculum, and social context are difficult to unravel. Therefore, as the authors say, variables cannot be easily isolated.

HIV/AIDS as SSI Content

Jiménez-Aleixandre: The SSI course was structured into three modules: death and dying, infectious disease, and HIV and AIDS. I think that the HIV/AIDS case offers a very productive context for SSI-based education. The case raised by the authors, in which scientists have disagreed about the causal mechanism of AIDS, provides on interesting avenue for science education. Another topic-related issue that we have used in biology education courses is the claim by Pope Benedict XVI in March 2009 about the AIDS epidemics in Africa. At the outset of his first visit to the continent as Pope, he claimed that condoms were not the answer to the continent's fight against HIV and AIDS and that condom use could even make the problem worse (Butt, 2009). He recommended sexual abstinence and fidelity as means of preventing HIV and AIDS. This claim, which was contested by health agencies as well as Catholic priests and nuns working in Africa, illustrate the interactions between beliefs grounded in ideological stances and health-care recommendations based on scientific evidence. In the more fundamentalist strands of Catholicism, the Pope is perceived to be infallible, and although this infallibility only affects his theological productions, the implication is that whatever he claims is true. On the other hand, the claim seems to support the reservations about the causal relationship between HIV and AIDS, casting doubts about the process of infection and how to prevent it. The Pope never offered scientific justifications for his position. This case could be a used as an example illustrating the difference science and religion and criteria for claims made in either domain. The situation becomes problematic when ideological positions, like the one made by the Pope, are interpreted as a claim based on scientific evidence, as is the case with many SSI.

The senior course also offers an interesting example of integration, not just of content, but also of action, engaging students in community service. It is an exemplary model for environmental education courses. Another feature of the paper that has potential to be useful for researchers is the rubric combining reasoning and perspectives presented in Tables 6.2, 6.3, and 6.4. The study may help us to understand and decide how SSI learning environments (and related communities of practice) should be structured.

References

Butt, R. (2009). *Pope claims condoms could make African AIDS crisis worse – guardian.co.uk.* Accessed from http://www.guardian.co.uk/world/2009/mar/17/pope-africa-condoms-aids

Sadler, T. D., & Zeidler, D. L. (2005). The significance of content knowledge for informal reasoning regarding socioscientific issues: Applying genetics knowledge to genetic engineering issues. *Science Education, 89*, 71–93.

Chapter 8
Discussing a Socioscientific Issue in a Primary School Classroom: The Case of Using a Technology-Supported Environment in Formal and Nonformal Settings

Maria Evagorou

Recent educational reports in the USA (Duschl, Schweingruber, & Shouse, 2007), the UK (Osborne, 2007), and elsewhere in Europe have called for a science education that places an emphasis on scientific literacy, and makes the connection between science and everyday life. The focus of this approach is on the social aspects of science, aiming to prepare young people for life beyond school. Aikenhead (2006) has attempted to define the term by explaining scientific literacy as acquiring knowledge for science—that is, both *knowledge of the content* and *knowledge about science,* which he sees as the social processes of science. Likewise, in national reform documents, the core of scientific literacy is related to understanding knowledge and processes of science, and the application of this knowledge (AAAS, 1993; National Research Council, 1996). For example, the most recent US science education reform document states that:

> Expectations of what it means to be competent in science and understanding science have also broadened. [...] Learners who understand can use and apply novel ideas in diverse contexts, drawing connections among multiple representations of a given concept. They appreciate the foundations of knowledge and consider warrants for knowledge claims. Accomplished learners know when to ask a question, how to challenge claims, where to go to learn more, and they are aware of their own ideas and how these change over time. (Duschl et al., 2007, p. 19)

This call for the emphasis in science education on understanding the evidence and claims of science and being able to assess them is associated with the shift from studying science as exploration and experiment to studying science as argument and explanation (Duschl et al., 2007). For instance, Duschl (1990) argues that if we do not present science as a process of revision and substitution of knowledge claims we run two risks. Firstly, to develop in students the perception that "scientific knowledge growth is governed by the addition of new ideas, facts and theories to old ones" (p. 54) and, secondly to portray science as an activity in which scientists always agree. Hence the emphasis on teaching argument and explanation can contribute to

M. Evagorou (✉)
School of Education, University of Nicosia (UNic), Nicosia, Cyprus
e-mail: evagorou.m@unic.ac.cy

T.D. Sadler (ed.), *Socio-scientific Issues in the Classroom: Teaching, Learning and Research*, Contemporary Trends and Issues in Science Education 39,
DOI 10.1007/978-94-007-1159-4_8, © Springer Science+Business Media B.V. 2011

students' appreciation of both the power and the limitations of scientific knowledge claims. Such an understanding is increasingly required within the context of a society where scientific and socio-scientific issues (SSI) dominate the cultural landscape, where social practices are constantly examined and reformed in the light of scientific evidence, and where the public maintain an attitude of ambivalence (Giddens, 1990) or anxiety about science (Beck, 1992).

The work presented in this chapter is informed by these recent trends in science education, and the goals of my research in the context of socio-scientific issues have been related to argumentation (a component of scientific literacy) and engaging students with problems from their local or national communities as a means to improve their science learning experiences. In the early stages of my work, I explored SSI contexts: (a) as a way to improve elementary school students' systems thinking (Evagorou, Korfiati, Nicolaou, & Constantinou, 2009), decision-making (Nicolaou, Korfiati, Evagorou, & Constantinou, 2009) and (b) as a way to support students' collaborative argumentation (Evagorou & Osborne, 2007, 2008) when supported by the use of technology and more specifically scaffolded with tools from the WISE platform (Linn, Eylon, & Davis, 2004). WISE (Web-based Inquiry Science Environment) is an online platform that was developed to scaffold teachers and learners as they were learning to take advantage and manage new Internet technologies (Cuthbert & Slotta, 2004). The tools within WISE are designed based on four metaprinciples: (a) making science accessible to all students; (b) making thinking visible by modeling students' ideas and evaluating how these ideas are transformed and synthesized to form new knowledge; (c) helping students learn from others by encouraging them to build on ideas presented by others and question peers or experts, and (d) promoting autonomy and lifelong learning (Linn et al., 2004). Early research in WISE focused on improving the platform, adding new tools or redesigning the existing tools that were integrated in the environment, in order to improve learning and scaffold students in more effective ways. Results from early studies were used as a means for refining the four metaprinciples (Linn et al.). The first projects implemented in WISE placed an emphasis on teaching heat and temperature through the use of simulations. Some of the projects implemented later included learning skills such as evaluating data gathered from websites, connecting claims to evidence or discussing findings with peers. More recent work has focused on using the tools within WISE to support students while collaboratively constructing arguments (Clark & Sampson, 2008). The choice of WISE is based on the evidence from previous research that the tools can support students as they collaborate with peers in their effort to investigate an issue (e.g., Bell, 2004).

For the first WISE project that was implemented in Cyprus, the problem that served as the curricular and instructional focus was not a real issue. Our group designed the project as a way to introduce systems thinking, and it focused on ways to control the population of the mosquitoes in a swamp. The elementary school students explored the issue through information provided to them; they used WISE tools to collect and organize information, and then reach a decision using their systemic knowledge to build computer models that would predict the long-term effects of their proposed solutions (e.g., use chemicals or introduce new species).

For the second WISE project we featured a problem that had emerged as a national issue in the UK. The focal issue related to the decline of the indigenous grey squirrel because of the introduction of the red squirrel. This project was implemented in the UK with middle school students and they used the WISE platform and tools to collect and organize information about squirrel ecology as well as prepare and discuss their arguments regarding ways to control the population of the grey squirrels online.

These learning environments helped students to engage with the different SSI and improve their systems thinking, to build their content knowledge related to the system and to strengthen their argumentation practices. The implementation of the two projects also helped to understand some deficiencies in the design. For example, the students did not have any personal experiences with the issues presented since they did not constitute real or authentic problems for them or their nearby area. Hence, the students did not have opportunities to collect evidence from the field to support their arguments, and were not motivated to further explore the problem. Another design problem was that the curriculum materials were designed by researchers, but not in collaboration with teachers that hold a more practical understanding of the classroom reality. Based on these experiences and the ideas and questions generated from my involvement in these projects I started developing a new research project that received funds from the Cyprus Research Promotion Foundation. This research project, the focus of the chapter, is called *Technoskepsi*, from the Greek words *technologia* (technology) and *skepsi* (thinking) and aimed to collaboratively develop (with a group of teachers) curriculum materials making use of technologies (the WISE platform and handhelds) in order to support elementary school students' argumentation through the context of an SSI. An additional element of the learning environment was the integration of formal and nonformal settings for students' investigations that would provide field experiences and an authentic aspect of the science learning process (Braund & Reiss, 2006) for the students.

Project Goals

The *Technoskepsi* project had four goals, both design and research goals: (1) to design, in collaboration with teachers curriculum materials that will support younger students' argumentation within an authentic SSI; (2) to explore elementary school students' arguments and collaborative argumentation, (3) to explore the use of WISE and handhelds and the effect they might have on students' learning; and (4) to explore supplementing formal with nonformal settings, and understand their impact on students' argumentation, decision-making, and emotions toward science.

One of the main goals of the *Technoskepsi* project was to utilize the knowledge and experience of a group of teachers in order to design curriculum materials that will support elementary school students' argumentation and decision-making within an SSI, making use of technology. The decision to involve teachers in the process of designing the curriculum is associated with results from the implementation

Table 8.1 Overview of teachers' qualifications and teaching experience

Name	Qualifications	Years of teaching experience
Anna	BA in Elementary Education	2
	MA in Science Education	
Mary	BA in Elementary Education	2
	MA in Educational Technology	
Costas	BA in Elementary Education	2
	MA in Educational Technology	
Aris	BA in Elementary Education	9
	MA in Science Education	

of curriculum materials from previous projects that indicated that teachers felt they did not have ownership of the materials because they were not involved in the design process. Furthermore, feedback from the same teachers states that researchers design curriculum materials for classes without having the experience from the classes. In order to address these issues I recruited four elementary school teachers to work collaboratively on the design of the curriculum. The teachers, two female and two male had special interest either in the use of technology or in science education, and were already working as elementary school teachers in Cyprus at the time. Table 8.1 shows the qualifications and years of teaching experience for the four teachers that participated in the design group.

Initially the four teachers participated in four, 3-h meetings aiming to familiarize the teachers with the goals of the research project and the main theoretical perspectives of the project: (a) argumentation, students' difficulties, and methodological issues, (b) socio-scientific issues and their importance for science learning, (c) scaffolding science learning with the use of technology, (d) sociocultural theories of learning, and (e) project-based learning. Between meetings, the teachers had to read papers, evaluate curriculum materials, and prepare short activities on the topics discussed. An online environment was designed in which the teachers could find all resources and also discuss issues with the other members of the group. In one of the meetings, Aris suggested the design of curriculum materials that would focus on the study of a socio-scientific problem that was causing problems to the community of his school. This problem was associated with a pig farm near this community. The issue was generating a lot of media coverage primarily because of complaints from community members related to the farm's smell. Following Aris's suggestion, we collected information regarding how pig farms function, what kinds of pollution they can potentially cause (water, soil, and air), the European legislation regarding farms and optimum waste management techniques, and all the latest techniques that can be used to minimize the smell and pollution issues. After having acquired this information, we worked collaboratively to design the structure for an instructional unit.

The second goal of the project was to explore elementary school students' arguments and collaborative argumentation. Argumentation is considered a major aspect of the resolution of scientific controversies (Fuller, 1997; Taylor, 1996). It is seen as "a social process, where cooperating individuals try to adjust their intentions and interpretations by verbally presenting a rationale for their actions" (Patronis, Potari, & Spiliotopoulou, 1999, pp. 747–748). Therefore, argument and argumentation have two different aspects, an individual and a social. The individual aspect of the argument refers to articulating a point of view (Jimenez-Aleixandre & Erduran, 2008), while the social aspect involves two or more people and aims to persuade others (Bricker & Bell, 2008). In science, arguments are commonly constructed to explain a phenomenon or to explain a theory or a new discovery, and argumentation is seen as part of the process of knowledge construction in science. More specifically,

> … argumentation in scientific topics can be defined as the connection between claims and data through justifications or the evaluation of knowledge claims in light of evidence, either empirical or theoretical. (Jimenez-Aleixandre & Erduran, 2008, p. 13)

However, argumentation is not a skill specific to science; on the contrary, it is central to people's ability to solve problems, make judgments and decisions, and formulate ideas and beliefs (Kuhn, 1991). Argumentation is essentially a thinking/reasoning skill. According to Kuhn (2005), thinking is the process that enables us to make informed choices between conflicting claims and understanding this leads a person to value thinking. Usually, when learners are constructing arguments, they need to evaluate alternative perspectives and opinions and select a solution that is supported by evidence and explanation (Cho & Jonassen, 2002). Hence, argumentation is an important skill for everyday life because we are frequently faced with situations in which we have to evaluate alternative solutions or scenarios and decide on a course of action based on evidence. The ability to identify alternative solutions—a skill associated with argumentation—can potentially help people move toward more informed decisions in their everyday life.

Even though argumentation in science education has been an emphasis of many studies (Erduran, Simon, & Osborne, 2004; Jimenez-Aleixandre, Rodriguez, & Duschl, 2000; Jimenez-Aleixandre & Pereiro-Munoz, 2002; Kuhn, 1991; Sandoval, 2003; Osborne, Simon, & Erduran, 2004a, b), little is yet known about how young students develop their arguments, especially in the context of socio-scientific issues, and how young students work collaboratively in order to construct their arguments. *Technoskepsi* explored the issues of construction of arguments and collaborative argumentation by younger students, and some of the findings are reported later in this chapter.

The third goal of the project was to explore the use of WISE and handhelds and the effect the combined use of these technologies might have on students' learning and development of argumentation. Results from previous studies agree that even with specially designed instruction, students do not construct the high quality arguments that might be desired of them (Erduran et al., 2004; Jimenez-Aleixandre & Pereiro-Munoz, 2002). Students' failure to construct high quality arguments can be explained partly by the dominance of a pedagogy which is authoritative and rooted in education as a form of transmission (Simon et al., 2005) which provides students with few opportunities to engage in the process of argumentation (Jimenez-Aleixandre et al., 2000). It has been suggested that online learning environments have the potential: (a) to scaffold the teaching process and help teachers move away from authoritative pedagogy and (b) to scaffold argumentation in more constructive ways (Andriessen, Baker, & Suthers, 2003; Bell & Linn, 2000; Clark & Sampson, 2008). Educators favoring the use of handhelds in education according to Zurita and Nussbaum (2004) suggest that handhelds "support constructivist educational activities through collaborative groups, increasing motivation, promoting interactive learning, developing cognitive skills, and facilitating the control of the learning process and its relationship with the real world" (p. 235). The choice of the WISE platform and the handhelds is explored further in the intervention section.

Finally, the last goal of the project was to explore supplementing formal with nonformal settings, and understand their impact on students' decisions, argumentation, and emotions toward the lesson. The decision to make use of nonformal settings is associated with the claim that when students engage in authentic practices that can provide a context that can potentially increase students' motivation (Edelson & Reiser, 2006). Blumenfeld, Kepler, and Krajcik (2006) add that motivation sets the stage for cognitive engagement and leads to achievement by increasing the quality of the cognitive engagement.

Setting

As described earlier, a group of four teachers proposed a set of activities they thought would be appropriate for 10–12 year old students for the socio-scientific issue proposed by one of the teachers, Aris. Three of the teachers changed school districts the following year and were no longer able to work on the project. Aris, on the contrary, stayed at the same school and offered to implement the curriculum with two of his classes. By that time, Aris was in his tenth year teaching at the elementary school level. Aris graduated from a prestigious 4-year Bachelor's degree program in Elementary Education with a specialization in Science Education. Immediately after graduation, he began teaching. During his career, he worked at three different elementary schools and taught all grade levels and subjects. Four years into his teaching career, Aris took two years off from teaching to pursue a Master's in Science Education and worked part-time as a researcher at the same university. During that time, Aris was involved in argumentation and computer modeling projects with elementary school students. After finishing his master's degree, he returned to teaching at the elementary school at which the current project was conducted. When we started working together on the *Technoskepsi* project, he was in his third year at that school which will be referred to as MA Elementary, and he was teaching sixth grade (student age: 9–10) language, mathematics, and science and fifth grade (student age: 11–12) science.

MA Elementary serves the local community of a small suburban town, with a total of 160 students (K-6) and 15 teachers. The curriculum materials were implemented in two of the classes that Aris was teaching science: a fifth grade class and a sixth grade class. The fifth grade, similar in terms of students' abilities to the sixth grade, had 17 students and served as a forum to pilot test the curriculum. Based on observations made during the fifth grade class implementation, we enacted curricular changes that were implemented with the sixth grade class. Hence, the sixth grade served as the class in which we collected the main data for the *Technoskepsi* project. Aris's sixth grade, with 18 students (10 boys and 8 girls), was a low achieving class. Since in the Cypriot educational system there are no national exams or formal grades until middle school, I asked Aris to describe his students' general abilities in mathematics, Greek language (the official school language), and science, and I also administered an argumentation questionnaire to identify their argumentation levels. According to the teacher, the majority of the sixth graders were low achievers in all three main subjects, with two of the students (immigrants) having Greek as a second language. Additionally, Aris reported that some of the students exhibited behavioral problems and it was difficult to include them in group activities. Furthermore, the argumentation questionnaire showed that most of the students had difficulties in either choosing or constructing the best argument (Evagorou, 2008).

Aris organized his students in groups at the beginning of the year, and the groups designed investigations and interacted with technology as part of their science curriculum throughout the year. Five desktop computers were available in the class, and students had already worked on collaborative computer programs prior to the initiation of the *Technoskepsi* project (e.g., they started a wiki describing the history of their village and also used modeling programs). More information regarding how the students worked during the implementation of the current project are presented in an upcoming section.

Teacher–Researcher Relationships

I had known Aris for several years, and we had worked together in the past as researchers in science education and technology projects. We shared similar ideas regarding teaching and learning and the same passion for the use of technology as a tool for learning. After the initial drafting of the curriculum with the group of teachers, we worked with a biologist and an educational technology expert from my institution to design the final version of the curriculum materials. The biologist assisted us with the content knowledge that was necessary to understand the air, water, and soil pollution that the waste from the pig farm could potentially cause as well as various waste management techniques. Implementation of the curriculum started in January and finished in March. Materials were implemented with the fifth grade class every Monday and then in the sixth grade class every Friday of the same week. Aris met with his students for two, 40-min sessions each week. During the implementation, I was a participant observer, sometimes coteaching with Aris, sometimes simply participating in the group discussions and offering technical support and other times simply observing the teaching process. After each implementation in the fifth grade, Aris and I held meetings during which we would discuss difficulties that the students had with the curriculum and possible changes to the activities to better meet student needs. During the week, we made the changes to the curriculum and implemented the modified learning experiences with the sixth grade class at the end of the week. After each implementation in the sixth grade, we held a shorter meeting in which we discussed the lesson.

Intervention

As described earlier, the first form of the curriculum was designed collaboratively by a group of teachers, and the final form was designed by myself and the teacher that implemented it with the help of a biologist and an expert in educational technology. The learning environment is partly designed within the WISE platform (Linn et al., 2004) and poses the following guiding question to the students: *What are the effects of the pig farm on your area and what course of action do you suggest?* The socio-scientific topic was chosen because it was relevant to the students' everyday lives and was an issue that could potentially engage these students in the investigation and challenge them to construct arguments considering all aspects of the topic (moral, financial, environmental, social). We designed the *Technoskepsi* curriculum materials to meet the following instructional objectives:

1. Help the students to develop an understanding of argumentation and how argumentation is different from simply expressing opinions.
2. Develop argumentation skills and be able to use evidence to justify their claims.
3. Engage in scientific investigations and collect evidence from the field.
4. Engage in investigations regarding an authentic socio-scientific issue and understand and appreciate the social and scientific factors that contribute to the controversy.
5. Understand the systemic nature of their environment and the short- and long-term effects that various decisions have on their environment.
6. Develop an understanding of waste management techniques and the impacts they can have on the environment.

In order to achieve these objectives we designed eight, 80-min lessons. We examined resources from the Cyprus Department of Agricultural with information regarding the pig farms and interviewed environmentalists and visited a pig farm to have a holistic approach to the problem. Our decision was to develop the curriculum based on project-based learning (Krajcik et al., 1998), sociocultural theories of learning (Rogoff, 2003), and what we already know regarding how people construct arguments. For example, Kuhn (1991, 2005) suggests that most people tend to be certain of their theories, even when they are using pseudoevidence; they tend to reason better on the subjects for which they have personal knowledge; and they assimilate new information in existing theories and they express considerable certainty that new evidence supports their theories even when it is contradictory.

Another decision was to use the WISE platform which incorporates knowledge representation and discussion-based tools that can potentially scaffold students in their effort to work collaboratively to construct arguments (Bell & Linn, 2000; Evagorou & Avraamidou, 2008). More specifically, a way to support students to construct higher-level arguments is through scaffolding provided by the use of computer-based tools (Cho & Jonassen, 2002). Several technology-based classroom interventions have been designed with the aim of argument construction in science classrooms (Sandoval & Reiser, 2004). Research findings on technology-based environments and argumentation suggest that technology has the capacity to support high quality argument construction within the classroom by taking account of the epistemic and social factors that support and promote argumentation (Bell, 2004). Such environments are designed to support students' argument construction in two ways: (1) by incorporating knowledge representation tools (Edelson, Pea, & Gomez, 1996) and (2) by incorporating discussion-based tools (Scardamalia, 2003).

Knowledge-representation tools have been designed to help students construct arguments by connecting evidence to the appropriate claim. For example, in these tools, evidence might be represented with a specific shape and color and claim, with a completely different shape and color. These tools address the difficulty that students have with evaluating evidence and claims and the fact that they usually tend to provide a claim with no evidence or with single rather than multiple pieces of evidence. Furthermore, as Suthers (1999) states, knowledge representation tools mediate discourse "by providing learners with the means to articulate emerging knowledge in a persistent medium, inspectable by all participants, where the knowledge then becomes part of the shared context" (p. 4). Suthers (1999) also explains how the visual presence of the knowledge unit in a shared representational context can serve as a reminder of the work that needs to be done by the learners. For example, a linear text, like an online discussion does not provide any hints to whether learners need to do something specific. On the contrary, a graphical representation tool illustrates how learners need to find connections between different bits of the knowledge. An example of a knowledge representation tool within WISE that we used to scaffold our students is Sense Maker (Fig. 8.2).

Discussion-based tools can facilitate communication, either through online asynchronous communication or face-to-face synchronous communication, with other learners, promoting dialogic argumentation. Discussion-based tools are based on the recognition that the construction of knowledge is not an individual process but rather a collective process of ideas and arguments that come together, such as in the work undertaken in scientific research groups (Scardamalia, 2003). Furthermore, according to (Lampert, Rittenhouse, & Crumbaugh, 1996) discussion-based tools allow students to take more time before formulating a contribution or an argument, something that usually does not happen within classrooms, and in that way help them contribute to discussions in more coherent ways. These tools can also reduce social and emotional obstacles of expressing opinions in public when lacking the appropriate representations, and, thus enable more students to take part in the discussions. Research associated with these environments has demonstrated that such technology-enhanced learning environments can be used to successfully scaffold dialogic argumentation (Sandoval & Reiser, 2004).

Finally, another form of technology we decided to use was handheld data collection devices. Recently, the use of mobile devices has taken the place of ordinary computers in the classrooms. What is of importance about the use of mobile technologies for education is that tools which first existed only on expensive desktop machines are now available on inexpensive handheld units (Soloway et al., 2001). In the case of *Technoskepsi,* the use of handhelds provided a number of affordances to support student learning experiences. Handhelds are highly portable, are relatively inexpensive (compared to other data collection systems), have a relatively long battery life (compared to laptops), and facilitate easy synchronization and data sharing. Given the opportunity for students to visit and collect data from a field site, the handhelds proved to be ideal tools for supporting the students' work.

The curriculum materials were made up of eight interconnected lessons that ranged from an introduction to argumentation, to an introduction to the problem, a visit to the nearby pig farm and a whole-classroom discussion of the decisions the groups reached. Table 8.2 presents the structure and content of the curriculum materials.

As shown in Table 8.2, the first lesson was an introductory lesson in argumentation, in which the students had to discuss in their groups a different socio-scientific

Table 8.2 Overview of the Technoskepsi curriculum

Lesson	Brief description of lesson	Arguments submitted by groups
Lesson 1: Introduction to argumentation	Students engaged in argumentation through the discussion of a socio-scientific issue: whether they agree with building a new zoo in their area or not (based on Osborne et al. IDEAS pack)	
	Presentation of students' arguments and models of a good argument	
Lessons 2 and 3: Introduction to the problem	Students were introduced to the problem of study: Environmental problems from a nearby pig farm and what cause of actions should be taken	Argument 1: Opinion
	Group work using the learning environment designed within the WISE (Linn et al., 2004) platform to understand the various aspects of the problem	Argument 2: Position after studying the evidence provided
	Supported by specially designed prompt windows and knowledge representation tools (SenseMaker)	

(continued)

Table 8.2 (continued)

Lesson	Brief description of lesson	Arguments submitted by groups
Lesson 4: Familiarizing with handhelds and water quality tools	Students were provided with the handheld devices and were asked to undergo some investigations, both indoors and outdoors, in order to familiarize with the handhelds Preparation of interview questions for the pig farm visit Familiarized with the water, air, and soil quality kit	
Lessons 5 and 6: Visit to the pig farm	Visit to the pig farm The students collected data from the pig farm regarding the waste management techniques, the position of the farm and distance from inhabited areas, and water quality and soil Interview with the farmer	
Lesson 7: Preparing the final argument	The students, in their groups, prepared their final argument based on the evidence from WISE (after an online discussion), and the data gathered during the visit to the pig farm The students made use of the online discussion tool to provide feedback to each other on the structure of their arguments	Argument 3: Final argument after the outdoors investigation
Lesson 8: Presentation of the final arguments and discussion	The groups presented their final arguments to the class and a whole-classroom discussion followed The students prepared a letter addressed to the local authorities with their suggestions regarding the pig farm	

issue, that of constructing a zoo in their area and then present their arguments. The material for this argument was adapted from the IDEAS pack (Osborne et al., 2004a). The teacher focused the discussion on what a good argument should look like and how an argument differs from an opinion. The emphasis during the whole classroom discussion was on the use of evidence to support one's claim and what kind of evidence we should trust.

The second lesson was an introduction to the problem—that of the excessive smell from the pig farm and the protests of the people in the community. The problem was presented through newspaper clippings and recorded interviews from people in the community, an environmentalist, and the owner of the pig farm. After studying the problem the students (working in groups of threes) were asked to state their opinion and whether they suggest closing down the pig farms. In Table 8.2, the argumentation task is shown in the third column as Argument 1. Figure 8.1 presents a screenshot from the learning environment showing a Note Window and the introductory page presenting the problem.

During the third lesson the students had to work in their groups in order to familiarize themselves with all aspects of the problem (e.g., environmental and financial issues) and understand the possible effects the waste from the pig farm could have on soil, air, and water, and the kind of solutions that the various waste management techniques could offer. The online platform (WISE) was used to scaffold students in two different ways during that time: (a) to collect all the available information to help them construct their arguments and (b) to scaffold students in the process of constructing and sharing their arguments. In order to achieve the first goal, various "note" windows were designed (see Fig. 8.1) which scaffolded students to collect information and evidence from the online learning environment. In order to achieve the second goal, knowledge representation tools and discussion-based tools were used. Figure 8.2 presents a screenshot of the knowledge representation tool (SenseMaker) that was used in the online learning environment.

SenseMaker allows students to coordinate their evidence with the appropriate claim, a function that addresses one of the difficulties that students face with

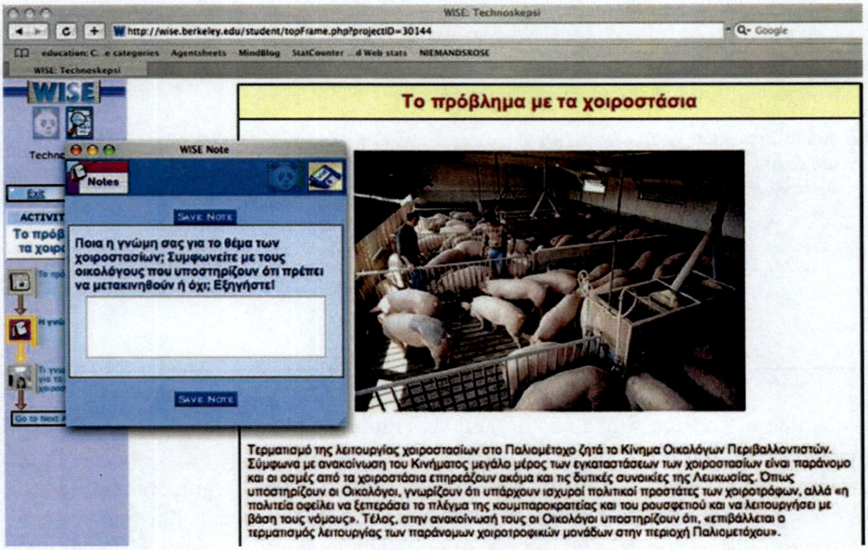

Fig. 8.1 The introductory page of the online part of Technoskepsi

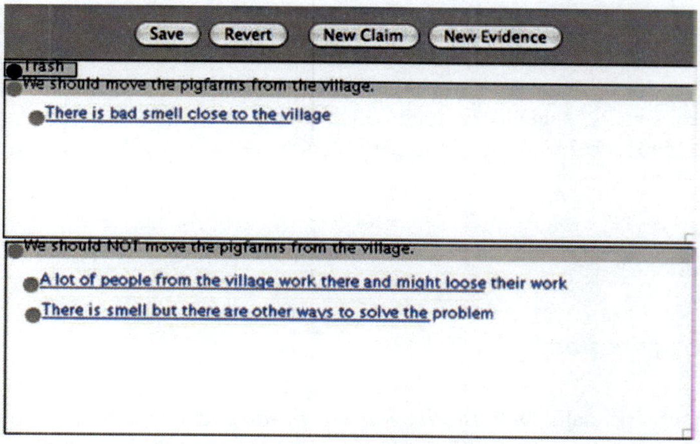

Fig. 8.2 The knowledge representation tool, SenseMaker

argumentation. More specifically, the rectangles represent the claims, and these are set in the learning environment; the students then have to type in their evidence (represented by the underlined text) and put them under the appropriate claim. At the end of the third lesson the groups had to submit online their argument as formed after studying all the available evidence within WISE. This is shown in Table 8.2 in the third column as *Argument 2*.

After carefully reading and discussing the online data, we explained to the students that they would visit the area of the pig farm to collect data. However, an important goal was to prepare them for the field investigation; hence after the teacher's suggestion we designed Lesson 4 in order to familiarize the students with the data collection techniques they would use on the field. During Lesson 4, the students developed familiarity with the handheld devices, and used the water, air, and soil quality kit in some investigations in the school yard, and prepared a list of questions for their visit to the pig farm.

Lesson 5 and 6 (two, 80-min lessons) consisted of visits to the pig farm. During that time, the students had to collect evidence to support their argument (e.g., water, soil, and air quality, interview with the pig farmer regarding his waste management techniques, location of pig farm relative to inhabited areas), and use their handhelds to store the data collected. After they returned to the classroom, students transferred their data from the handheld devices collected in the field to their computers.

The aim of lesson 7 was to help students unpack the experiences from farm visit and to scaffold students in using the evidence collected from the field to further support or dispute their arguments. After revisiting their arguments, the groups had to submit a new argument online and share it with the other groups using the discussion-based tool. The groups then commented on another group's argument. The purpose of this activity was to help the students strengthen their arguments based on the feedback from another group. The final outcome of this lesson was *Argument 3* (as shown in Table 8.2, third column) that was submitted by each group.

Finally, during the last lesson the groups presented their arguments during a whole-class discussion, and students engaged in a debate. Additionally, the class talked about the different kinds of justifications and how to rank them based on importance. Finally, students decided which points to include in a letter that was addressed to the local authorities and presented the outcomes of their investigations.

Research

Research Questions

The research associated with the *Technoskepsi project* explores how various types of technologies can be used to support argumentation and decision-making within a socio-scientific issue, both in formal and nonformal settings. More specifically the research questions guiding this study are: (a) Is the specially designed curriculum material (combining investigations in formal and nonformal settings, and the use of technology) successful in engaging 11–12 year old students in argumentation? (b) How do students' arguments and decisions develop/change after the outdoors visit? (c) What is the contribution of the learning environment on 11–12 year old students' attitudes and emotions toward science?

This study is significant because it describes how different technological tools could be used in order to support investigations and argumentation in formal and nonformal settings, aiming toward argumentation. Even though a lot of studies in science education place an emphasis on argumentation, previous studies in argumentation have not identified how students' arguments develop and change (and why) especially after a nonformal investigation of an authentic problem. Although some research has been published on the argumentation practices of relatively young learners (e.g., Neylor, Keogh, & Downing, 2006), most of the work in this area has focused on older students (e.g., Jimenez-Aleixandre et al., 2000; Osborne et al., 2004b; Sandoval & Reiser, 2004). Therefore, the current study's focus on elementary students' argumentation has potential to offer new insights for the field. Furthermore, students' attitudes after participating in such argumentation projects supported by technology have not been previously documented.

Methods

The students worked in groups of three both indoors and outdoors for a period of eight, 80-min lessons. Furthermore, students' artifacts including online submissions, online discussions, final presentations were also recorded. After the end of instruction, all students were interviewed in order to identify their emotions and

Table 8.3 Research questions and corresponding data

Research question	Data collected
Is the specially designed learning environment successful in engaging 11–12 year old students in argumentation both in formal and nonformal settings?	Students' written arguments (level of argumentation, number of pieces of evidence, socio-scientific aspect of argument) Video from whole-classroom discussion (levels of arguments)
How do students' arguments and decisions develop/change after the outdoors visit?	Students' written arguments (change in decision after the outdoors visit)
What is the contribution of the learning environment on 11–12 year old students' emotions and attitudes towards science?	Interviews with students after the instruction

attitudes from the implementation of the learning environment. The video interactions from the implementation and the interviews were transcribed, and a qualitative case study research approach (Creswell, 1998; Merriam, 2002) was used in order to analyze students' construction of arguments and their responses to the interview. Table 8.3 presents the research questions and corresponding data.

Data Collection and Analysis

Analyzing Students' Written Arguments

In order to analyze students' written arguments, a modified version of Toulmin's (1958) Argumentation Pattern (TAP) devised by Erduran et al. (2004) was applied in order to assess the structure of the arguments. In Toulmin's framework, the essential elements are claims, data, warrants, and backings. According to this framework, data are "the facts we appeal to as a foundation for the claim" and warrants "general hypothetical statements, which can act as bridges" (pp. 97–98). According to TAP, data are the facts that those involved in the argument appeal to in support of their claim. A claim is the conclusion whose merits are to be established. Warrants are the reasons that are used to justify the connections between the data and the conclusion, and backings are the basic assumptions that provide the justification for particular warrants. Additionally, in more complex arguments, Toulmin identifies two more features in his framework; the qualifiers that specify the conditions under which the claim is true—and rebuttals—which specify the conditions in which the claim may not be true. The elements of argument are also presented in Table 8.4, with an explanation of each of the terms.

Summarizing, in terms of Toulmin's framework, quality means having all the different components that Toulmin suggests. However, how can you decide which argument is better quality than the other? In order to address this methodological issue, Erduran et al. (2004) devised five argumentation levels to "measure" or explain the quality of argumentation, especially as a measure of interactive discourse, since

Table 8.4 Definitions of the elements in Toulmin's framework of argumentation

Element	Definition
Claim	The conclusion whose merits are to be established
Warrant	The reason that is used to justify the connections between the data and the conclusion
Backing	The basic assumption that provides the justification for particular warrants
Qualifier	Specifies the conditions under which the claim is true and are phrases that show what kind of degree of reliance is to be placed on the conclusions, given the arguments available to support them
Rebuttal	Specifies the conditions in which the claim is not true

the main identifier of quality in their levels is the presence or not of rebuttals (Erduran, 2008). These levels are based theoretically on Toulmin's framework and are informed from empirical evidence on how young students construct arguments (e.g., Osborne et al., 2004a,b). The authors suggest the following levels of argumentation, which were used in the analysis of students' artifacts in this study:

- Level 1: Arguments that are a simple claim versus a counter-claim or a claim versus a claim.
- Level 2: Consist of a claim versus a claim with either data, warrants, or backings but which does not possesses any rebuttals.
- Level 3: Consists of a series of claims or counter-claims with either data, warrants, or backings with the occasional weak rebuttal.
- Level 4: Arguments with a claim with a clearly identifiable rebuttal. Such an argument may have several claims and counter-claims.
- Level 5: An extended argument with more than one rebuttal. (Erduran et al., 2004).

According to these levels, a sophisticated argument is one that consists of more than one rebuttal (Level 5) which points to the circumstances under which the claim would not hold true, and an argument that consists of only a claim is a Level 1 argument. The value of this modified version of Toulmin's framework lies in the fact that it enables an identification of the level, or what might be termed the quality of argumentation, and can be used to evaluate both interactive or oral argumentation, and written arguments, even though the presence of rebuttals in written arguments should not be expected to be as frequent. This modified version of TAP by Erduran et al. (2004) is the main framework that guides the analysis of the data in this study, and the choice of this framework is based mainly on the fact that is has been previously applied for the analysis of students' dialogs for a similar age group as the one in the current study (e.g., Osborne et al., 2004a,b), and it has been widely used by science education researchers (e.g., Jimenez-Aleixandre et al., 2000; Osborne et al., 2004a,b). However, we appreciate that this is not a sufficiently elaborated representation of the levels of argumentation that can occur in a science class. A person, when constructing an argument, for example, can propose one that consists of a claim and a single piece of datum and another person might propose the same claim but support it with multiple data. Are these two arguments at the same levels of sophistication or should they be placed in different levels? Is the presence of more data an indication of the quality of the argument? The Erduran et al. (2004) framework does not discriminate between the two (Evagorou et al., 2009), hence an additional measure of the quality of the students' written argument was the number of pieces of evidence in each one of the arguments. Furthermore, the arguments were also analyzed in terms of the *socio-scientific nature*, using the following coding categories: social, environmental, moral, and financial. Finally, all written arguments (Argument 1, Argument 2, Argument 3; see Table 8.2) were also analyzed in terms of the decision that the groups made (e.g., to move or not to move the pig farm) in order to identify the impact of the outdoors visit on the decision.

Analyzing Video Interactions and Interviews

The video interactions from the whole-classroom discussion were transcribed and analyzed in order to identify any arguments constructed and presented by the students, using the Erduran et al. (2004) argumentation framework. The interviews were transcribed and open coded in order to identify students' emotions and attitudes from the implementation of the learning environment.

Results and Discussion

Students' levels of Arguments and Changes in Decision

The analysis of students' artifacts as presented in Table 8.5 indicates that the students engaged in argumentation during the implementation of the curriculum materials with some of them providing higher level arguments by the end of the instruction. More specifically the table presents the levels of arguments for all six groups, for their three arguments during the lesson, and their decision in each one of the arguments (to move or not the pig farm). The first argument is the opinion that the students offered at the beginning of lesson 2, the second argument is the one offered by the groups after familiarizing with the problem and ways of solving

Table 8.5 Levels of arguments and decision

	Argument 1		Argument 2		Argument 3	
Groups	Level	Decision	Level	Decision	Level	Decision
1	2	Yes	3	No	2	Yes
2	2	Yes	2	No	2	No
3	1	Yes	2	No	2	Yes
4	2	Yes	2	Yes	2	Yes
5	2	Yes	2	No	3	Yes
6	2	Yes	2	No	4	No

it using the WISE platform and all available information during the indoors investigation. The third argument is the final argument presented by the groups after the visit to the pig farm.

As shown in the table above, during the first lesson, all the students could provide an argument, but some were unsupported claims (Level 1 arguments) or claims supported by a single piece of evidence (Level 2), usually based on everyday experience. By the end of the implementation as shown in the table only two of the groups improved their final argument in terms of the level of argumentation. However, looking into details into the arguments offered by the groups it was evident that even though there was no improvement in the levels of the arguments, there was improvement in the content of the argument, and the number of pieces of evidence offered by the groups. An example is the arguments constructed by Group 1 in the first and last lesson.

"We think that the pig farm should be removed from the area because it is causing problems" (Group 1, Level 2 argument/Lesson 1).

"We should not close the pig farm because: if we do so many people will lose their jobs, we will have no meat to eat, people will lose their jobs, it might smell but there are various ways of waste management that can help reduce the smell." (Group 1, Level 2 argument, Final Lesson).

Based on the issue identified above, and the fact that the Erduran et al. (2004) levels of argumentation cannot always capture the improvement in students arguments, especially the written ones (Evagorou et al., 2009), the data were also analyzed based on the number of pieces of evidence provided by the groups (Table 8.6), and the socio-scientific nature of the argument (social, S; environmental, E; financial, F; moral, M) as shown in Table 8.7.

Comparing the first and the second argument constructed by the groups, it is evident that even though only Groups 1 and Group 3 improved their levels of argumentation, all six groups improved in terms of the number of pieces of evidence they included in their arguments. This finding suggests that the learning environment supported the students in collecting and including new pieces of evidence in their argument, even though the structure of the argument (e.g., inclusion of rebuttals) did not necessarily change. Comparing the first argument, with the argument submitted after the outdoors visit, it is evident that only three

Table 8.6 Levels of argumentation and number of pieces of evidence for each group

	Argument 1		Argument 2		Argument 3	
Groups	Level	Evidence	Level	Evidence	Level	Evidence
1	2	1	3	4	2	2
2	2	2	2	2	2	2
3	1	0	2	3	2	2
4	2	1	2	2	2	2
5	2	1	2	2	3	3
6	2	1	2	5	4	7

Table 8.7 Levels of argumentation and socio-scientific nature of argument

	Argument 1		Argument 2		Argument 3	
Groups	Level	Nature	Level	Nature	Level	Nature
1	2	E	3	S, E, F	2	E
2	2	E, M	2	S	2	E, M
3	1	E, S	2	S, F	2	E
4	2	E	2	E	2	E
5	2	E	2	F, S	3	E
6	2	E	2	F	4	E, M, S

groups improved their arguments, both in terms of the level of argumentation and the number of pieces of evidence, something that suggests that the outdoors visit did not necessarily help the students to improve their argument. This finding can be explained by looking (a) into how their decision (to move or not the pig farm) changed after the outdoors visit, and (b) the socio-scientific nature of the groups' arguments. Table 8.7 presents the socio-scientific nature of the arguments for each one of the groups.

As shown above, for the first argument all groups offer an argument which focuses on the environmental aspect of the problem (that the smell is bothering the people in the village), an aspect that they were experiencing in their everyday lives. Only two of the groups (Group 2 and Group 3) offer additional moral and social aspects for their arguments. For the second argument the socio-scientific aspect is more complex since the students offer evidence that link to the social, environmental, and financial aspect of the problem as well. Examples of arguments that were offered by the students and accommodate those aspects are presented below:

The pig farms should close because there is a lot of bad smell in the air. They should build the pig farm away from inhabited areas. (Group 5, Argument 1)

We believe that the pig farm should close because the smell is very bad and influences the people at the village. On the other hand though the people need the meat (Group 6, Argument 1)

For the last argument, the one constructed after the pig farm visit, the nature of the arguments for all groups again focused on the environmental aspect of the problem—the smell—an aspect that is associated with the experience they had when visiting the pig farm. The analysis of the data above supports that all the groups improved their arguments in terms of the number of pieces of evidence they included after the use of WISE (argument 2), but returned back to their original argument after the pig farm visit. This finding suggests that the experience in the visit, and the excessive smell pushed students to ignore the evidence they had collected from WISE, a finding that is supported by previous studies in argumentation (e.g., Kuhn, 1991).

The analysis of the students' arguments in terms of the decisions (see Table 8.5) show that the WISE learning environment supported students in collecting all the available information and changing their initial decision which was to move the pig farm to a different area, to not moving the pig farm. More specifically, for the first argument all groups decided that the pig farms should be moved to a different area, whilst for the second argument only one group (Group 4) supported the same idea. An example of how groups changed from the first to the second argument is that of Group 6:

We believe that the pig farm should close because the smell is very bad and influences the people at the village. On the other hand though the people need the meat (Group 6, Argument 1)

The pig farm should not be moved because then we will not have meat, a lot of people in our area will stay without a job, we can use the waste to produce energy, and we can find ways to minimize the bad smell [...] (Group 6, argument 2)

However, what is more interesting is the change in decision after the visit to the pig farm. After the pig farm visit, four groups reverted back to their original decision to move the pig farm to a different area, and only two groups insisted on their decision constructed after studying the evidence explaining the problem with the pig farm. All these arguments focused on the environmental aspect of the problem—the bad smell—based on the students' experience from the visit, and their experience from living in the community close to the pig farm. These findings support findings from previous studies showing that students easily ignore evidence if these are not in accordance with their own claims (e.g Kuhn, 1991). In the case of the issue under study, a problem of personal interest in the area, especially after experiencing the bad smell during the visit, the students ignored the evidence they had previously collected, and the evidence from the field regarding the water and soil pollution. Furthermore, the analysis of the whole classroom discussion shows how the students during their presentations focused on a specific aspect of the problem—the bad smell, and even though they would recognize that there are solutions to the problem, they insisted on moving the pig farms because of the smell.

Students' Attitudes and Emotions Regarding Project

After the implementation of the curriculum, all students were interviewed either by their teacher or the researcher. Students were asked to express how they felt about the research project, what they liked, what they did not like, and whether the experience was different from what they usually do in their class. All of the students who participated in the interviews offered positive appraisals of the project. They expressed excitement in reflecting on the use the handheld devices, enjoyed interactions and experiences in the WISE platform, and appreciated the opportunity to visit a field site.

The excerpt below, taken from a postintervention interview, provides an example of a typical student reaction to the use of handhelds. This excerpt is representative of many student comments offered relative to their experiences with the handhelds.

Researcher: How did you feel when you first used the handhelds?
Erena: I was happy because it was the first time that I had seen such a thing and I wanted to use it. I was smiling, I was happy.
Researcher: Do you remember similar feelings from school?
Erena: Yes, when we went to the pig farm.
Researcher: Other than from the learning environment, do you remember having similar feelings?
Erena: Yes, every time I have my birthday.
Researcher: Do you think that it matters that you were happy?
Erena: Yes, because when I am happy it means that I want to learn more.
Researcher: What other feeling do you remember having?
Erena: I was anxious to go and visit the pig farm.

This excerpt highlights an interesting pattern that emerged across the data set. All of the students indicated that they had seen handheld devices (including Erena despite her statement in the beginning of this excerpt), but they had never considered using them for the purposes of science. Most students discussed handhelds as something their parents used for business and a device that they may be able to use to play games. The idea of using these devices for school and science was clearly novel to the students but also very well received.

The excerpt below provides an example of a typical student reaction to participating in the research project, with references to the aspects of the project that the student enjoyed the most. This excerpt is representative of many student comments offered relative to their experiences with the project.

Researcher: What did you like about the learning environment?
Kyriaki: That we visited the pig farms and someone explained the process. I also like that we did research and presented the outcomes to the other groups.
Researcher: What do you mean when you say you did research?
Kyriaki: We searched online for information, we interviewed the pig farmer, we collected information from other resources, we visited the pig farm to see what is happening. And at the end we presented the outcome to our class.

This excerpt highlights an interesting pattern that emerged across the data set with most of the students indicating that they enjoyed participating in the *Technoskepsi* research project because they engaged in research (searching for information from various resources). They were asked to express their opinion, they visited the pig farm and had the chance to talk with the farmer, and to see whether the information they collected from the other resources were trustworthy.

Finally, the expert below is representative of students expressing that they enjoyed working in groups, and expressing their opinions.

Researcher: What did you like about the learning environment?
George: That we could work in groups for so long and ask questions and talk to each other about this topic. We could express our opinion.

The analysis of the interviews and classroom observations suggest that important aspects of the learning environment were the positive feelings that students expressed both during and after the instruction, especially about the use of the handhelds, working collaboratively in groups, expressing their opinion, and visiting the pig farm to collect data for their research.

Implications for ...

Teaching and Learning

An important implication of this study is that students can improve their written arguments, when supported by an online learning environment as the one in *Techoskepsi* even though the quality of argument, and hence improvement in argumentation is an issue that needs to be further explored and is discussed in the implications for research section. Associated with this issue is the question of how we can enable teachers to evaluate students' arguments and provide feedback. One of Aris's concerns during the instruction, even though he was familiar with argumentation frameworks, was how to evaluate his students' SSI arguments and provide feedback during the lessons. He was concerned with what was "wrong" and what was "right" in the discussions, and how to frame that for the class. Hence, one of the implications from this study is associated with finding ways to support teachers, not only to teach argumentation, but also to find consistent ways to evaluate argumentation, especially in socio-scientific contexts.

Another important finding in this study that has implications for teaching is that students can easily revert back to their original argument even though they have opposing evidence. It is interesting how most of the groups changed back to their original argument/decision to move the pig farms because of the bad smell after the visit to the pig farm. This finding suggests that the teacher's role during the instruction should be to scaffold students to weigh the evidence and decide based on not only selected but all evidence. Furthermore, the specially designed inquiry-based instruction supplementing formal and nonformal studies seems to have the potential to support students' argumentation while concurrently contribute to increasing student "motivation" for participation in science activities. Based on findings from the students' interviews, various characteristics of the research project were considered positive and as the students stated helped them engage in the learning process. These characteristics (collaborative work, engaging with authentic problems, meeting the "actors" of an issue, use of novel technology) could be incorporated by teachers when they design their lessons in order to help them feel happy during the lesson and motivate them. According to Blumenfeld et al. (2006), motivation sets the stage for cognitive engagement and leads to achievement by increasing the quality of the cognitive engagement.

Research and Methodology

An important aspect of the *Technoskepsi* project was supplementing formal with nonformal settings when trying to engage students with an SSI. The findings from this study suggest that there is a great impact on students, both in terms of learning, and emotions when using nonformal settings. Most of the groups changed their decision after the visit to the pig farm, something that suggests that the visit (non-formal setting) had a greater impact on how they talk about the SSI, and what kind of evidence they use to support their arguments. Furthermore, most of the students when interviewed stated that they enjoyed the visit to the pig farm and the opportunity to talk to the people who are involved in the issue they were studying (e.g., farmer). Future research should focus explicitly on the impact that nonformal visits have on students' SSI arguments, and the kinds of evidence they choose to use, and try to explore possible reasons for the change in the decision/arguments, based on the students' experiences or personal identities.

A methodological implication that derives from the analysis of the data is what counts as quality of argument in SSI, especially in the case of the short, written arguments (*artifacts*) that were the main data sources of this study? In the methods section I explain how I decided to use the Erduran et al. (2004) modified levels of argumentation as the framework to guide "measuring" the quality of students' arguments and argumentation, even though I was aware of the limitations. Two of the limitations of the framework identified by previous studies are (a) It is not easy to distinguish between warrants, data, and justifications (e.g., Duschl, 2008; Erduran et al.), something that has an effect on the inter-rater reliability of the coding. However this issue was addressed through the coding of part of the data by a second researcher, (b) it does not account for the content of the argument but only for the structure (e.g., Clark & Sampson, 2008; Osborne et al., 2004a,b), something that was evident in the analysis. Two additional limitations of the Erduran et al. framework identified through this analysis is that it fails to account for the number of pieces of evidence in the students' arguments as an additional characteristic of the *Levels*, and it is a framework designed to evaluate *dialogic* argumentation, and not *written arguments*, since the focus is on the rebuttals. According to TAP, rebuttals specify the conditions in which the claim is not true, and are more easily found in *dialogic* argumentation, in which claims are challenged by someone else, hence the person who is arguing offers rebuttals to justify and protect their argument. But how easy is it to include rebuttals in written arguments? This position is also supported by Carey (1985) who argued that both children's conceptual change and their growth in scientific reasoning are fundamentally driven by a growth in domain-specific knowledge. More specifically, Kuhn (1991), who studied the skill of argumentation, found that people who have knowledge of the subject seem to be more able to provide an alternative theory and that they tend to reason better on the subjects for which they have personal knowledge. Based on the above, I suggest that the number of pieces of evidence should be a measure of the quality of argumentation since they indicate an improvement in knowledge, and knowledge and skills are interrelated. Hence, a research implication from this study is the need for a framework that is designed to evaluate written arguments that should have an intermediate level between Level 2 and Level 3, in which the emphasis is on the number of pieces of evidence, before evaluating the use of rebuttals.

Conclusions

The *Technoskepsi* project was designed to explore young students' argumentation within a socio-scientific context, and explore how students argue when their study concerns an authentic problem and with formal and nonformal investigations. The analysis produced evidence regarding how elementary school students change their arguments during their investigations regarding a socio-scientific, authentic problem, and raised questions on how to design these learning experiences to support more integrated arguments that include all aspects of the problem. Supplementing formal with nonformal investigations also raised questions about the affordances of nonformal experiences, and how they can change students' attitudes and emotions toward science. An important aspect of the project was the collaboration of teachers and researchers for the development and implementation of the project, and it allowed the design on a curriculum that was based on the needs of the teacher and the students and not on the needs of the researchers. One of the limitations is that the specific curriculum was implemented in collaboration with a teacher and a researcher, an opportunity that is not given to many teachers, but I believe that this is still a good example of how a socio-scientific context can be used as a means to improve students' argumentation and decision-making.

References

AAAS. (1993). *Benchmarks for science literacy: A Project 2061 report*. New York: Oxford University Press.

Aikenhead, G. S. (2006). *Science education for everyday life: Evidence-based practice*. New York: Teachers College Press.

Andriessen, J., Baker, M., & Suthers, D. (2003). Argumentation, computer support, and the educational context of confronting gognitions. In J. Andriessen, M. Baker, & D. D. Suthers (Eds.), *Arguing to learn: Confronting cognitions in computer-supported collaborative learning environments*. Dordrecht: Kluwer.

Beck, U. (1992). *Risk society: Towards a new modernity*. London: Sage.

Bell, P. (2004). The educational opportunities of contemporary controversies in science. In M. Linn, E. Davis, & P. Bell (Eds.), *Internet environments for science education*. Hillsdale: Lawrence Erlbaum and Associates.
Bell, P., & Linn, M. (2000). Scientific arguments as learning artifacts: Designing for learning from the web with KIE. *International Journal of Science Education, 22*(8), 797–781.
Blumenfeld, P. C., Kepler, T. M., & Krajcik, J. S. (2006). Motivation and cognitive engagement in learning environments. In Sawyer, R. K. (Ed.), *The Cambridge handbook of the learning sciences* (pp. 575–488). New York: Cambridge.
Braund, M., & Reiss, M. (2006). Towards a more authentic science curriculum: the contribution of out-of-school learning, *International Journal of Science Education, 28*(12), 1373–88.
Bricker, L., & Bell, P. (2008). Conceptualizations of argumentation from science studies and the learning sciences and their implications for the practices of science education. *Science & Education, 92*(3), 473–498.
Carey, S. (1985). Are children fundamentally different thinkers and learners from adults? In S. F. Chipman, J. W. Segal, & R. Glaser (Eds.), *Thinking and learning skills* (2nd ed., pp. 485–517). Hillsdale: Lawrence Erlbaum Associates, Inc.
Cho, K., & Jonassen, D. (2002). The affects of argumentation scaffolds on argumentation and problem solving. *Educational Technology Research and Development, 50*(3), 1042–1629.
Clark, D., & Sampson, V. (2008). Assessing dialogic argumentation in on-line environments to relate structure, grounds and conceptual quality. *Journal of Research in Science Teaching, 45*(3), 293–321.
Creswell, J. W. (1998). *Qualitative inquiry and research design: Choosing among five traditions*. Thousand Oaks: SAGE Publications.
Cuthbert, A., & Slotta, J. (2004). Designing a web-based design curriculum for middle school science: the WISE "Houses in the desert" project. *International Journal of Science Education, 26*(7), 821–844.
Duschl, R. (2008). Quality argumentation and epistemic criteria. In S. Erduran & M. Jiménez-Aleixandre (Eds.), *Argumentation in science education: Perspectives from classroom-based research*. Dordrecht: Springer.
Duschl, R., Schweingruber, H., & Shouse, A. (Eds.). (2007). *Taking science to school: Learning and teaching science in grades K-8*. Washington, DC: The National Academies Press.
Duschl, R. A. (1990). *Restructuring science education: The importance of theories and their development*. New York: Teachers College Press.
Edelson, D., Pea, R., & Gomez, L. (1996). The collaboratory notebook. *Communications of the ACM, 39*(4), 32–33.
Edelson, D. C., & Reiser, B. J. (2006). Making authentic practices accessible to learners: Design challenges and strategies in K. Sawyer (Ed.), *The Cambridge handbook of the learning sciences* (pp. 335–354). Cambridge University Press.
Erduran, S. (2008). Methodological foundations in the study of argumentation in science classrooms. In S. Erduran & M. Jiménez-Aleixandre (Eds.), *Argumentation in science education: Perspectives from classroom-based research*. Dordrecht: Springer.
Erduran, S., Simon, S., & Osborne, J. (2004). TAPing into argumentation: Developments in the application of Toulmin's Argument Pattern for studying science discourse. *Science & Education, 88*(6), 915–933.
Evagorou, M., & Avraamidou, L. (2008). The role of technology in supporting the process of argument construction in science learning. *Educational Media International, 45*(1), 33–45.
Evagorou, M., Korfiati, K., Nicolaou, C., & Constantinou, C. (2009). An investigation of the potential of interactive simulations for developing system thinking skills in elementary school: A case study with fifth-graders and sixth-graders. *International Journal of Science Education, 31*(5), 655–674.
Evagorou, M., & Osborne, J. (2007). Argue-WISE: Using technology to support argumentation in science. *School Science Review, 89*, 103–110.
Evagorou, M. (2008). *Technoskepsi project report* (in Greek). Cyprus.

Evagorou, M., & Osborne, J. (2008). *Identifying features of young students' construction of arguments in the science classroom*. New York: Annual American Educational Research Association.

Fuller, S. (1997). *Science*. Buckingham: Open University Press.

Giddens, A. (1990). *The consequences of modernity*. Cambridge: Polity Press.

Jiménez-Aleixandre, M., & Erduran, S. (2008). Argumentation in science education: An overview. In S. Erduran & M. Jiménez-Aleixandre (Eds.), *Argumentation in science education: Perspectives from classroom-based research* (pp. 3–27). Dordrecht: Springer.

Jiménez-Aleixandre, M., Mortimer, F. E., Silva, C., & Diaz, J. (2008). *Epistemic practices: An analytical framework for science classrooms*. New York: American Educational Research Association.

Jimenez-Aleixandre, M., Rodriguez, A. B., & Duschl, R. (2000). "Doing the Lesson" or "Doing Science": Argument in high school genetics. *Science & Education, 84*(6), 757–792.

Jimenez-Aleixandre, M., & Pereiro-Munoz, C. (2002). Knowledge producers or knowledge consumers?Argumentation and decision making about environmental management. *International Journal of Science Education, 24*(11), 1171–1190.

Krajcik, J., Blumenfeld, P. C., Marx, R. W., Bass, K. M., & Fredricks, J. (1998). Inquiry in project-based science classrooms: Initial attempts by middle school students. *The Journal of the Learning Sciences, 7*, 313–350.

Kuhn, D. (1991). *The skills of argument*. Cambridge: Cambridge University Press.

Kuhn, D. (2005). *Education for thinking*. Cambridge, MA: Harvard University Press.

Lampert, M., & Rittenhouse, P. (1996). Agreeing to disagree: Developing sociable mathematical discourse. In N. Torrance (Ed.), *The handbook of education and human development* (pp. 731–764). Cambridge, MA: Blackwell.

Linn, M., Eylon, B., & Davis, E. (2004). The knowledge integration perspective on learning. In M. Linn, E. Davis, & P. Bell (Eds.), *Internet environments for science education* (pp. 29–46). Hillsdale: Lawrence Erlbaum Associates.

Merriam, S. (1998). *Qualitative research and case study applications in education*. San Francisco: Jossey-Bass.

Neylor, S., Keogh, B., & Downing, B. (2006). Argumentation and primary science. *Research in Science Education, 37*, 17–39.

Nicolaou, C., Korfiati, K., Evagorou, M., & Constantinou, C. P. (2009). Development of decision-making skills and environmental concern through computer-based, scaffolded learning activities. *Environmental Education Research, 15*(1), 39–54.

National Reasearch Council (NRC). (1996). *National science education standards*. Washington, DC: National Academy Press.

Osborne, J. (2007). Science education for the twenty first century. *Eurasia Journal of Mathematics, Science & Technology Education, 3*(3), 173–184.

Osborne, J., Erduran, S., & Simon, S. (2004). Enhancing the quality of argumentation in school science. *Journal of Research in Science Teaching, 41*(10), 994–1020.

Osborne, J., Erduran, S., & Simon, S. (2004a). *Ideas, evidence and argumentation in science (IDEAS) project*. London: King's College London.

Osborne, J., Erduran, S., & Simon, S. (2004b). Enhancing the quality of argumentation in school science. *Journal of Research in Science Teaching, 41*(10), 994–1020.

Patronis, T., Potari, E., & Spiliotopoulou, S. (1999). Students' argumentation in decision-making on a socio-scientific issue: Implications for teaching. *International Journal of Science Education, 21*(7), 745–754.

Rogoff, B. (2003). *The cultural nature of human development*. New York: Oxford University Press.

Sandoval, W. A. (2003). Conceptual and epistemic aspects of students' scientific explanations. *The Journal of the Learning Sciences, 12*(1), 5–51.

Sandoval, W. A., & Reiser, B. (2004). Explanation-driven inquiry: Integrating conceptual and epistemic scaffolds for scientific inquiry. *Science & Education, 88*(3), 345–372.

Scardamalia, M. (2003). Knowledge forum (Advances beyond CSILE). *Journal of Distance Education, 17*, 23–28.
Simon, S., Erduran, S., & Osborne, J. (2005). Learning to teach argumentation: Research and development in the science classroom. *International Journal of Science Education, 28*(2–3), 235–260.
Soloway, E., Norris, C., Blumenfeld, P., Fishman, B.J., & Marx, R. (2001). Devices are ready-at-hand. *Communications of the ACM, 44*(6), 15–20.
Suthers, D. D. (1999). *Effects of alternate representations of evidential relations on collaborative learning discourse*. Third Conference on Computer Supported Collaborative Learning, Stanford.
Taylor, C. (1996). *Defining science*. Madison, WI: University of Wisconsin Press.
Toulmin, S. (1958). *The uses of argument*. Cambridge: Cambridge University Press.
Zurita, G., & Nussbaum, M. (2004). A constructivist mobile learning environment supported by a wireless handheld network. *Journal of Computer Assisted Learning, 20*, 235–243.

Chapter 9
Metalogue: Assessment, Audience, and Authenticity for Teaching SSI and Argumentation

Maria Evagorou, Troy D. Sadler, and Tali (Revital) Tal

Assessing Argumentation

Sadler: This chapter raises several interesting issues associated with the assessment of argumentation. There is obviously a great deal of support throughout the science education community for featuring argumentation as a fundamental scientific practice that ought to be featured in science learning experiences. However, the tools available for assessing argumentation both for research and teaching purposes remain somewhat limited. Toulmin has had an enormous impact on how science educators think about argumentation and the assessment of arguments, and modifications of the Toulmin argument pattern have been used extensively for assessment purposes (Erduran, Simon, & Osborne, 2004). As discussed in the chapter, Toulmin's model can be useful but it has a number of drawbacks.

I thought that Maria's use of multiple dimensions to assess argumentation was a definite strength for this study. In addition to argumentation levels defined by the Toulmin scheme, Maria documents the number of pieces of evidence used and characterizes the "socio-scientific nature" of the arguments. She also documented the students' decisions (i.e., to move the pig farm or allow it to remain in its current location). I strongly agree with the notion that a unidimensional system for assessing argumentation can limit our understanding of student argumentation practices and the nature of argumentation development that may occur in response to learning opportunities. However, assessing multiple dimensions of argumentation creates potential challenges as well. It is important for researchers and the audiences of that

M. Evagorou (✉)
School of Education, University of Nicosia (UNic), Nicosia, Cyprus
e-mail: evagorou.m@unic.ac.cy

T.D. Sadler
School of Teaching and Learning, University of Florida, Gainesville, FL, USA

T. Tal
Department of Education in Technology and Science, Technion–Israel Institute of Technology, Haifa, Israel

T.D. Sadler (ed.), *Socio-scientific Issues in the Classroom: Teaching, Learning and Research*, Contemporary Trends and Issues in Science Education 39,
DOI 10.1007/978-94-007-1159-4_9, © Springer Science+Business Media B.V. 2011

research to acknowledge fundamental differences in the data sets. For example, assessment of the argumentation levels provides ordinal-level data; the number of pieces of evidence is ratio-level data; and categories of socio-scientific nature, as used here, are nominal-level data. I think that it is important for us to remember these differences as well as the descriptive and inferential procedures that can be used with each of the data sets.

Creating ways to represent these various forms of data such that they can be understood in a comprehensive manner is another challenge. Maria's results tables helped in offering an integrated perspective, but I would like to see more sophisticated approaches to the analysis and representation of multidimensional data particularly for the assessment of complicated constructs such as scientific practices. This is less a critique of Maria's work and more of a call to all of us working in these areas to pay more attention to assessment issues.

In the implications section, Maria raises issues associated with using a single scheme for assessing argumentation. One of her recommendations is to expand the Erduran et al. (2004) framework by incorporating the number of pieces of evidence. My initial reaction to this suggestion is that doing so has the potential to inappropriately conflate different dimensions of argumentation. This would introduce ambiguity to the scale making it difficult to interpret results. For instance, improvements in student practice may be due to the use of more evidence and/or the incorporation of counter-claims and rebuttals.

Evagorou: The assessment of argumentation has become an important issue in the field of argumentation, but one that as Troy suggests is still unresolved. Even though an array of different frameworks exist for evaluating argumentation (see Sampson & Clark, 2008; Erduran, 2008 for a review) there is no consensus as to what counts as successful argumentation or quality of argument, as different studies use different frameworks. I agree with Troy's comment that we should be careful about using different kinds of data sets when trying to address this issue of structure versus content versus nature of arguments. As Sampson and Clark (2008) suggest we need to start thinking about assessment in argumentation in a more synergistic fashion.

Troy also raises the issue of examining the number of pieces of evidence as a measure of improvement of arguments. None of the written argument frameworks explicitly addresses the issue of multiple pieces of evidence as a criterion for best quality of an argument. However, is it true that more is not necessarily better, especially if more refers to pieces of evidence? According to Duschl, Schweingruber, and Shouse (2007), students should "[...]understand both the body of knowledge and the process by which this knowledge is established...." (p. 26). Furthermore, Kuhn (1991), who studied the skill of argumentation, found that people that have knowledge of the subject seem to be more able to provide an alternative theory and that they tend to reason better on the subjects for which they have personal knowledge. The decision to add an intermediate level to Toulmin's Argumentation Pattern (TAP) is based on these ideas. I suggest that the number of pieces of evidence should be a measure of the quality of argumentation since they indicate an improvement in knowledge, and knowledge and skills are interrelated. In later work (Evagorou & Osborne, in preparation) we analyzed data by adding an intermediate level to the TAP between Level 2 and Level 3. This is Level 2B which is defined as an argument consisting of a claim (as opposed to a claim with more pieces of data, warrants, or backings) but does not possess any rebuttals. With the use of an intermediate level we were able to document improvement in students' arguments in terms of the structure. This modified version can be used with younger students who find it more difficult to improve their arguments in terms of rebuttals, and easier to improve in terms of evidence.

Tal: I think that while discussing whether the number of pieces of evidence should be considered in determining argumentation quality, we should not forget the context of SSI-based teaching in general and the relevant SSI in particular. With the small sample featured in Maria's study, I wanted to read more examples of student work that could add to the quantification presented in the tables. I was quite curious, for example, about possible alignment between the number of pieces of evidence and the different aspects brought up by the students. Looking at Tables 6.6 and 6.7, I noticed, for example, that in Argument 2, group 1 provided four pieces of evidence from three aspects (social, environmental, and financial), while group 6 provided five pieces of evidence – all financial. I really wondered how one group of sixth graders provided five different financial reasons. I doubt we learn anything from this because of the overall small number of groups, but the point I want to make is that while investigating argumentation very carefully, we should not forget the context. I, for example, do not see the uniformity in Argument 3 as disturbing. It is true that while studying in school using the well-structured WISE platform, students improved their argumentation. However, the concrete experience at the farm, smelling and observing the pigs, seemingly convinced them that that environmental aspects should be prioritized in making their decision. As there are no examples of students' arguments and as Maria reports, the overall argumentation level was rather low, it is hard to determine anything about whether and how the students explained the shift.

Sadler: The implications section of the chapter raises the issue of teacher assessment of argumentation. Maria discusses the challenges associated with evaluating student argumentation for research purposes versus teaching purposes. I would be interested in hearing about any suggestions for actually addressing this issue. The significance of formative assessment of student learning and practice is overwhelming. So it stands to reason that strategies for supporting teacher assessment of student argumentation would be very valuable in terms of improving argumentation.

Evagorou: In the case of *Technoskepsi*, the teacher was familiar with the TAP framework and felt comfortable evaluating written arguments both in terms of the structure and SSI aspects. However, not all teachers are familiar with analytic frameworks for argumentation. Furthermore, even for this teacher, evaluating oral arguments (argumentation) in the context of the class was very challenging. In these contexts, he struggled in determining whether and what students were learning and what kinds of supportive actions he ought to be taking. His question during the implementation was, "what does research tells us about assessing argumentation, especially in SSI, and how can we transfer that in the classroom?" Unfortunately, I did not have a good answer for him.

Tal: Maria indicates that the first lesson was dedicated to teaching about argumentation and modeling good argumentation by the teacher. McNeill and Krajcik (2008) highlight the significance of aligning teacher practices with expectations for student argumentation. They coded four instructional practices: modeling scientific explanation, making the rationale of scientific explanation explicit, defining scientific explanation, and connecting scientific explanation to everyday explanation. In their study, McNeill and Krajcik found that teachers differed a great deal in discussing the rationale behind argumentation and in connecting to real life. Although it is apparent that Aris, the teacher in Maria's chapter, is an exemplary teacher in terms of his practices, I wondered whether continuous support of the teacher and being explicit about what makes a good argument throughout the learning process could yield better arguments in stage 3.

SSI and Argumentation for Young Learners

Evagorou: SSI and argumentation have been an emphasis of many studies; however, most of these studies focus on high school or university students. One of the challenges of the *Technoskepsi* project was to implement the curriculum with younger students (10–12 year old). Working with elementary school students can be informing and inspiring. However, a reservation I would like to share and discuss is: What kind of SSI can be relevant for younger students (that can promote scientific knowledge within the issue)? How can we help the students to understand the multiple aspects of these issues? And finally, how can we support their teachers to support them.

Sadler: These are great questions; questions for which we do not have good answers (at least to my knowledge). As reflected in this volume, most of the work being conducted in the area of SSI is done in secondary contexts. Maria is one of the few researchers that I know of who are looking at how SSI can be incorporated in classrooms for younger learners. Theoretically, the complexity of SSI and connections to real-world issues can serve to motivate learning, but for young students, delving into the complexity of socially-contentious issues may not be developmentally appropriate. In general, I think that SSI are more easily situated in secondary and college level classrooms. That is not to say that SSI cannot have a place in elementary education; however, I think that teachers and researchers will have to be very careful in terms of which issues they choose and how they frame these issues.

Authenticity

Tal: I wish to highlight some similarities and a few differences between Maria's project and the study conducted by my group, which is described in Chap. 2. Like Maria, we intended to use the WISE platform and to increase authenticity by anchoring the instructional unit to real events and people. In our case, we engaged students in a cystic fibrosis (CF) fundraising event and a visit to the CF unit of a local children's hospital. Because of challenges associated with scaling a field trip model, we offered an online interaction with a young CF patient instead. Similar to Maria, we found that the actual visit to the hospital did not contribute to better performance of the students in pre/post tests. Yet, the students who visited the hospital and had a diversified experience were more enthusiastic and showed higher motivation than their counterparts. Unlike what Maria found, we concluded that the field trip to the hospital broadened the students' views and contributed to their holistic understanding of the SSI. Taking our study into account, it could be that the visit to the pig farm, given its overwhelming smell, was not sufficient in this case. Perhaps meeting with other stakeholders such as an employee of the farm, a town council member or an environmental engineer could have contributed to a more balanced view of the issue, and consequently, to better argumentation following these experiences.

Evagorou: The *Technoskepsi* curriculum was designed around a problem that students and their families were experiencing during their everyday lives – that of the smell from the pig farms. The problem was in the local news for months. Environmentalists and members of the parliament visited the area on a regular basis to discuss possible solutions, and the residents organized several protests. One difference between *Technoskepsi* and other SSI projects presented in this book is that all students in our sample experienced the problem and protests first hand. Most students had strong opinions on the issue before engaging with the curriculum. When we started designing the curriculum we were aware of this situation and were prepared for students to resist solutions different from the one proposed by the local community (to move the pig farm away from the inhabited area). Hence, our emphasis in Lessons 2 and 3 (see Table 6.2 and subsequent description) was on introducing the problem and presenting all aspects and possible solutions. For example, students explored waste management techniques that can minimize air and water pollution, talked about financial costs of moving the pig farm (including local job losses), and learned about ways to process compost from the pig farm to produce electricity (and the costs).

Table 6.5 summarizes what I consider an important aspect of the implementation of *Technoskepsi*: All groups' original decision was to move the pig farm. Engaging with the materials in WISE supported changes to their decision-making, but after the visit they changed back to their original decision. This chain of changes in decisions suggests that experiencing the problem as it is (authentic learning) can be more powerful than any specially designed learning environment. López-Facal and Jiménez-Aleixandre (2008) explored how students identify with the "actors" in SSI and how this defines their decisions. In the case of *Technoskepsi*, the students identified with their parents and local authorities ignoring all other evidence and solutions. In the case of Tali's project the students probably identified with the doctors and the patients. Hence, the two projects even though similar in some regards, are different as to the affordances and the implications of the visit to an authentic site.

References

Duschl, R., Schweingruber, H. A., & Shouse, A. W. (2007). *Taking science to school: Learning and teaching science in grades K-8*. Washington, DC: The National Academies Press.

Erduran, S. (2008). Methodological foundations in the study of argumentation in science classrooms. In S. Erduran & M. Jimenez-Aleixandre (Eds.), *Argumentation in science education: Perspectives from classroom-based research* (pp. 47–69). Dordrecht, the Netherlands: Springer.

Erduran, S., Simon, S., & Osborne, J. (2004). TAPing into argumentation: Developments in the application of Toulmin's argument pattern for studying science discourse. *Science Education, 88*(6), 915–933.

Evagorou, M. & Osborne, J. (2011). Towards improving the measurement of quality of argument using Toulmin's framework: A methodological contribution. *Presented at the annual conference of the National Association of Research in Science Education (NARST)*, April 3–6, Orlande, FL.

Kuhn, D. (1991). *The skills of argument*. New York: Cambridge University Press.

López-Facal, R., & Jiménez-Aleixandre, M. P. (2008). Identities, social representations and critical thinking. *Cultural Studies in Science Education, 4*(3), 689–695.

McNeill, K. L., & Krajcik, J. (2008). Scientific explanations: Characterizing and evaluating the effects of teachers' instructional practices on students learning. *Journal of Research in Science Teaching, 45*, 53–78.

Sampson, V., & Clark, D. (2008). Assessment of the ways students generate arguments in science education: Current perspectives and recommendations for future directions. *Science Education, 92*(3), 447–472.

Chapter 10
Decision Making and Use of Evidence in a Socio-scientific Problem on Air Quality

Shirley Simon and Ruth Amos

The study reported here arises from the overlapping interests of the two authors as we came together as supervisor and student for a master's dissertation. As an experienced researcher, Shirley Simon had been studying argumentation in school science over many years, in particular focusing on the ways in which teachers develop their pedagogical approach to argument and the challenges they experience when trying to change their practice (Simon, Erduran, & Osborne, 2006; Simon & Maloney, 2007). One feature of Simon's work with teachers was to study how they organised and managed small group discussion, role play, and class debates, and how students engaged with scientific evidence or socio-scientific issues (SSI) to construct arguments in different contexts (Osborne, Erduran, & Simon, 2004a). The research, conducted in schools in the United Kingdom, led to the development of activities and guidance for teachers in argumentation (Osborne, Erduran, & Simon, 2004b), which coincided with changes in the science component of the English national curriculum for 14–16 year olds, and the emergence of a course aimed to enhance scientific literacy (SL) called *Twenty First Century Science* (OCR, 2005). The new national curriculum places more emphasis on the nature of science (NOS), and *Twenty First Century Science* includes activities for students to debate and construct arguments on issues related to science topics, with a focus on relevance to everyday life; it also includes pedagogical guidance for teachers in how to organise and manage such activities.

As an experienced school teacher of chemistry, Ruth Amos had included debates on SSI in her classroom with students aged 16–17 years and found that students often had difficulty formulating convincing arguments. The new *Twenty First Century Science* course appeared to offer potential for making debates on

S. Simon (✉) and R. Amos
Institute of Education, University of London, London, UK
e-mail: s.simon@ioe.ac.uk

T.D. Sadler (ed.), *Socio-scientific Issues in the Classroom: Teaching, Learning and Research*, Contemporary Trends and Issues in Science Education 39,
DOI 10.1007/978-94-007-1159-4_10, © Springer Science+Business Media B.V. 2011

SSI more relevant to students, while building on their science learning. At the time the new course was being piloted by teachers in England, Amos became a lecturer for preservice teachers in Simon's university department, and aware of the issues faced by the teachers in Simon's studies. She also became involved in working with practicing teachers who were engaged in the pilot scheme to trial the *Twenty First Century Science* activities. Amos decided to undertake a small-scale piece of research for her dissertation in the context of a *Twenty First Century Science* SSI, which was the focus of a debating activity that came at the end of a module on a substantive scientific topic. Her study looked in depth at how one teacher interpreted the activity, used the resources, and organised and managed small group discussion and argument. The study also focused on the sources of evidence used by the students to see how they built on the ideas within the science topic and on the quality of their argumentation (Erduran, Simon, & Osborne, 2004).

The choice of module on which to focus the study was determined by Amos's interest in chemistry-related issues and how the *Twenty First Century Science* course would provide students with an opportunity to discuss SSI related to chemistry. A meta-analysis of research into the use of small group discussion in science lessons showed that few studies had been undertaken in chemistry contexts, or using computer simulations (e.g. Bennett, Lubben, Hogarth, Campbell, & Robinson, 2005; Hogarth, Bennett, Campbell, Lubben, & Robinson, 2005; Sadler, 2004). The selection of a suitable activity to be the focus of the study was influenced by a desire to extend this research base, in particular to see whether chemistry understanding could be utilised in discussion-based activities, so a decision was made to observe a debate set in a chemistry context that involved a computer simulation. The module called 'Air Quality' was chosen as this has a substantial focus on chemicals in the air, air pollution, and choices we make personally, locally, and nationally to improve air quality. The debating activity, called Bleaksville, involves the use of a computer simulation to engage the students' interest and model the outcomes of decision making on different policies relating to air quality. The activity was also chosen because it includes role-play, an approach where students take on roles that represent different positions and which can stimulate argumentation as students engage with their opposing roles. It was of interest to see how the teacher would organise and manage the role-play approach, given that it has proved challenging for many teachers (McSharry & Jones, 2000). Indeed, the most common types of role-play used in science lessons have tended to be 'action' scenarios with minimal accompanying dialogue between students; for example, modelling the mechanisms in the human kidney (Johnson, 1999). Simon's research on the teaching of argumentation (Simon et al., 2006) had also indicated that only teachers who are confident in the use of argumentation will include role-play in such activities. It was of interest therefore to investigate how an example of *Twenty First Century Science* pilot materials enabled the teacher to facilitate discussion and argumentation in a role-play activity.

Background and Goals

The changes in the national curriculum for science in England were influenced by ongoing debates of the time, including the review of school science reported in *Beyond 2000* (Millar & Osborne, 1998). The new curriculum demonstrates a commitment to enhancing SL through the inclusion of a more epistemic focus in the teaching of science; in addition to science content, the curriculum requires that students are taught about the NOS, including the evidential basis for making scientific claims. The change emphasises the importance of educating students about *how* we know and why we believe in the scientific world view, to see science as a distinctive and valuable way of knowing. To appreciate the origins of scientific knowledge and thus develop an epistemological understanding of science, students need to explore reasons why accepted ideas have become established and why alternative theories are considered to be 'wrong'. The ability to comprehend and follow arguments of a scientific nature is perceived as a crucial aspect of SL, the importance of which has been highlighted in documents and debates within science education worldwide (Braund, Lubben, Scholtz, Sadeck, & Hodges, 2007; Goodrum, Hackling & Rennie, 2001; Norris & Phillips, 2003; OECD, 2006; Roberts, 2007; UNESCO, 1999). The changes in the national curriculum also aimed to make school science more relevant, coherent, and engaging, and help to prepare adult citizens for decision making on scientific issues that have social consequences. If students are to learn how to debate and use evidence to construct arguments in socio-scientific contexts, and thereby take part in future decision making, then science education needed to provide opportunities for these skills to develop (Ratcliffe & Grace, 2003). Argumentation and decision making thus became an important feature of the *Twenty First Century Science* course.

The study reported here on the Bleakesville debate in *Twenty First Century Science* was informed by previous research on enhancing the quality of argument in school science and the theoretical basis for that earlier work in which Simon was coinvestigator (Erduran et al., 2004; Osborne et al., 2004a; Simon et al., 2006). The perspective on argument informing this research was based on the premise that argument encompasses both individual and social meaning, this dual meaning comprising an inner chain of reasoned discourse (individual) and a dispute or debate between people holding contrasting positions (social) (Jiménez-Aleixandre & Erduran, 2008). The internal and social aspects are linked (Kuhn, 1993) in that social dialogue offers a way to externalise internal thinking strategies embedded in argumentation. The classroom research of Osborne et al. thus focused on the value of small group discussion in promoting students' argumentation skills. In their study of argumentation as it occurred between teachers and students, and between students engaged in small group discussion, Osborne et al. (2004a) draw a useful distinction between 'argument' and 'argumentation'. Argument refers to the claim, data, warrants, and backings that form the substance or content of an argument (Toulmin, 1958), so constructing an argument through presenting a claim with supporting data, warrant, and backing can be undertaken by an individual working alone.

Argumentation, however, refers to the process of arguing, essentially between two or more people, in which the construction, justification, and refutation of arguments take place as individuals externalise their thinking in a social setting. By engaging collaboratively in argumentation activities that make reasoning public, students can gain collective experience of constructing arguments, justifying arguments with evidence, evaluating alternative arguments, and reflecting on the outcomes of argumentation. Though the role of argumentation has now become more highly valued in science education, research has shown that only if it is specifically addressed in the curriculum and explicitly taught through task structuring and modelling will students gain the skills needed to explore its use in science (Jiménez-Aleixandre & Erduran, 2008; Osborne et al., 2004a). Such pedagogical approaches to promote the process of argumentation can be developed in a range of contexts and activities, though there are thought to be differences in the kinds of reasoning needed to construct arguments in scientific and socio-scientific contexts. Erduran (2008), for example, focuses on the nature of claims that can be made in each context, pointing out that in socio-scientific debates, where social and scientific factors have to be taken into consideration, claims can be made from multiple perspectives, whereas when arguments are constructed in scientific debates, claims take the form of explanatory conclusions or descriptive frameworks. SSI, therefore, make different demands in terms of reasoning, as students are required to weigh up the pros and cons of different positions in complex situations, a process Sadler (2004) terms informal reasoning. The Bleakesville debate in *Twenty First Century Science* has been designed for students to draw on scientific evidence in constructing arguments; however, it is presented in a decision-making context with social, economic, and environmental factors to consider. It is of interest therefore to investigate the nature of evidence used by students in constructing their arguments to inform the policy debate.

Twenty First Century Science is presented as a high school balanced science course with core science and additional units dedicated specifically to biology, chemistry, and physics. The core science component aims to provide students with a broad, qualitative grasp of major scientific explanations, sufficient to give them the basic understanding that would be needed for informed decision making about scientific and SSI, that is, draw on evidence to provide explanatory conclusions in a scientific context, or draw on evidence while engaging in informal reasoning to argue for a position in a socio-scientific context. The course also provides students with opportunities to develop an understanding of the nature of scientific knowledge and of the ways in which it is established through arguments that draw on evidence to support claims to assess different contested views (Millar, 2006). *Twenty First Century Science* includes activities that focus on discussion of open-ended issues, debate, and role-play, all of which involve argumentation and the use of evidence. Thus one aim of the course is to enable students to consolidate their knowledge and understanding of science through debating issues that draw on scientific evidence related to core content.

Our concern in the undertaking of this research was that many teachers and students would be unfamiliar with *Twenty First Century Science* activities as part of 'science'. A previous study by Newton, Driver, and Osborne (1999) showed that as little as 5% of the available time in typical UK school science lessons was given over to student discussion. Reasons teachers gave for such low levels of discussion included difficulties in managing discussion, low levels of confidence in leading discussion, limited perceptions of the value of discussion in science, and a view that discussion was for English lessons, not science. Therefore, if the *Twenty First Century Science* course was to be effective in enabling students to develop decision-making skills, teachers would have to adopt unfamiliar pedagogical strategies to help students engage in social contexts (Levinson & Turner, 2001). Simon et al.'s (2006) research into teachers' skills in the use of argument in science had shown that scaffolding argument processes during small group discussion, particularly using role-play and encouraging higher order processes of counter-argument and debate, could be challenging for many teachers.

The purpose of the research project featured in this chapter was to explore how teachers and students responded to decision-making scenarios in the chemistry component of the *Twenty First Century Science* course, through an in-depth observation of the Bleaksville scenario in the module on Air Quality. Our aim was to complement the structured evaluation of the *Twenty First Century Science* pilot course which was undertaken by independent researchers between 2003 and 2006 (UYSEG & Nuffield Foundation, 2007). One of three strands of this evaluation focused on classroom teaching in the Core Science course, and in particular such aspects that might be considered novel by many science teachers. The focus included the management of classroom discussions of science-related issues, which may involve a range of social, economic, political, and ethical ideas, as well as scientific ones (Millar, 2006). Evaluation focusing on changes in classroom practices highlighted some familiar concerns associated with school science education (Ratcliffe & Osborne, 2007). For example, much of the classroom discourse observed during the evaluation included closed questions, which offered little opportunity for extended teacher-student dialogue or student-student dialogue in small groups. Few collaborative activities were observed. These findings suggested that teachers were not changing their practice readily during the pilot project. The *Twenty First Century Science* project team responded by stating that many of the findings were inconclusive, in terms of the impact on changes in teachers' pedagogical strategies and in student learning outcomes (Burden, Campbell, Hunt, & Millar, 2007), and they called for replication and development of the research. Surveys and sample observations such as those carried out by the evaluation team only painted a broad picture of what was going on. To gain a more comprehensive insight into a complex situation, there was a need for detailed case studies to be carried out alongside larger, more quantitative studies (Lunn, 2002).

Given the aims of *Twenty First Century Science*, the difficulties associated with implementing the debating activities, and the limitations of a quantitative evaluation, this research aimed to provide a detailed study in the context of the Bleaksville activity, the over-arching goals being a) to see how students engaged with and used the resource materials in their decision-making task, b) to evaluate the quality and focus of argumentation arising from the implementation of discussion activities and c) to see how an experienced teacher facilitated discussion and argumentation in a *Twenty First Century Science* activity based on a SSI.

The Setting and Teacher-Researcher Relationship

Initially, three schools in the Greater London area that were taking part in the *Twenty First Century Science* pilot were approached to see whether they would be willing to be involved in the research. The aim was to work with ordinary teachers who were grappling with the demands of the new course and trying to change their practice to use more discussion in the science classroom. Two of these three schools were known to both Simon and Amos as they were working as partnership schools with the university's preservice teacher education programme. Of the three schools, two were unable to take part, but in the third school a teacher called Jackie, who was committed to *Twenty First Century Science,* was able and willing to use the Bleaksville activity. Jackie knew Amos as she had been a preservice teacher in Amos's school science department.

Jackie's school is a nonselective, comprehensive school with a very mixed student intake in terms of social, ethnic, and cultural background (mainly white British, white European, Asian, and African-Caribbean). Jackie was teaching the pilot of *Twenty First Century Science* for the third year running, although this was her first time teaching the Air Quality unit. She had received only 2 days training in managing small group discussions, 1 day as part of Continuing Professional Development provided by the *Twenty First Century Science* project and one as in-school training, and had access to the pilot guidance material. The first day of training had taken place at a university, led by the professional development team working with the designers of the course materials. This training included a focus on facilitating small group discussion and developing students' argumentation skills, and was attended by teachers from the pilot schools in the southern region of England. The training included lectures and interactive workshops to help teachers become familiar with strategies for small group discussion. The second training day took place within the school, and was led by an external consultant who also focused on facilitating small group discussion. In spite of the training days, Jackie found discussion activities challenging. She also discovered there was limited opportunity to share her concerns with colleagues in her science department as they were sceptical of the value of the *Twenty First Century Science* course.

The class of 19 students that took part in the study were aged 14–15 years, and all were of average to low ability according to their national test results in science at age 14 years, assessed 6 months before the study. In England, students are expected to fall into five levels of achievement (levels 3–7) by the time they reach the age of 14. These 'levels' broadly pertain to students' abilities to use scientific language, which ranges from using 'everyday' language at level 3 and using scientific language appropriately at level 4 to forming simple explanations for phenomena at level 5, using scientific theory to explain phenomena at level 6, and using and applying more complex understanding at level 7. The students in the study had all achieved between levels 3 and 5 on this test. The *Twenty First Century Science* pilot evaluation noted that the student sample following the course may be skewed toward the lower end of the attainment range (Millar, 2006) and so these case study students were representative of the pilot cohort.

The teacher-researcher relationship between Jackie and Ruth Amos continued over a 12-month period, beginning just prior to the Bleaksville activity. After the use of Bleaksville, they continued having discussions of pedagogy and reviewed the findings as Amos analysed her data. Jackie added insights on the capabilities and achievements of her learners. Their discussions also focused on how Jackie was developing her approach to using small group discussion in science activities, and her attempts to encourage colleagues to redesign the school science scheme to include more discussion-based SSI that would enhance students' skills in argumentation and decision making.

The Bleaksville Simulation

Twenty First Century Science provides opportunities for students to engage in debate about a wide variety of current SSI, including several with a chemical theme. In the pilot, the unit on Air Quality focused on the nature of air pollution and the challenges of dealing with its effects in modern towns and cities. The resources for the unit included both written materials and computer simulations for use in discussion and debate. The unit begins with a review of students' knowledge of the composition of unpolluted air and an introduction to the concept of 'air quality' and how this can be investigated. The unit explores which air pollutants are regularly monitored and how the concentrations of certain pollutants vary with time and location. Students carry out activities that investigate the particulates produced when a fuel burns and collect their own data on the concentration of an atmospheric pollutant. There is a strong focus on understanding the products of combustion when hydrocarbon fuels and carbon burn. The unit then explores what happens to atmospheric pollutants and how air quality can affect health.

The Air Quality unit finishes with the role-play debate called the Bleaksville Simulation. This simulation attempts to engage learners by enacting the scenario of an imaginary town struggling to improve the quality of air and thus of people's day-to-day lives. To that end, students take on the role of advisors to the mayor of Bleaksville and consider a number of possible solutions to a series of environmental air pollution problems. A key requirement for the successful running of such an activity is students' acceptance of the relevance of the SSI involved. If they were unimpressed with or untouched by issues arising, then meaningful debate and decision making would be unlikely (Aikenhead, 2006). In the simulation, students are in charge of Bleaksville, trying to improve its air quality. The scenario shows that unregulated industrialisation has made Bleaksville an unpleasant place to live. High pollution has caused respiratory disease and destroyed much of the city's environment.

Citizens are now leaving. The task for students is to advice the new mayor on policies to improve air quality and make the citizens happier. They evaluate different policy options through structured group discussion. A computer model of the city data accompanies the simulation giving students feedback on their decisions. The activity's position, at the end of the teaching unit on Air Quality, made it a good choice in terms of exploring whether students would draw on scientific ideas that they had encountered in the unit.

The Bleaksville sequence of lessons begins with students observing the air pollution situation in Bleaksville and the detrimental effect it is having on the citizens in terms of health and welfare. They are presented with photographic images of these effects, including people with respiratory difficulties, cars and chimneys emitting smoke, and a written scenario describing the effects and data on pollution levels. Graphical representations of levels of sulfur dioxide and nitrous oxides are actively displayed over 5-year periods as the simulation progresses and as students enter the decisions that they have made about how to deal with pollution levels into the computer simulation, they can observe the impacts resulting from different policies (Fig. 10.1).

Alongside these data, people's happiness levels and environmental damage are used as impact factors for assessing the successful implementation of anti-air pollution policies.

The Bleaksville Simulation presents students with four policies for improving air quality, which are investigated sequentially. Each policy involves a choice of options for students to discuss in order to decide which option is the best, and once a decision is made this policy option is then implemented in the town for 5 years. Thus the four policies represent a 20-year period in the life of the town. The four policies in the activity are shown in Table 10.1. The table displays three options for implementing each policy. The policies and options are more fully described on the students' policy sheets.

The policies are presented individually at whole-class 'policy' meetings, where the teacher outlines the policy options to the whole class. These options, shown in

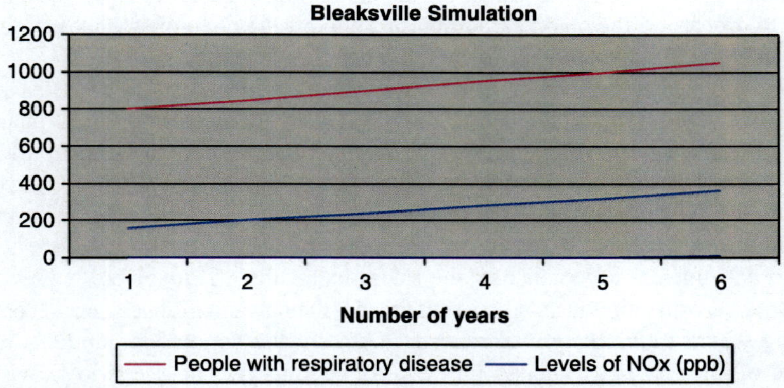

Fig. 10.1 Example chart from the Bleaksville Simulation showing levels of NOx and number of people suffering from respiratory diseases

Table 10.1 Policies and options for the Bleaksville Simulation

Policy	Option 1	Option 2	Option 3
1: Improving car technology to reduce car emissions	Fit all cars with catalytic converters	Develop ultra low sulfur fuel	Make people convert their engines to lean burn[a]
2: Improving power station technology to reduce sulfur dioxide emissions	Switch the power station fuel from coal to gas	Fit 'flue gas desulfurization' technology	Introduce low nitrogen oxide burners
3: Encouraging the use of alternative forms of transport	Introduce a congestion charge	Ban cars with big engines	Build a new tram system
4: Encouraging the use of alternative energy resources	Introduce a tax on using energy inefficiently	Invest in renewable forms of energy	Build another power station

Note: [a]A lean burn engine typically uses an air: fuel ratio of up to 23:1, as compared to a normal engine where the ratio is 14.7:1

Table 10.1, are examined by the students in small groups to help them decide which one they think will make the citizens happiest. Each group discusses the pros and cons of each option and presents their recommendations to the mayors (another group of students). The decision is entered into the computer model, which works out what happens to the city for the following 5 years before the next policy is considered. Running the computer simulation shows the impact on the various parameters as shown in Fig. 10.1; however, decisions about improving air quality can have different consequences, for example, measures to control pollution can hurt the economy or restrict people's lives, so the simulation alerts the students to the impact of their decisions. Thus the aim is for students to pay attention to each area to keep the citizens happy.

To structure discussions, students were organised by Jackie into large groups of six, which were then further subdivided into two small groups of three, referred to as listening triangles by Jackie. The work strategy for this group was that each student in the group takes on a different role, one as questioner, one as respondent, and the other as recorder, a strategy with which the students were familiar. Two other students acted as the mayors of Bleaksville. Students were able to consult data concerning the implementation of each policy option, as these were described on the policy sheets provided. The guidance for the activity suggested that all groups consider all three options for a policy and make decisions about which they would choose. However, in this case, Jackie chose to modify the activity by assigning just one option to each of the three large groups. Students were initially asked to discuss the advantages and disadvantages of their given option in their listening triangles and then confer with their 'partner' triangle to compare the points they had raised. At the end of each policy discussion, a spokesperson from each of the three large groups presented their main argument to the class and the mayors, who were responsible for deciding which option to adopt for each policy. Once a decision was made, Jackie selected the chosen option and ran the computer simulation with the whole class to show the impact of the chosen option (see Figs. 10.2 and 10.3).

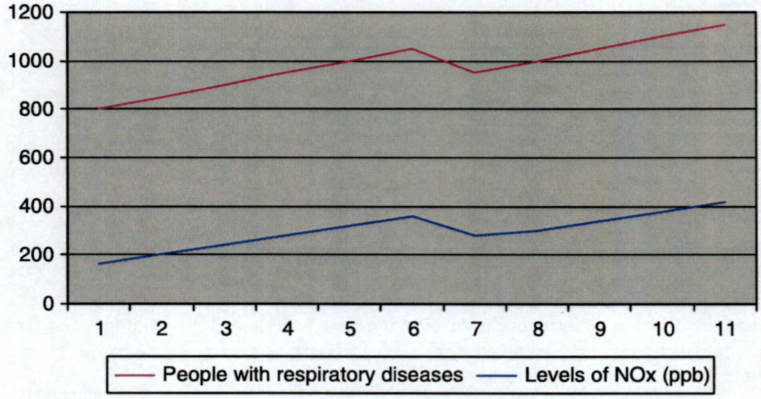

Fig. 10.2 Impact of implementing option 1, policy 3, on respiratory diseases

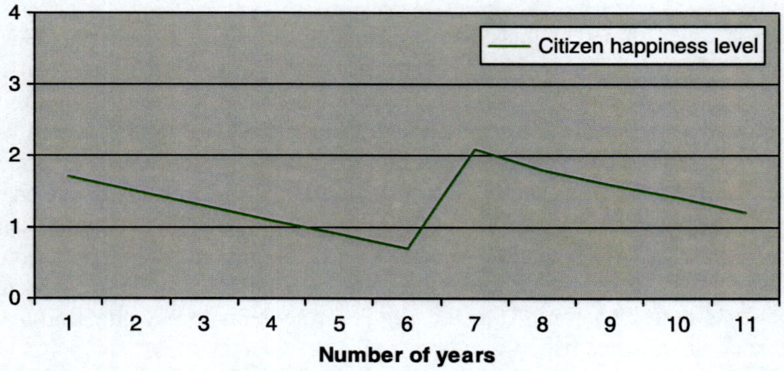

Fig. 10.3 Impacts of implementing option 1, policy 3, on happiness level

Students therefore received immediate feedback on the consequences of the mayor's decision after each policy.

Research

The research design was that of a case study, the unit of analysis being Jackie's implementation of the Bleaksville Simulation. The study drew on multiple sources of data as the Bleaksville Simulation was played out over the four policies, which took two lessons. The research questions were:

1. Do students build on the scientific content knowledge of the unit in their debating activity by drawing on scientific evidence from the resources in their arguments, or from what they have learnt during the teaching unit?

2. Does the combination of resource and teaching approach provide opportunities for good quality argumentation?
3. How does Jackie scaffold argumentation during a discussion/role-play debating activity?
4. How do teacher and students view the discussion-based activity?

Data Sources

To address these research questions data sources included audio and video-recordings of the sequence of Bleaksville lessons. During whole class and small group discussions, the video-recordings were centred on two of the small groups (three students in one and two in the other) that would come together to form the larger advisory group (group A) for making the decision of which option to recommend to the mayors. The five students were selected by Jackie, and were representative of the class (in terms of the national test levels mentioned previously, one student was assessed at level 3, two at level 4, and two at level 5). Discussions were audio-recorded using a double microphone system, one microphone pointing at each of the two small groups. Together with the video-recording, it was possible to establish who said what in the series of parallel discussions. The audio-recordings were fully transcribed. The presentations by the group spokespeople were also video-recorded at the end of each policy debate along with the mayor's decision. The two other advisory groups were named B and C, the mayors were named Group D.

Jackie was also audio-recorded using a separate recorder to capture her discussions with a variety of students in the class, as well as all her discourse with the whole class. Once again the video-recording helped to locate the students with whom she was interacting during the lessons. The researcher recording observational data was as close to being a nonparticipant as is possible within all the constraints of being present in a classroom and attempting to minimise influence on events.

To supplement the observational data, further data to ascertain views and perceptions of the simulation, role-play, and group discussion activity were collected from Jackie and the students. After the final lesson, Jackie was interviewed to ascertain her views about the Bleaksville Simulation and discussion activities in *Twenty First Century Science*. She was interviewed again at the end of the academic year (9 months later) to explore whether her approaches to such activities had changed at all. To explore their perceptions of group discussion and their interest in the activity, all the students were asked to complete a short attitudinal questionnaire about the Bleaksville Simulation. This questionnaire was designed to see whether the students had a positive affective response, one aim of the course and of the debates being to engage and interest the students. Other questions in the survey focused on issues relating to classroom interactions and familiarity with small group discussion, given that previous research undertaken in similar classrooms had demonstrated low levels student interactivity (Newton et al., 1999). The survey also explored how the students felt about making contributions, their experience of small group organisation, and their perception of its value, given the importance of small group discussion in developing reasoning through argumentation.

Data Analysis and Results: Students' Use of Evidence

To evaluate the students' use of evidence in constructing arguments, their discourse from all the recorded sources was coded in terms of the kinds of evidence they used for constructing arguments for all the policies. One aim of the Bleakesville activity was for students to draw on the science in the Air Quality topic in making decisions, so it was of interest to see how much scientific content was incorporated into the statements of objectives for Bleaksville (i.e., reflecting the aims of the topic). The objectives included reference not only to science concepts but also to other aspects of content that could be categorised as environmental, economic, and social. The policy texts were then examined to see whether these categories of information were incorporated for students to use as evidence. Analysis of the students' argumentation in terms of evidence could then be set against the objectives and materials provided for them. In summary, we could see whether the proportion of scientific evidence presented in the materials was the same as that which featured in the students' arguments. The following codes were assigned to evidence statements or arguments: Sci = scientific, Env = environmental, Ec = economic, S = social. There is some overlap between the scientific and environmental codes, but for the purposes of this analysis, 'scientific' is used to describe statements or arguments containing specific scientific terminology, for example, *cars with big engines burn more fuel*. The environmental category was used to describe less scientifically precise statements with a generalised reference to pollution such as *power stations will generate more pollution*. Clearly other codes could be used to describe evidence, for example, statements with 'legal' connotations, but we decided to adhere to a four-category system derived partly from the learning aims of the course unit.

The Bleaksville objectives suggest that students should be able to use and develop the following concepts introduced in the Air Quality unit. The objectives have been coded using the system described above:

- Pollution from fossil fuel power stations can be reduced by using less electricity (Env), removing sulfur dioxide from the flue gases of coal-burning power stations (Sci) and sulfur from natural gas fuel (Sci)
- Pollution from motor vehicles can be reduced (Env) by burning less fuel, by having more efficient engines (Sci), using low sulfur fuels, and using catalytic converters (Sci)

- Governments can take action to adjust the balance between public and private transport (S), through legal limits to emissions (Sci), taxes (Ec), and providing more public transport (S)
- People can take action to reduce their energy consumption (S, Env), by using energy more efficiently (Ec)

The highest proportion of statements in these objectives are scientific in nature, more so when combined with those pertaining to environmental impacts. This categorisation enabled us to see whether the kinds of evidence provided in the resources for students were selected when they constructed their arguments. In order to illustrate how the evidence has been analysed, the statements or arguments provided in the policy 3 options are shown below, the categories are shown in brackets:

Option 1: 'Congestion charge: people will have to pay for driving in the polluted city centre (Ec). This will encourage them to use public transport more often (S). The tax will be £5 per day (as at the time the resources were created). It will not operate at weekends. (Note: this is the scheme introduced in London in 2003). The charge will have a bigger effect on people with low incomes (Ec)'.
Option 2: 'Cars with big engines burn more fuel (Sci). The new regulation will ban engines bigger than 1.6 l. People will have to sell their sporty cars and buy smaller models (S). If people cannot buy more expensive cars, the profits of the car industry will decrease (Ec)'.
Option 3: 'We will improve the poor public transport in the city with a tram system (S). This will link the city centre with people's homes. The trams produce no exhaust emissions (Env). It will be cheap to travel (Ec), and encourage people to leave their cars at home (S). The cost will be funded by an extra tax of £50 (Ec). The power station will generate a bit more electricity (and therefore pollution Env) to power the trams (Sci)'.

An example of how the codes were applied to students' arguments is shown below, again for policy 3, which involved encouraging the use of alternative forms of transport (see Table 10.1). The *underlined* text shows where students quoted the evidence from the activity resources word for word. The application of the codes to the policy texts and to students' arguments was undertaken by Amos, with these extracts from policy 3 reviewed by Simon for agreement, with discussion of how Amos applied the coding scheme:

Group A: um we think *trams will be cheap (Ec) and encourage people to leave their cars at home (S)*... and *trams produce no exhaust emissions (Env)*. Option 2. Um we think option 2 is bad because if people don't buy new cars, *the profits of the car* companies *will decrease (Ec)* and bigger cars will not be allowed. Option 1 is bad because *the charge will affect people in low incomes (S)* and people will use their cars more at weekends because there is no charge (S, Ec).
Group B: Option 1. Because *people will have to pay to drive in the city centre (Ec)*, this will *encourage people to use public transport more often (S)*. If people used more public transport, less pollution will be made which means less cars on the road (Env). The people in the towns' happiness will go up due to all this

(S). Option 2 is bad because people have to sell their original cars which will take time to sell and too much chaos because they have the latest model big engine cars (S). Whereas in our one, you won't have to do anything except pay £5 per day to use public transport (Ec). Option 3 is bad because you have to pay £15 per person extra tax (Ec).
Group C: Ok, our option is... *ban new cars with big engines*. And the good thing is smaller engines will need smaller amounts of fuel (Sci), so there'd be less pollution (Env). And the bad things are option 1 will pay with more fuel. People might not want to use trams because it's noisy and inconvenient (S). And... and how would you get on the tram with all your shopping? (S)
Group D (the mayors): we've chosen option 1, introducing a congestion charge. We've chosen option 1 because there'll be less cars in the town centre and *it will encourage people to use public transport more (S).*

Using the four derived codes to identify different kinds of evidence, the full analysis of the text from the policy documents in all policies reveals the frequency of types of argument as shown in Table 10.2.

Table 10.2 shows that over the whole of the Bleaksville Simulation, 29% of the *presented* evidence can be categorised as 'scientific', with a further 14% describing environmental impact. Policy 1 and policy 2 presented more scientific evidence than 3 or 4, which were more focused on economic and social issues. The types of argument that the students *chose* to include in their final arguments presented in the whole class debate revealed a much lower proportion of scientific evidence, as shown in Table 10.3.

Table 10.2 Categories of evidence presented to students in policy texts

Policy	Scientific	Environmental	Economic	Social
1	6	3	5	0
2	7	1	5	1
3	2	2	4	4
4	2	2	9	5
Total	*17*	*8*	*23*	*10*
% of presented evidence statements	*29*	*14*	*40*	*17*

Table 10.3 Categories of evidence selected by students when presenting arguments for each policy

Policy arguments	Scientific	Environmental	Economic	Social
Total for policy 1	2	7	13	2
Total for policy 2	5	3	5	2
Total for policy 3	1	3	5	8
Total for policy 4	2	4	10	4
Overall total	*10*	*17*	*32*	*16*
% selected evidence by students	*13*	*23*	*43*	*21*

The relationship between the presented evidence and that selected by students during the Bleaksville Simulation appears to confirm that students will tend to choose economic and social types of evidence readily. Indeed, the frequency with which the case study students selected these two types of evidence slightly exceeded that of the presented evidence. They did use some scientific evidence in their arguments but this was far outweighed by their use of environmental evidence. In other words, what seemed to happen is that students were more inclined to talk about 'pollution' in a generalised way, rather than to use more specific scientific terminology or examples, when building their arguments. They diluted the scientific content in favour of using less precise terms.

Data Analysis and Results: Quality of Argument

To ascertain the quality of students' argumentation, a framework was used based on the level-system developed by Erduran et al. (2004). Derived from the work of Toulmin (1958), Erduran et al. identify low levels of argumentation as simple claims and counter-claims, whereas higher levels include the use of rebuttals. Rebuttals occur where the data or warrants of an argument are opposed and are distinguished from counter-arguments where an opposing claim is presented. The focus in Erduran et al.'s research was on those episodes of student-student dialogue where there was a clear opposition between students, and the nature of this opposition was assessed in terms of the strength of rebuttals offered. The ability to use rebuttals is 'the most complex skill' (Kuhn, 1991); thus, rebuttals are an essential element of arguments of better quality and demonstrate a higher-level capability with argumentation. Arguments that include rebuttals force students to evaluate the validity and strength of arguments. The framework of levels devised by Erduran et al. (shown in Table 10.4) was applied to opposition episodes identified in the data these researchers recorded from small group discussions. The method of analysis

Table 10.4 Codes for analysing episodes of opposition in students' discourse

Level of argumentation	Characteristics of argumentation
Level 1	Consists of arguments that are a simple claim versus a counter claim
Level 2	Consists of arguments that have a claim versus a counter claim with either data (evidence), warrants or backings (justifications) but do not contain rebuttals (anticipation of counter argument)
Level 3	Consists of arguments that are a series of claims or counter-claims with either, data, warrants or backings and the occasional weak rebuttal
Level 4	Consists of arguments that have a claim with a clearly identifiable rebuttal. Such argument may have several claims and counter-claims as well, but this is not necessary
Level 5	Consists of an extended argument with more than one rebuttal

using this level system enabled the researchers to perform comparisons pre- and post-intervention, and for different contexts (see Osborne et al., 2004a). Though the level system enabled comparisons to be made based on assumptions of quality, the nature of grounds and rebuttals remained unexplored, thus the definition of quality was confined to argument structure rather than content and strength of evidence.

The students' episodes of argumentation in this study were coded according to the five-level system of Erduran et al. in two ways. First, the recorded discourse of the two small groups within group A was analysed using the level system. Second, the level system was applied to the arguments presented by each of the large groups (A–D) in the whole-class debate. This application of the level system diverges from its original use for oppositional episodes in student-student discourse, however the descriptors for each level category can be applied to written arguments for comparison with respect to justifications and anticipation of rebuttals, hence were used in this analysis. This latter analysis could then be coupled with the analysis of evidence sources, to see whether there was any pattern in the kinds of evidence used when students presented arguments at different levels.

In their small group discussion the Group A students constructed their arguments using the data available, but much of their small group discussion consisted of conferring with one another to clarify their understanding of that data. Analysis of the quality of argumentation using the level system demonstrated that these students mostly reached low levels of argumentation; there were few rebuttals. As the activity progressed through policy discussions 1–4, students' argumentation in the small group format improved and the beginnings of weak rebuttals were found in policy 4 discussions:

S5: we need to build a new wind farm which produces no emissions which is a good thing.
S4: but it's also a bad thing because you have to build it and it costs
S5: they don't need to know that…need a new wind farm…which produces no emissions.

Students S4 and S5 were anticipating a counter-argument, recognising evidence that might damage their case. In terms of being representative of the class as a whole, group A was the most consistent at presenting an argument for their option with justifications, and in trying to counter the arguments from the other two groups' options in the whole class debate. Group A's argumentation in each debate typified the argumentation of the group during their listening triangle discussion (in preparation for debate) and it also represented their normal level of interaction, according to Jackie.

In presenting arguments in the debate, students from all the groups responded well to Jackie's requests for justification, for example:

S: OK. Our option is…ban new cars with big engines. And the good thing is smaller engines will need smaller amounts of fuel, so there'd be less pollution. And the bad thing about option 1 is people will pay for more fuel. And people might not want to use trams because it's noisy and inconvenient.

The quality of argumentation in the presentations did vary from group to group. The arguments of group A were mostly at level 2 across all four policies, as they made claims and justified these with evidence from the data sheets, whereas the other groups were less consistent. In policy 1, group B came close to a level 3 argument by anticipating a counter-argument, albeit rather tentatively:

S: And, um, even though it costs, like £500, you know, it makes the air really clean and it's much safer.

In the other policies, group B consistently argued at level 2. Group C constructed arguments at a slightly lower level than the others. Three of their arguments in policies 1 and 2 struggled to exceed level 1, although the remainder were clearly at level 2, suggesting that they did improve their argumentation skills during the course of the activity. The mayors were the most effective in terms of developing higher-level arguments, displaying elements of level 3 in policies 1 and 4, with the rest being level 2s.

Drawing on both analyses of student discourse, a comparison was made between the types of evidence selected by each group of students as they presented their arguments to the whole class, and the levels of argument demonstrated in these presentations. The majority of arguments constructed by the students were at level 2, as described previously. However, it was of interest to see whether the students arguing occasionally at higher levels would select scientific evidence to build their arguments, or would argue from a socioeconomic standpoint. Table 10.5 shows the levels of argument for different types of evidence used.

Table 10.5 shows that level 3 arguments were uncommon (3 in total) and of these, one argument drew on scientific evidence, one on economic evidence, and one on social evidence. Thus there is insufficient evidence from this small data set to draw conclusions about whether students presenting arguments at higher levels might draw on scientific evidence or not. When in small groups, the analysis suggests that the two groups (from A) who sometimes reached level 3 in small group discussion did draw on scientific evidence more than is represented in the final group presentations of Table 10.5. They did this more in policy 4, which could be either because they were becoming more adept at rebuttal, or their arguments were stronger as they became more confident with the science. The finding suggests that this comparison between nature of evidence and quality of argument may be of

Table 10.5 Levels of argument in each evidence category for each policy debate

Policy	Scientific	Environmental	Economic	Social
1	2 level 2	7 level 2	1 level 1; 11 level 2; 1 level 3	1 level 1; 1 level 2
2	5 level 2	3 level 2	2 level 1; 3 level 2	2 level 2
3	1 level 2	3 level 2	5 level 2	8 level 2
4	1 level 1; 1 level 3	4 level 2	10 level 2	3 level 2; 1 level 3

interest to pursue with a larger data set involving more students over different SSIs, and using more data sets from small group discussion.

Data Analysis and Results: Scaffolding Argumentation

To determine the extent to which Jackie perceived it necessary to support the process of argumentation while the students were involved in decision-making discussions, the framework developed by Simon et al. (2006) for evaluating argumentation processes was applied to transcripts from the audio-recording of Jackie's interactions. This analytical framework that was chosen as teachers' oral contributions demonstrate their implicit goals, in this case to facilitate the processes needed for successful small group discussion. The processes identified by Simon et al. that are needed for argumentation include talking and listening, defining argument, taking a position, using justification, constructing and evaluating arguments, counter-arguing, and reflecting on argument. In their work, Simon et al. found that the processes of counter-argument, evaluating argument, and reflecting on argument were not always present in argumentative interaction, or scaffolded by the teachers' contributions, and so were considered to be 'higher order' processes that require careful planning and teacher awareness for facilitation. Jackie's interactions with students reveal how she has conceptualised the processes involved for decision making through discussion and argument and how she acts to facilitate these processes. Table 10.6 shows how these processes were used to code Jackie's talk, the extracts in the table are indicators of how Jackie's oral contributions encourage each process.

Jackie adopted some of the scaffolding processes identified by Simon et al. but was limited in her encouragement of higher-order argumentation. She encouraged talking and listening by using a 'listening triangles' strategy. Students in the triangle were to assume the roles of developing arguments 'for' and 'against' the anti-air pollution policy that they had been assigned to, with a scribe to record their argument. This approach enabled the students to position themselves in the argument process. Jackie used the phrase 'pros and cons' repeatedly throughout

Table 10.6 Codes for identifying the scaffolding of argumentation processes

Argumentation processes
Talking and listening – 'You need to be able to speak, but you also need to be able to listen'
Knowing meaning of argument – 'what is an argument and why is it a valuable thing?'
Positioning – 'what's your argument 'for' the flue gas desulfurization?'
Justifying with evidence –'why is that a good thing?'
Constructing arguments – 'your argument needs to follow these points'
Evaluating arguments – 'what sorts of things will make a good argument?'
Counter-arguing/debating – 'Can anyone think of anything that somebody might say to oppose that?'
Reflecting on argument – 'would you like to explain how you persuaded Sally?'

her interactions with the whole class, and with groups of students. Students in the main observed group, group A, are designated 'S1, 2, 3, 4, and 5'. All other students are designated 'S'. For example:

T: Right so…you've done…you are option 3… you don't want to say the cons of your thing, you want to say the cons of their thing, so that the mayor chooses your thing and not their thing… yeah?
S5: oh so these are cons

Jackie occasionally used the terms 'good thing/bad thing' and also 'advantage/disadvantage' to define argument. She rarely made any other reference to the structuring of an argument and did not model examples of arguments. However, by encouraging the students to identify 'pros' and 'cons' for their arguments, she helped them to focus on pieces of evidence that supported a particular position in the argument:

T: So, what's one argument 'for' the flue gas desulfurisation?
S3: …um, removes most of the sulfur dioxide from the waste gases given off by the power station.

Jackie strongly scaffolded the justification of arguments on many occasions by asking questions such as:

T: Very good, and why is that a good thing?
S3: well…um…it's pollution from the power station
T: What problems is the pollution causing for the people…? What diseases is it giving them?… Asthma and things like that. Chest infections and things… yes. So what's an 'against' the other people's answer?

Jackie continually asked the students to read and check the evidence on the policy sheets and on the relevant pages of the textbook resource. There were clear examples of Jackie playing devil's advocate with the focus group, a stance taken to promote further justification. These occurred during policy 3, which sets up three options for improving air quality by making changes to public and private transport. The first example came during a discussion about the 'pros' and 'cons' of introducing a congestion charge in the city centre:

S3: Er..er..I don't know…I don't know what to do for 'againsts'
T: Against the other people's…?
S3: Yeah
T: Right so… Well it says…what does it say here?…The charge will have a bigger effect on people with low incomes… so…it's alright for people who can afford to pay the congestion charge but what about the people who can't afford the congestion charge? What are they going to do?…Is that fair?
S3: no
T: So maybe that's something against this, yeah?
S2: yes
S3: I don't understand it…

T: 'Low incomes'... that means how much they earn. If people do jobs and those jobs don't pay very much...and those people need to go into central London for whatever...for their job...or if they want to go shopping or whatever...they won't be able to afford it. That's not very fair. Those people who earn loads and loads of money...do you think they're going to be bothered if they have to pay congestion charge or not? They don't care...they don't care if it's £5 here and there. They're really rich. But people who don't earn very much, they do care...and maybe it's not fair on those people who don't earn very much.

A further example followed on immediately from this:

T: And what about this one? 'If people cannot buy new cars, the profits of the car industry will decrease'. So ban new cars with big engines, so we say, look, you've got a big 4-wheel drive...you have to get rid of it. But what if I can't afford to buy a new car?
S2: well you sell the car so you get money for the car.
T: But it's been banned! Who's going to buy a car that's been banned?
S4: umm!
S5: sell it to another country!
S4: yeah, we bought ours from Japan
T: Fair enough but maybe it's been banned there as well!
S4: well actually it was from England.

Both examples related the arguments to issues and experiences in the students' everyday lives, and in the second, the students seemed to be engaging and seeing personal links. Jackie was creating relevance for the students and it had the desired effect. Policies 1 and 2 did not elicit any discussions of this nature, so an important aspect of 'authenticity' (Aikenhead, 2006) was missing for the greater part of the observed interactions.

Jackie strongly encouraged students to make written records of their arguments and to prepare for the presentations at the decision-making stage of the debate, hence focusing strongly on constructing arguments. Each policy sheet contains the same introductory sentence about making the citizens happy, and three statements to provide a writing frame. Jackie referred to these prompts during the whole activity. Jackie did not encourage the evaluation of argument to any great extent; however, she did ask the students who were role-playing as decision-making mayors to judge how persuasive their advisors' arguments were, but did not give explicit evaluation criteria:

T: So, you've got a basic idea of what's a good option and what's a bad option?
M: Yeah...
T: So try to keep an open mind and listen to their arguments...see how persuasive they are...ok?

Jackie did not encourage counter-argument or reflection on argument with the whole class, and subsequently these processes were limited in the students' discourse. Although the activity is set in the context of a role-play debate, Jackie did not encourage debate at all between spokespeople and the mayors during the presentation of arguments after each policy. She simply thanked each spokesperson and moved on to the next one:

T: Ok, thank you…right…are we going to give her applause? Ok, option 3.

There was a real opportunity to open up the arguments and to promote debate at each stage but Jackie did not do so, despite having attended the training provided by *Twenty First Century Science* which encourages teachers to take this approach. Jackie's experience reinforces our previous work with teachers (Simon et al., 2006) in that the changes required to bring about sustained development in teaching argumentation through small group discussion require careful planning, clear teaching goals, and strategies for interaction needed to achieve those goals.

Data Analysis and Results: Teacher and Student Views

After the Bleaksville activity, Jackie commented on the challenges she faced while trying to facilitate the role-play debate. She had been worried that the students 'would not care enough about the subject' to take part in discussions enthusiastically. However, by the end, she was delighted with their general engagement and enjoyment. Jackie also thought that the role-play debate appealed to the students but that they 'didn't see it as science'. In commenting on the nature of discussion she had been pleasantly surprised that students had actually discussed ideas with one another in their listening triangles. She had experienced success with the strategy in other lessons, observing that it ensures students are all involved and have a responsibility. She did express disappointment about the number of questions from the students concerned with the definition of technical terms. She had hoped that students would draw upon ideas from previous lessons, but she was asked to define several times the terms that she felt they should be familiar with, for example, many students asked 'what's an emission'? She felt that the students had covered this issue thoroughly in the unit and so she was not sure whether or not they linked the concept as it emerged in the options to the chemical reactions they had studied in the unit. These concerns were reflected during the debates by the students, who ignored scientific evidence in favour of more generalised environmental issues.

Overall, Jackie felt pleased with the participation and engagement of the students but felt unsure about how much they had gained from the activity. She had not seen evidence of the students relating the role-play to their own everyday lives, suggesting that they fail to make links with their own experiences as one would hope for concerning many SSI. Indeed, group A only made one reference to anything close to this during the whole activity, as shown previously in the discussion on banning four-wheel drives. Such links with students' lives are an important outcome of *Twenty First Century Science,* given the aim for students to become more informed citizens, taking responsibilities for their own life choices.

Nine months after teaching the Bleaksville Simulation, Jackie had not used any further such activities with her class, citing lack of time and the need for 'effective' preparation for examinations as the main constraints. Moreover, she revealed that none of her departmental colleagues had even used the Bleaksville Simulation while teaching parallel classes that year. However, she had recently convened a departmental meeting to explore ways of developing teacher confidence in using argumentation and decision-making activities in the future and to embed them more explicitly in teaching sequences.

Since small group discussion was new to many teachers involved in *Twenty First Century Science* and an approach that many teachers find challenging, one aim of this research was to explore how students, who were also new to small group discussion, would perceive and respond to this strategy. The success of the debating activities within *Twenty First Century Science* would rest on how students responded, in terms of their interest and also their acceptance of small group discussion as having purpose and value.

To evaluate the students' affective response, the 19 student questionnaires were analysed to determine students' feelings and thoughts about discussion in science lessons, as outlined earlier, and specifically the Bleaksville Simulation. The results are summarised as follows:

- All but two students agreed with the statement 'science has got loads of things to discuss', hence acknowledging that discussion has a role to play in science.
- All but one student agreed (and two were unsure) with the suggestion 'discussing things in groups in science helps me to understand', which even if they were not confident, suggests they perceive discussion to have value in learning.
- Ten students felt that they would 'worry about giving answers to questions in science in case they got it wrong', revealing the lack of confidence felt by this group of weaker students.
- Six students were unsure or disagreed with the suggestion that they would 'say to other people if they thought they had got something about science wrong'.
- Just over half agreed that the teacher (Jackie) did 'not often put them into different groups in science' and five were not sure, revealing that small group formats were usually the same.
- Six students were unsure about, or disagreed with statements about listening well to one another.

Reinforcing the concern of Jackie, it would seem that several students were not confident about contributing scientific ideas in lessons, though it is hoped that through small group discussion such confidence can grow. However, with the support of the policy sheet resources, and structuring from Jackie, there was evidence to show that even the weakest students had managed to learn how to construct a basic argument and to present it in front of their peers. The students did manage to discuss ideas among themselves in positive ways during the activity. Eleven of the students said that they 'do not ask (my) teacher questions about science in most lessons' but by policy 4, several had started to do so more readily, moving beyond 'what am I supposed to do?' Seventeen out of 19 students felt positive about the Bleaksville activity itself (one 'didn't mind it' and one thought it 'boring'). Favourable comments included:

> It was fun to discuss and see an outcome;
> I learnt a lot about how to save the air from getting polluted;
> It was useful for future life'; 'it was interesting to hear what everyone had to say about it;
> I liked it better than most lessons as we got to discuss stuff.

Jackie thought that the students had not particularly enjoyed this first unit of the course, Air Quality, as it was rather dry and lacked opportunities for practical work. The latter negative feedback was common in the pilot evaluation (Millar, 2006). However, if their positive responses to the Bleaksville Simulation are a true reflection of the students' feelings, then they bode well for one of the main goals of *Twenty First Century Science,* that is, to stimulate students' engagement with, and understanding of, SSI.

Implications

The aim of the research was to select a representative context from the pilot course and see how Jackie used the *Twenty First Century Science* resources to generate argumentation in a typical classroom and to evaluate the students' use of evidence and argumentation. Jackie encountered many familiar challenges reported by science teachers when trying to adopt new pedagogical strategies, in particular, concerns about student engagement and meaningful learning via discussion and argumentation were raised. This finding is reflected in more recent studies using small group discussions in science teaching (Bennett et al., 2010). Jackie was positive about the Bleaksville experience overall, agreeing that the students had seemed to engage in and enjoy the activity. Later in the year, her department made a commitment to include more discussion activities in their course in the following year.

This case study has highlighted some implications for teachers trying to implement discussion and argument in science lessons in more detail than is possible in a large-scale evaluation. The engagement in the Bleaksville role-play enabled students to exercise some argumentation processes, as some students' argumentation progressed to include weak rebuttals, but the results suggest that the guidance needs to include more explicit teaching strategies and learning outcomes for higher levels of argumentation to be achieved. Jackie did not encourage higher-order argumentation processes such as evaluating and reflecting on argument and, as Maloney and Simon (2006) note, unless teachers make these skills part of their learning objectives, students will not achieve them. Students were primarily satisfied to read and use the data in the option policies, without questioning or referring to their own every day experiences, or pay attention to opposing positions, a finding evidenced in other studies (Sadler, 2004). The frameworks used for this analysis could provide a means for making those links and for enhancing the guidance to include strategies for evaluating argument, counter-argument, and reflection on argument. The results show that engagement with the materials enabled students to evaluate their arguments in a minimal way, but if more students are to benefit from opportunities to develop decision-making skills with *Twenty First Century Science,* then the guidance needs to be more clearly focused and the materials enhanced to prompt these higher order processes. Without practice in counter-argument and evaluation, students are unlikely to gain sufficient grounding in those processes that would enable them to fully engage in reasoned debate as citizens.

Students' abilities to use and understand scientific concepts and terms during discussion activities were questionable. The research suggests that students engage more readily with socio-scientific evidence that is relevant to their general understanding. In designing the materials, one possibility is to ensure that initial engagement is based on socio-scientific evidence, to which students can then be introduced and encouraged to use. The framework for identifying different kinds of evidence can be used to help support teachers by exemplifying and modelling arguments that build on scientific evidence. Support materials can be adapted to model the use of key terms and ideas for students so that they improve upon this aspect. By building up students' skills of argumentation using different kinds of evidence, their engagement and reasoning in SSI can provide a basis for science learning (Sadler, Barab, & Scott, 2007). Though the comparison between sources of evidence used and quality of argument was attempted by applying both analytical frameworks to the presentations made in each policy, the results (Table 10.5) show no conclusive pattern linking any particular source of evidence with higher levels of argument using this data set. More data for students in small group discussion would be useful to ascertain whether the indications from group A, that the higher level argumentation in groups achieved in policy 4 using scientific evidence, was an indication that these two factors may be in any way linked.

Conclusions

The research has contributed to the ongoing evaluation of a novel, demanding, course by looking in depth at a real-life situation in a typical non-selective London school. It has highlighted the complex nature of the expectations of such a course, including students' ability to draw on scientific evidence, an assumption made in the course design, and their ability to produce high quality arguments within small group discussion and whole-class debate. In terms of teacher organisation and management, the research has reinforced our knowledge of the challenges faced by teachers in changing their practice to adopt more discussion activities, such as those written into the course. Similar problems continue to exist as more recent evaluation of the course has shown (Ratcliffe & Millar, 2009).

As SSI are located firmly in everyday contexts, surely their inclusion in school science lessons should improve engagement and foster a stronger sense of relevance for students. Aikenhead (2006) has drawn together many important issues pertaining to the use of everyday contexts and the relevance of students' experiences of school science, thus providing a useful lens for examining the *Twenty First Century Course* approach. He argues that the issue of relevance is at the heart of a humanistic science curriculum, and that while many teachers state that they subscribe to a humanistic approach to learning science, they actually have great difficulty in making changes to their own natural orientation toward traditional school science. A humanistic science course requires a wider range of approaches than those with a traditional bent, including small-group work and cooperative learning. Decision making, and the process of doing so, is often central to a humanistic science curriculum, as it serves as a vehicle to transport students into their everyday world. Therefore, if teachers struggle to adopt appropriate pedagogical approaches toward the teaching of argumentation and decision making, students will fail to develop the necessary skills.

References

Aikenhead, G. (2006). *Science education for everyday life: Evidence-based practice*. New York: Teacher College Press.

Bennett, J., Hogarth, S., Lubben, F., Campbell, B., & Robinson, A. (2010). Talking science: The research evidence on the use of small group discussions in science. *International Journal of Science Education, 32*(1), 69–95.

Bennett, J., Lubben, F., Hogarth, S., Campbell, B., & Robinson, A. (2005). A systematic review of the nature of small-group discussions aimed at improving students' understanding of evidence in science. In *Research evidence in education library*. London: EPPI-Centre, Social Science Research Unit, Institute of Education, University of London. Retrieved July 9, 2007, from http://eppi.ioe.ac.uk/cms/Default.aspx?tabid=326

Braund, M., Lubben, F., Scholtz, Z., Sadeck, M., & Hodges, M. (2007). Comparing the effect of scientific and socio-scientific argumentation tasks: Lessons from South Africa. *School Science Review, 88*(324), 67–76.

Burden, J., Campbell, P., Hunt, A., & Millar, R. (2007). A project response. In *UYSEG and Nuffield Foundation (2007): Twenty first century science pilot evaluation report*. Retrieved June 26, 2007, from http://www.21stcenturyscience.org/data/files/c21-evaln-rpt-feb07–10101.pdf

Erduran, S. (2008). Methodological foundations in the study of argumentation in science classrooms. In S. Erduran & M. Jiménez-Aleixandre (Eds.), *Argumentation in science education* (pp. 47–69). Dordrecht, the Netherlands: Springer.

Erduran, S., Simon, S., & Osborne, J. (2004). TAPping into argumentation: Developments in the application of Toulmin's argument pattern for studying science discourse. *Science Education, 88*(6), 915–933.

Goodrum, D., Hackling, M., & Rennie, L. (2001). *The status and quality of teaching and learning of science in Australian schools*. Canberra, Australia: Department of Education, Training and Youth Affairs.

Hogarth, S., Bennett, J., Campbell, B., Lubben, F., & Robinson, A. (2005). A systematic review of the use of small-group discussions in science teaching with students aged 11–18, and the effect of different stimuli (print materials, practical work, ICT, video/film) on students' understanding of evidence. In *Research evidence in education library*. London: EPPI-Centre, Social Science Research Unit, Institute of Education, University of London. Retrieved July 9, 2007, from http://eppi.ioe.ac.uk/cms/Default.aspx?tabid=324

Jiménez-Aleixandre, M. P., & Erduran, S. (2008). Argumentation in science education: An overview. In S. Erduran & M. Jiménez-Aleixandre (Eds.), *Argumentation in science education* (pp. 3–27). Dordrecht, the Netherlands: Springer.

Johnson, G. (1999). Kidney role-plays. *School Science Review, 80*(292), 93–97.

Kuhn, D. (1991). *The skills of argument*. Cambridge, UK: Cambridge University Press.
Kuhn, D. (1993). Science as argument: Implications for teaching and learning scientific thinking. *Science Education, 77*, 319–337.
Levinson, R., & Turner, S. (2001). *Valuable lessons: Engaging with the social context of science in schools*. London: Wellcome Trust.
Lunn, S. (2002). What we think we can safely say – Nature of science. *British Educational Research Journal, 28*(5), 649–672.
Maloney, J. F., & Simon, S. (2006). Mapping children's discussions of evidence in science to assess collaboration and argumentation. *International Journal of Science Education, 28*(15), 1817–1841.
McSharry, G., & Jones, S. (2000). Role-play in science teaching and learning. *School Science Review, 82*(298), 73–82.
Millar, R. (2006). Twenty First Century Science: Insights from the design and implementation of a scientific literacy approach in school science. *International Journal of Science Education, 28*(13), 1499–1521.
Millar, R., & Osborne, J. (1998). *Beyond 2000. Science education for the future*. London: King's College London, School of Education.
Newton, P., Driver, R., & Osborne, J. (1999). The place of argumentation in the pedagogy of school science. *International Journal of Science Education, 21*, 553–576.
Norris, S. P., & Phillips, L. M. (2003). How literacy in its fundamental sense is central to scientific literacy. *Science Education, 87*, 224–240.
OCR. (2005). *Twenty first century science*. Oxford, Cambridge and RSA Examinations (OCR) retrieved July 26, 2007, from http://www.ocr.org.uk/qualifications/type/gcse
OECD. (2006). *Assessing scientific, reading and mathematical literacy. A framework for PISA 2006*. Paris: OECD.
Osborne, J., Erduran, S., & Simon, S. (2004a). Enhancing the quality of argument in school science. *Journal of Research in Science Teaching, 41*(10), 994–1020.
Osborne, J., Erduran, S., & Simon, S. (2004b). *Ideas, evidence and argument in science. In-service training pack, resource Pack and video*. London: Nuffield Foundation.
Ratcliffe, M., & Grace, M. (2003). *Science education for citizenship teaching socio-scientific issues*. Maidenhead, UK: Open University Press.
Ratcliffe, M., & Millar, R. (2009). Teaching for understanding of science in context: Evidence from the pilot trials of the Twenty First Century Science courses. *Journal of Research in Science Teaching, 46*(8), 945–959.
Ratcliffe, M., & Osborne, J. (2007). Part 3 changes in classroom practices. In *UYSEG and Nuffield Foundation (2007). Twenty first century science pilot evaluation report*. Retrieved June 26, 2007), from http://www.21stcenturyscience.org/data/files/c21-evaln-rpt-feb07–10101.pdf
Roberts, D. (2007). Scientific literacy/science literacy. In S. K. Abell & N. G. Lederman (Eds.), *Handbook of research on science education* (pp. 729–780). Mahwah, NJ: Lawrence Erlbaum.
Sadler, T. D. (2004). Informal reasoning regarding socioscientific issues: A critical review of research. *Journal of Research in Science Teaching, 49*(5), 513–536.
Sadler, T. D., Barab, S. A., & Scott, B. (2007). What do students gain by engaging in socioscientific inquiry? *Research in Science Education, 37*, 371–391.
Simon, S., Erduran, S., & Osborne, J. (2006). Learning to teach argumentation: Research and development in the science classroom. *International Journal of Science Education, 28*(2–3), 235–260.
Simon, S., & Maloney, J. (2007). Activities for promoting small group discussion and argumentation. *School Science Review, 88*(324), 49–57.
Toulmin, S. (1958). *The uses of argument*. Cambridge, UK: Cambridge University Press.
UNESCO. (1999). *Science for the twenty-first century. A new commitment.* Retrieved July 6, 2007, from, http://www.unesco.org/science/wcs/abstracts/I_7_education.htm
UYSEG and Nuffield Foundation (2007). *Twenty first century science pilot evaluation report.* Retrieved June 26, 2007, from http://www.21stcenturyscience.org/data/files/c21-evaln-rpt-feb07–10101.pdf

Chapter 11
Metalogue: Engaging Students in Scientific and Socio-scientific Argumentation

Victor Sampson, Shirley Simon, Ruth Amos, and Maria Evagorou

Evagorou: In this chapter, Shirley and Ruth raise several interesting issues related to (socio-scientific issues) SSI, argumentation, and decision-making. A major question in this study was the quality of students' arguments, and if there is a link between the nature of the evidence (e.g., scientific, environmental, financial) and the quality of the arguments. The results are not conclusive as to this point, and the authors suggest that more evidence is necessary. The issue of quality of arguments has been prominent in discussions in the science education community lately (e.g., Erduran, 2008) and reading this study made me think about the following issues:

(a) Is the quality of the arguments also connected to the quality of the evidence that is presented in the learning environments? If so, what more do we need to learn to inform the design of SSI and argumentation curriculum?
(b) Are the arguments supported by evidence collected/produced by the students bound to be of higher quality?

Another issue that arises from this chapter is the level of engagement with the learning environment and what kind of affordances different curriculum materials might have, especially when SSI are involved. The closer the connection between the issue under study and students' identities, the more students' beliefs systems are affected, making it more probable that students will ignore evidence and provide weaker justifications that are mostly based on personal values

V. Sampson (✉)
School of Teacher Education and the FSU-Teach program, Florida State University, Tallahassee, FL, USA
e-mail: vsampson@fsu.edu

S. Simon and R. Amos
Institute of Education, University of London, London, UK

M. Evagorou
School of Education, University of Nicosia (UNic), Nicosia, Cyprus

T.D. Sadler (ed.), *Socio-scientific Issues in the Classroom: Teaching, Learning and Research*, Contemporary Trends and Issues in Science Education 39,
DOI 10.1007/978-94-007-1159-4_11,

(Simonneaux & Simonneaux, 2009) or their personal and cultural identities (López-Facal & Jiménez-Aleixandre, 2008). This trend supports that development of learning experiences that are not authentic. On the other hand, other researchers argue that the issue is not whether students' decisions are more value-based than knowledge-based, but on what kind of knowledge is regarded as relevant by the students (Kolstoe, 2006). The Bleaksville debate presented to Jackie's class engaged her students in discussions, but I am wondering whether a more authentic learning experience (e.g., measuring and discussing air quality in London) would afford better quality arguments, and what kind of evidence the students would choose (scientific, environmental, financial) to support their arguments.

Sampson: The study described in this chapter, which examined how students made decisions and used evidence as they discussed various ways to improve air-quality in the imaginary town of Bleaksville and how the classroom teacher attempted to promote and support this process, provided many key insights into how students engage in socio-scientific argumentation when they are given an opportunity. It also, like all good research, raises new questions and opens up some potential avenues for future work.

The first question that came to my mind as I read Shirley and Ruth's study concerned the relationship between argumentation that is scientific in nature, which is the focus of my research, and socio-scientific argumentation is: How much of an overlap is there between scientific argumentation and socio-scientific argumentation? Shirley and Ruth briefly touched on this issue when they described argumentation in general and compared it to the nature of socio-scientific argumentation. It seems to me that there is some obvious overlap between the two because both processes require people to construct, justify, and refute arguments. However, there are also some major differences. For example, Shirley and Ruth describe how the students in their study were trying to determine which policy to endorse in order to improve air quality without making the citizens unhappy. This type of activity, where people need to consider different courses of action related to a complex problem from multiple viewpoints, seems to be a hallmark of socio-scientific argumentation but it is very different from the purpose of argumentation that is more scientific in nature. In scientific argumentation, people are often attempting to explain or describe a natural phenomenon or develop a valid and acceptable answer to a research question. This is a different goal and thus the types of claims that can be made in each context will be different. There are also several differences in the nature of the supports and challenges that people can use in these two contexts. In socio-scientific contexts, a wide range of reasons are viewed as an acceptable way to support or challenge the viability of a course of action. These reasons include, but not are limited to, social, economic, moral, and empirical. In scientific contexts, in contrast, the reasons that tend to be used to support or challenge a claim are often limited to those that are empirical, theoretical, methodological, or analytical in nature. There are other differences to be sure. I therefore think science educators working in the field of socio-scientific argumentation will need to help students understand the similarities and differences between argumentation that is scientific and socio-scientific in nature.

My second question, which stems from my first, concerns the issue of transfer. If there are significant differences between scientific and socio-scientific argumentation such as the ones I described earlier, can we expect students to transfer what they have learned about participating in one type of argumentation to the other? This issue has not been well investigated but I think it is important. I do not think we can assume that students will be able to participate in these two forms of argumentation in a desired manner without learning about both of them. This conjecture is based, in large part, on the difficulties that students face when they are first asked to participate in scientific forms of argumentation that are so well documented in the literature. Yet, it is important to note that the difficulties that students have when they first participate in scientific argumentation do not seem to stem from a lack of natural ability. Most students just have never had an opportunity to participate in scientific argumentation and do not understand the "rules of the game" and are therefore forced to rely on "everyday" forms of argumentation. That is one reason why short interventions often lead to substantial improvements in students argumentation skills (e.g., Venville & Dawson, 2010); students just need to be introduced to what counts as quality in a given context. Therefore, if socio-scientific argumentation and scientific argumentation are different but related to forms of argumentation, perhaps it would be better for science educators to treat the ability to participate in each type of argumentation as distinct but equally desirable outcomes of a high quality science education.

The third question that I have is: How should we, as field, define evidence in a socio-scientific context and should all types of information be considered evidence? Shirley and Ruth, for example, describe how the students were given "evidence statements" that they classified as scientific, environmental, economic, and social. Yet, as I read the samples they included in the chapter, the statements appeared to be different types of reasons rather than evidence. I tend to define evidence in science as observations, measurements, or findings from other studies that have been collected, analyzed, and interpreted by researchers (Sampson & Gerbino, 2010). I use this definition to help students understand the difference between evidence, data, and unsubstantiated inferences when I ask them to construct evidence-based argument in response to a research question. However, this definition is not the only one in the literature and it is perhaps not even the most useful definition. Other authors, for example, describe evidence in science simply as data that is used to support a claim (Berland & Reiser, 2009; McNeill, Lizotte, Krajcik, & Marx, 2006). This is a much more general definition than the one I use in my research.

I raise this question because I think it is important for students to understand what does and does not count as genuine evidence in science. Do we also need to help students understand the difference between the various types of reasons that can be used during an episode of socio-scientific argumentation? Some reasons such as intuitive and emotive ones or appeals to the greater good are often used to persuade people but these types of reasons might not be as strong or convincing as others (such as ones that are economic, political, empirical, and ethical or moral in nature). Therefore, it might be a productive strategy to help students learn how to identify the various types of reasons people use to support a viewpoint or course of action and how to challenge these reasons in an appropriate manner if we want them to learn how to participate in better socio-scientific argumentation.

Fourth, should we, as field, expand our assessments of argumentation to include the nature or types of criteria students use to evaluate claims, answers to research questions, or alternative courses of action proposed by others? Shirley and Ruth provided an interesting analysis of the nature of the argumentation that students engaged in during the Bleaksville activity. However, this analysis was structural in nature and focused on the absence or presence of various components of an argument. I cannot help but wonder what we would have learned about the students' socio-scientific argumentation if Shirley and Ruth had also examined the nature of the criteria that students used to evaluate the different policies or the nature of the rebuttals these students privileged in this context. For example, did the students rely on economic reasons more than scientific reasons when they evaluated the different policies or to challenge an alternative idea? I think this type of analysis would tell us a great deal about the students' thinking during an episode of socio-scientific argumentation and would give us a measure of how often students tend to use scientific explanations to evaluate different perspectives.

Finally, Shirley and Ruth's description of how the teacher, Jackie, attempted to promote and support student participation in socio-scientific argumentation during the Bleaksville activity was extremely interesting. I think the field, as a whole, needs to focus more on how teachers modify and adapt curricula and structure classroom instruction in different contexts and the underlying reasons for their decisions. This study, for example, made me wonder about the underlying goals of the classroom teacher. It seems to me that Jackie's main goal was to increase the likelihood that the students would discuss the policies and not for them to learn how to engage in better socio-scientific argumentation. I think this is one reason why we tend to see teachers, such as Jackie, scaffold student engagement in this type of activity so much; teachers often want to make sure their students do it "right" the first time. However, it might be better to let students make mistakes and allow them to learn from them (along with more productive strategies and techniques) if the long-term goal is better argumentation skills, especially if the teacher plans to engage students in argumentation repeatedly over the course of the semester. I also wondered about how much the "unwritten rules of school" influenced the students' actions during this activity. The Bleaksville activity is clearly different from typical science classroom activities and it often takes students a long time to learn how to participate in unfamiliar activities. Students also need to see the value of this type of activity such that it makes sense and they have a reason to construct and evaluate arguments with their peers in the context of school science (Berland & Reiser, 2009). Is there a need for more longitudinal studies of how teachers scaffold socio-scientific argumentation inside the classroom over time and how students learn to participate in this type of complex activity? I think we could learn a great deal from this type of research.

Maria also raises some interesting questions about the quality of arguments generated by students and the nature of the activities used by science educators to promote and support students in socio-scientific argumentation. Overall, I think there clearly is a relationship between the nature of the arguments crafted by students and the amount and type of information available to students and the nature of the topic. For example, if we want students to construct an evidence-based argument, then students need to have access to data gathered through empirical research (or findings from empirical studies) that they can analyze, interpret, and transform into evidence. In socio-scientific contexts, students also need access to information about political motives, economic realities, and other factors. Students, in other words, need to have access to a great deal of information before we can expect them to look at an issue from multiple perspectives and construct two-sided arguments.

In the context of an imaginary scenario, such as the Bleaksville activity, the responsibility for finding or creating this information lies with the developers of the activity. Yet, students do not have to engage in socio-scientific argumentation around an imaginary scenario. Students can be asked to weigh the pros and cons of a proposed tax on beverages with high sugar content or evaluate the merits of a proposed cap and trade policy as a way to control carbon emissions. In this type of activity, students could use the available literature to develop their arguments and critique the arguments of their peers. I think the more important question is how to structure an activity in the appropriate manner in light of the topic, students involved, and the student learning objectives. The number of questions educators must consider when designing an activity, curriculum, or learning environment is vast. For example, what do we want students to be able to do during an episode of socio-scientific explanation? Should students be supplied with information or should we expect them to find their own? If we supply students with information should it all be relevant or should we expect students to determine what is and what is not important? Is it better to start simple for students and get more complex or is better to start with a complex issue and let students learn from their mistakes? There is a great deal of research that needs to be done before we can begin to develop tentative answers to these types of questions. I think we also need to learn more about student thinking in these various situations before we can begin to take advantage of the potential benefits of engaging students in socio-scientific argumentation. Last, but certainly not least, we also need to determine if the answers to these various questions are context specific or broadly applicable.

Simon and Amos: Maria raises two questions in her response to the chapter. First she asks whether the quality of arguments is connected to the quality of evidence provided to the students in the learning environment and second, whether quality of argument would be higher when evidence is sourced by the students themselves. The answer to the first question is clearly yes; in a recent study of students engaged in argumentation in socio-scientific contexts we have found that if information is provided with an activity then the students tend to use it in their argumentation in addition to, or in preference to, their own ideas and knowledge, thus quality of argumentation is connected to the kinds of evidence available. Exploring the links between quality of socio-scientific argument and the kinds of evidence both provided and used (i.e., scientific, social, economic) could be a subject of further research using different socio-scientific contexts.

With reference to Maria's second question, it is interesting to note that prior to the Bleaksville debate, the *Twenty First Science* teaching unit did include activities where students measured local air quality. However, neither they nor the teacher made specific links between the practical investigation and the debating activity. To make the Bleaksville debate more authentic and relevant, with the possibility of higher quality argumentation, guidance could be provided to help teachers scaffold the links between the activities of the scientific components and the socio-scientific debates. Indeed such links could be a vehicle for helping students to develop scientific argumentation as well as socio-scientific argumentation, through drawing out the kinds of arguments needed in both contexts and evaluating the differences. Such a process would address the distinction between the two forms of argumentation that is the basis of Vic's first question, where he raises the issue of overlap and difference between scientific and socio-scientific argumentation.

Vic also raises the question of how we define "evidence" in a socio-scientific context. The course materials for Bleaksville use the term "evidence statements," even though, as Vic suggests, the statements provide different types of reasons rather than evidence. The term evidence is clearly problematic if its definition is based on observations, measurements or findings collected, analyzed, and interpreted by researchers. In our work on argumentation, "evidence" is defined as information used in arguments to support claims, either as data, warrants or backings. In designing materials for teachers to use in socio-scientific argumentation, curriculum developers are clearly specifying statements that could be used as evidence to support claims, which is helpful to teachers inexperienced in teaching argumentation as well as the students themselves. We agree with Vic's suggestion that strategies to help students identify the reasons used to support different viewpoints and how to challenge these reasons in an appropriate way are important aspects of learning how to engage effectively in socio-scientific argumentation, but such strategies can be achieved through examining the nature and strength of reasons or "evidence" irrespective of how these are labeled. This point leads to Vic's fourth question about the criteria students use in evaluating arguments. In our study on quality of argumentation we did focus on the structural nature of students' arguments and it would have been interesting to explore the criteria students used. However, our study was undertaken in an authentic classroom where *Twenty First Century* activities were being piloted. We would have needed to take a more interventionist stance to explore students' ideas about how they judged evidence. Jackie's intervention was restricted to identification of pros' and "cons" rather than strength of evidence and argument. In our other work with teachers using IDEAS (Osborne, Erduran, & Simon, 2004), we have tried to emphasize the importance of developing criteria for evaluating knowledge claims, but teachers find this unfamiliar and need tools for supporting such a process with students.

Vic's final point is that the field needs to focus more on how teachers modify and adapt curricula and structure reasons for their decisions. Since the publication of IDEAS, we have been re-examining the criticality of guidance and how it relates to teachers' interpretation and enactment of argumentation activities (Simon & Richardson, 2009). There are many layers of interpretation involved, and enactment is driven by underlying goals and values. Our recent work in developing argumentation practice in whole school science departments has revealed the need for both teachers and students to practice argumentation activities, and our current analysis building on earlier work on teachers' scaffolding of argumentation (Simon, Erduran, & Osborne, 2006) aims to understand teachers' development in this practice.

References

Berland, L., & Reiser, B. (2009). Making sense of argumentation and explanation. *Science Education, 93*(1), 26–55.

Erduran, S. (2008). Methodological foundations in the study of argumentation in science classrooms. In S. Erduran & M. P. Jiménez-Aleixandre (Eds.), *Argumentation in science education* (pp. 47–70). Dordrecht, the Netherlands: Springer.

Kolstoe, S. (2006). Patterns in students' argumentation confronted with a risk-focused socioscientific issue. *International Journal of Science Education, 28*(14), 1689–1716.

López-Facal, R., & Jiménez-Aleixandre, M. P. (2008). Identities, social representations and critical thinking. *Cultural Studies in Science Education, 4*(3), 689–695.

McNeill, K. L., Lizotte, D. J., Krajcik, J., & Marx, R. W. (2006). Supporting students' construction of scientific explanations by fading scaffolds in instructional materials. *The Journal of the Learning Sciences, 15*(2), 153–191.

Osborne, J., Erduran, S., & Simon, S. (2004). *Ideas, evidence and argument in science. In-service training pack, resource pack and video*. London: Nuffield Foundation.

Sampson, V., & Gerbino, F. (2010). Two instructional models that teachers can use to promote and support scientific argumentation in the biology classroom. *American Biology Teacher, 72*(7), 427–431.

Simonneaux, L., & Simonneaux, J. (2009). Socio-scientific reasoning influenced by identities. *Cultural Studies in Science Education, 4*(3), 705–711.

Simon, S., Erduran, S., & Osborne, J. (2006). Learning to teach argumentation; research and development in the science classroom. *International Journal of Science Education, 28*(2–3), 235–260.

Simon, S., & Richardson, K. (2009). Argumentation in school science: Breaking the tradition of authoritative exposition through a pedagogy that promotes discussion and reasoning. *Argumentation, 23*, 469–493.

Venville, G., & Dawson, V. (2010). The impact of a classroom intervention on grade 10 students' argumentation skills, informal reasoning, and conceptual understanding of science. *Journal of Research in Science Teaching, 47*(8), 952–977.

Chapter 12
Different Music to the Same Score: Teaching About Genes, Environment, and Human Performances

Blanca Puig and María Pilar Jiménez-Aleixandre

> *What argument against social change could be more effective than the claim that established orders exist as an accurate reflection of innate intellectual capacities?*
>
> Stephen Jay Gould, The Mismeasure of Man

Introduction: Biological Determinism, When Science Meets Ideology

There is agreement within the science education community on the contributions of argumentation about socio-scientific issues (SSI) to scientific literacy and to the development of critical thinking (Kolstø, 2006). SSI involves scientific arguments in addition to political, personal or ethical questions about what action to choose (Kolstø, 2006). It is suggested that argumentation about SSI makes scientific learning meaningful, as it provides a context that connects science with everyday problems where citizens are expected to make decisions, and requires taking an active role to solve controversies. Argumentation in these contexts involves not only applying scientific knowledge, but also developing an independent opinion in order to critically examine scientific claims and arguments, in other words, becoming a critical thinker (Jiménez-Aleixandre & Puig, 2010).

Biological determinism, which is the focus of this chapter, has social relevance because determinist views have been used, and still are used, to support political agendas challenging the notion that all humans are equal. But it differs from other SSI, as for instance cloning (Jiménez-Aleixandre & Federico-Agraso, 2009) where

B. Puig (✉) and M.P. Jiménez-Aleixandre
Department of Science Education, University of Santiago de Compostela (USC), Galicia, Spain
e-mail: blanca.puig@usc.es

T.D. Sadler (ed.), *Socio-scientific Issues in the Classroom: Teaching, Learning and Research*, Contemporary Trends and Issues in Science Education 39,
DOI 10.1007/978-94-007-1159-4_12,

the issues at stake relate to genetics but the controversies themselves are ethical in nature. In our unit about gene expression, students need to appeal primarily to causal explanations about genetics, rather than to values, although social representations play a role in the arguments about this issue. The relevance of different contexts (for instance, causal explanations or decision-making) for argumentation and the use of evidence are discussed elsewhere (Jiménez-Aleixandre & Puig, 2011).

The focus of the current study and the embedded teaching unit has ethical, social, and political implications: causal explanations (interaction versus determinism) about gene expression in the phenotype, in particular, about human performances. The current model of gene expression explains the relationships among genotype and phenotype, as for instance human traits and performances, as a consequence of gene-environment interactions. Causal explanations are constructed through social processes of questioning, evaluation, and revision (Berland & Reiser, 2009), meaning that the scientific practices of argumentation and explanation are interconnected. Educational research has examined students' difficulties in the genetic domain (Duncan, Rogat, & Yarden, 2009; Knippels, 2002). Nevertheless, to our knowledge there are no studies that explore the difficulties in the construction of the phenotype notion in connection with human traits and performances and the determinist representations that may be associated with it.

This chapter discusses the design and implementation of a teaching sequence, through a pilot study and two teaching cycles, in five high school classrooms. The topic is the causal model of gene expression and gene-environment interactions. The goals are to engage students in modeling gene expression and in using evidence to build explanations about human performances.

Our aim is to analyze teacher-students interactions or, as Tiberghien, Vince, and Gaidioz (2009) call them, joint productions, to increase our understanding about the challenges in teaching the model of gene expression and its potential interference with determinist views. We seek to explore how meanings about gene-environment interactions are constructed in the class. This analysis is framed in the model of didactical transposition (Chevallard, 1991), characterizing knowledge transformations, from scientific community (reference knowledge) to curriculum and teaching resources (knowledge to be taught), and from these designed instructional sequences to taught knowledge. The objectives, related to these steps in knowledge transformation are:

1. To analyze the process of design of the learning tasks with the purpose of making explicit assumptions and decisions guiding it, that is, the first step in didactical transposition, from reference knowledge to knowledge to be taught.
2. To examine two teachers' actions during the unit about gene expression, in particular how they dealt with the gene-environment interactions, and to characterize the didactical contracts created in both classrooms, that is, the second step in didactical transposition, from designed to taught knowledge.
3. To examine students' difficulties related to the construction of explanations that acknowledge gene-environment interactions.

Rationale: Determinism and Genetics Learning

In this section we will first review the notion of biological determinism and its current resurgence in the context of particular political agendas; then we turn attention to science education studies about genetics learning, in particular about the model of gene expression.

Determinism from Mainstream Science to Support for Racism

Biological determinism is the view that genes entirely determine all individual traits and performances, including intelligence, criminality or academic achievements. In daily life it is commonly expressed in racist and sexist opinions. Determinists attribute social and economic differences among different races or genders to hereditary; they consider these differences to be innate distinctions. These views were part of mainstream science; for instance, a hierarchy was established distributing human races from "superior" (white) to 'inferior', according to features such as skin color. Cuvier (1817) "studied" the body of Saartjie Baartman, known as the Hottentot Venus, concluding that she was a proof of why these (black) races were "condemned to eternal inferiority".

Nowadays "human race" is not accepted as a biological notion, much less as a hierarchy of "superior" and "inferior" people. Determinist views have been replaced by a consensus on the interaction between genes and environment. As Lewontin (1991) points out, although there is a large amount of variation from one individual to another from the same ethnic group, accounting for 85% of all genetic variation, there is remarkably little variation on average among major groups (7% of all genetic variation). The remaining 8% of variation is found between ethnic groups within a race. However, as with other socially constructed representations (Moscovici, 1961–1976) determinist views, explicit or implicit, continue to exist in society. The persistence of these views is reflected in the media, literature or jokes. Biological determinism cannot be justified on the basis of current scientific evidence, such as genome research. Then, why is this view still circulating? Who legitimizes it?

A particularly disturbing occurrence is that of political discourses that relate issues such as alcoholism, violence or suicide to genetic determination. We illustrate this trend with two statements of European politicians: first, the French president Sarkozy, in an interview: "Every year about 1,200 to 1,300 young people commit suicide in France. It is not because their parents do not care about them, but because, genetically they had frailty, a previous pain" (*El País*, September 10, 2009). The second example is taken from a newspaper article by Mariano Rajoy, current leader of the conservative party (Partido Popular), which has been alternating in the Spanish government with the socialists since the late 1970s. The article, entitled *Human equality and models of society* questions the "cliché of human equity":

> Natural inequality among men is written on the genetic code, where the roots for all human inequalities are found: in it are transmitted all our conditions, from physical, health, color

of eyes, hair, corpulence... to the ones that we call psychical, as intelligence, disposition for arts, studies or business. (Faro de Vigo, March 4, 1983)

Although this text was written years ago, this author has never distanced himself from the positions offered in this and a later article on the same issue.

Determinism explains social inequalities as a result of biology. Its message, as point out, is that all social phenomena are rooted in human nature. This reductionist view on human beings may be comforting for individuals because it provides an explanation for inequalities. If genes were responsible for determining exactly who each person would become and individuals do not have control over the outcome, then there would be no social responsibilities. The resurgence of determinism has been related to conservative proposals for reducing investments in social programs (Gould, 1981). It is used to support political agendas seeking scientific justification for reducing support for deprived segments of society; poverty, unemployment, and educational exclusion are interpreted as a result of innate features rather than social conditions (Kaplan & Llomovatte, 2009). In other words, responsibility is placed on individuals and on genetic traits, not on society (Lewontin, 1991).

Contemporary determinism asserts for instance that the differences in intelligence between blacks and whites are due to genetics (Herrnstein & Murray, 1994; Jensen, 1969). On October 14, 2007, in an interview for the *Sunday Times*, the Nobel laureate James Watson, talking about Africa, said that "all our social policies are based on the fact that their intelligence is the same as ours – whereas all the testing says not really." Watson went on to argue that people who employ black workers challenged the notion of equality. Watson's claim is a statement with political implications, which may be used to justify the reductions of investments in African countries. Racism is a target for educators, but we have not located any studies, besides Levy, Selles, Ferreira's (2008) exploration of textbooks, about determinism in the science classroom. In this study we focus on the students' positions between acknowledgment of interactions and determinism.

Teaching and Learning About Phenotype-Genotype Relationships

Research shows that genetics is one of the most difficult topics for students (Duncan & Reiser, 2007; Knippels, Waarloo, & Boersma, 2005; Lewis & Wood-Robinson, 2000), so the challenge for science educators is to develop learning environments that promote students' scientific literacy in this topic. This chapter deals with the model of gene expression and its application requires an appropriate understanding of the phenotype notion and of the influence of environment in gene expression. Given the complexity of these ideas and the lack of resources, as evidenced by the analysis of textbooks below, our goal was to develop a sequence supporting students' appropriation of this model. We agree with Gelbart and Yarden (2006) about the need to provide students with a context giving opportunities to apply genetics concepts, develop new knowledge, and present it in different ways. Tasks set in real life and SSI constitute appropriate contexts for this purpose.

Why is it important to understand this model? There are two primary reasons: first, it explains the relationships among genotype and phenotype accounting, for instance, for human traits as height and performances such as athletic achievements or intelligence. Without understanding these relationships students may not be able to reason about how the environment influences some phenotypic traits, for instance, why nowadays people are taller than several generations ago. Although phenotype is sometimes defined as the "visible" manifestation of genes, it needs to be noted that some traits, like blood type, are detectable rather than "visible."

Second, understanding this model is necessary for a critical evaluation of determinist views. Although there are studies showing that students have poor understanding of the relationships between genotype and phenotype (Tsui & Treagust, 2007; Venville & Donovan, 2005), they do not deal with biological determinism. For instance, Lewis and Kattmann (2004) have reported students' difficulties in distinguishing between genotype and phenotype, and as a result, in considering the microscopic and molecular causal mechanisms of genetic phenomena. Duncan et al. (2009), in their proposal of a learning progression in genetics, suggest that one of the core ideas of students' understanding in genetics is related to the interaction between genes and environment, but that this set of ideas is often entirely missing from the genetics curriculum. They indicate that the risk of developing a deterministic view is greater when students lack explanatory mechanisms that link genes to traits, being unaware of what organization level the genetic information specifies.

In a previous study (Puig & Jiménez-Aleixandre, 2010a), we suggested that students' difficulties in identifying data related to the model of gene expression and using them to evaluate a determinist claim may be influenced by social representations of human races. Artistic skills, human performances, and health conditions are sometimes presented as genetic traits without any environmental influence, misinterpreting an individual's genetics endowment as predestination. The fact that these ideas are still circulating makes it difficult for students to understand what phenotype actually is and what influences it. We think that understanding the model of expression is a necessary but not sufficient condition to evaluate determinist claims and that students need to develop critical thinking. In our characterization of critical thinking (Jiménez-Aleixandre & Puig, 2010), there is a component related to social emancipation and the capacity to develop one's own opinion as opposed sometimes to the mainstream ideas of a community or society.

Methods and Educational Context

Methodological Framework: Didactical Transposition

Our methodological approach is framed by the theories of didactical situations (Brousseau, 1998) and *didactical transposition* (Chevallard, 1991) based on the assumption that there are social conditions required for knowledge to exist, as knowledge can only stay alive if it is studied, used or both.

Didactical transposition characterizes the process of transformation of knowledge from one community, scientists (reference knowledge), to another, classrooms (taught knowledge). As Tiberghien et al. (2009) point out, there are two steps in the transposition: (1) from the reference knowledge to the knowledge to be taught and (2) from the knowledge to be taught to the taught knowledge. The knowledge to be taught consists of official curricula, textbooks, and other resources and the taught knowledge is related to the way a teacher enacts it in a particular class.

For objective 1, the analysis of the design process is based on Tiberghien et al. (2009), who propose a framework to develop research-based design, relating decisions to theories about knowledge, teaching, and learning. For objective 2, the analysis of the teachers' actions is based on the work of Sensevy (2007). Sensevy characterizes the "didactical action" (*action didactique*) as a reciprocal action based in communicative processes between the teacher and the students. This approach is based on the assumption that knowledge shapes the teaching and learning practices and that didactical action is a joint action between teacher and students. For analyzing these interactions between teacher and students, we adapted Mortimer and Scott's (2003) tool.

The analysis of teachers' actions is framed in the notion of *didactical contract* (Brousseau, 1998), which characterizes the teacher's expectations about students and the students' expectations about the teacher. As Tiberghien et al. (2009) discuss, the didactical contract constitutes a system of norms, some of which are generic and will be lasting, while others are specific to elements of knowledge and need to be redefined with the introduction of new elements.

For objective 3, the examination of students' difficulties in explaining phenotype as a result of the gene-environment interactions, instead of seeing it as depending only on genes (biological determinism), the students' wrote reports and the transcriptions of their talk were analyzed, and categories constructed in interaction with the data.

Context and Data Collection

This is a multi-case study conducted through a pilot study and two research cycles in five classrooms from three public (state) secondary schools in Galicia, the northwest region of Spain. All of the students involved in the research ($N=127$) were native Spaniards. The teaching sequence was carried out in several (ranging from three to five) sessions, each lasting 50 min except the pilot study that was developed in two sessions. The full study including the pilot and research cycles extended over a 3-year period. All the teachers, identified by pseudonyms, are male and hold a degree in Biology. In this section, we outline the specific context of each case study and the data collected. Table12.1 summarizes this information.

Students worked in small groups and each group was audio and video taped. The researcher (first author) took notes about teacher-students interactions. All drawings

Table 12.1 School context in the pilot study and research cycles

		First research cycle	Second research cycle	
	Pilot study	Case study 1	Case study 2	Case study 3
School and location	School A city	School A city	School B town	School C small village
Grade and age	9th grade (14–15)	10th grade (15–16)	10th grade (15–16)	11th grade (16–17)
Course	Biology & geology	Biology & geology	Biology & geology	Science for the modern world
Number of classes and students	1 (24 students)	2 (50 students)	1 (18 students)	1 (35 students)
Number of sessions	2	3	5	4

and diagrams used on the blackboard or electronic board were also registered. Sometimes, at the request of the teachers, the researcher engaged in classroom activities to offer guidance with the tasks.

Pilot Study

The pilot study was developed with a group of 9th graders (14 to 15 years old). The students were from middle class backgrounds, and the teacher was a professional with more than 20 years of experience and involvement in inquiry approaches.

Case Study 1

The setting was the same school as the pilot study with two tenth grade biology classrooms (including some of the students from the pilot study). The teacher was a novice, with only 2 years of experience. All tasks were discussed with him, although, due to lack of experience, he offered little input.

Case Study 2

The second case study, conducted in the second research cycle, was developed with tenth grade students from a working class background in a small town. The teacher had more than 15 years of teaching experience, most of it in this particular school. The tasks were discussed with him. Issues related to data collection and students' distribution in groups were also negotiated with the teacher. The scenario was quite different from the pilot study and case study 1 because many of the students involved were not high achieving or expressed little interest in school activities. Seven of the 18 students were repeating the course.

Case Study 3

The third case study was conducted in a rural school located in a remote village where agriculture and roofing slate quarries constituted the main socioeconomic activities. The students and teachers shared particularly strong and trusting relationships in this school. The teacher with whom we worked had been teaching for 10 years. He was involved in research projects related to inquiry and was pursuing a Ph.D. in science education at the time of the intervention.

From Reference Knowledge to the Knowledge to Be Taught: Design of the Teaching Sequence

In this section, we analyze the design process of a teaching sequence (TS) about the model of gene expression for tenth grade. It is a developmental research approach, drawing from theory and practice (Knippels, 2002), and consists of a process of testing the adequacy of learning tasks and design in a classroom setting.

The concept of *didactical transposition* helps to make explicit the process of the development of the teaching sequence. The first step of the didactical transposition consists of how the reference knowledge is transformed into the knowledge to be taught. Figure 12.1 summarizes the different elements, both theoretical (goals, design principles, and literature review) and empirical (textbook analysis, pilot study, and time constraints) that shape this first step.

Reference Knowledge

The reference knowledge is composed of two strands: (1) Genetics, in particular, the model of gene expression and gene-environment interactions, a topic connected to a social issue and biological determinism, and (2) Scientific practices, including

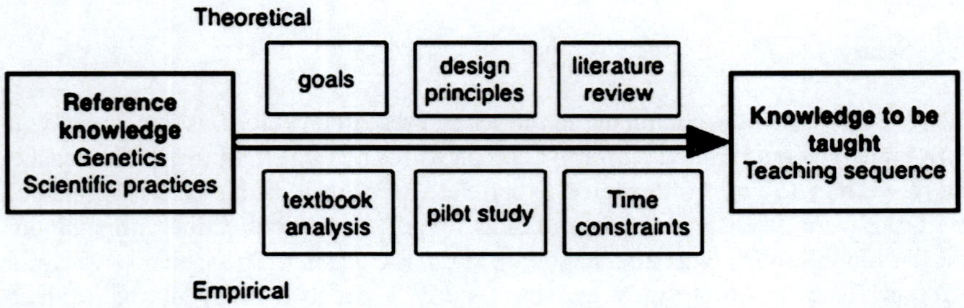

Fig. 12.1 First step in the didactical transposition: from reference knowledge to knowledge to be taught

modeling, argumentation, and the use of evidence. Argumentation and modeling are two central practices in science that are connected in that argumentation is a process that underlies the examination, evaluation, and revision of models (Berland & Reiser, 2009).

Goals and Design Principles

The transformation of reference knowledge into knowledge to be taught does not involve a simplification. It is a process of knowledge transformation expressed in the choice of learning tasks for students supporting the appropriation of the model of gene expression and its application to different contexts. As Tiberghien (2008) indicates, elaborating the knowledge to be taught that leads to scientific culture and to citizenship education, necessarily goes beyond disciplinary goals. Therefore, this sequence has three goals for students, the first two related to science education and the third to citizenship: (1) to be able to apply the model of gene expression to real life contexts; (2) to develop the competency of using evidence and building arguments; and (3) to be able to develop a critical stance toward biological determinism.

How did we use these goals and design principles in order to transform the reference knowledge? The consequences of the goals were first that tasks 2 and 3 required students to apply the model of gene expression to everyday contexts. Secondly, these tasks engaged students in the practice of selecting and using evidence in order to support a claim. Finally, the tasks, particularly numbers 2 and 3, addressed biological determinism. It has to be noted that this is not the standard way of transformation of reference knowledge, as illustrated by the textbook analysis reported below.

The choice of the design principles related to these goals draws from Jiménez-Aleixandre's (2008) proposal for learning environments that support argumentation. We briefly discuss how three of these design principles, role of students, of teachers, and learning tasks, influenced the decisions made in the design.

- *Active role of the students*: The tasks were designed in order to actively engage students in the scientific practices of modeling and use of evidence. For instance, in task 1 adapted from "Take two people" (Dixon, 1982), students worked on the construction of a model about relationships between genotype and phenotype. In task 2, students needed to relate pieces of evidence to claims about causes for the achievements of black sprinters.
- *Learning tasks set in real life contexts*: The learning tasks were drawn from real life. For instance, the second task ("Athletics"), related to the use of evidence, provided a context where students were asked to build arguments about black athletes' achievements. In task 3 ("Watson"), students had to evaluate a determinist claim. In terms of classroom organization, the tasks used cooperative learning strategies and required students to pay attention to different points of views within their groups.

- *The role of the teacher*: According to Sensevy (2007) didactical action is a joint production of teacher and students, which means that the teacher should act as a knowledge mediator. The teachers' actions and interactions with students are discussed in the section about taught knowledge.

Literature Review

The literature review about genetics learning, discussed in the rationale, was used first to identify learning problems that needed to be addressed such as understanding the genotype – phenotype relationship and second to locate proposals for tasks that were incorporated into the sequence such as the modeling activities designed by Dixon (1982) and Johnson (1991). This first step in the didactical transposition was also influenced by empirical elements: the analysis of school textbooks and the development of a pilot study as well as the time constraints, which were a chief concern for the teachers.

Analysis of Textbooks

We wanted to check how school textbooks introduced the model of genotype expression and whether they dealt with biological determinism. For this purpose, the contents of genetics chapters in five tenth grade textbooks were analyzed. Tenth grade is the first year in which genetics is a part of the Spanish curriculum. The four teachers involved in the study used two textbooks (TB1 and TB2), which are the most widely used in Spain. We decided to examine three more textbooks in order to have a representative sample, providing an overview of how this topic is typically presented in schools. Four dimensions of textbook presentations were analyzed: (a) the definition of phenotype; (b) examples of the influence of the environment on gene expression; (c) questions and activities requiring the application of the phenotype notion; (d) references to determinism or "race." The findings are summarized in Table 12.2.

Table 12.2 Summary of the textbooks analysis. N=5 textbooks, TB1 to TB5

Dimensions	Number of textbooks $N=5$
Definition of phenotype	
Phenotype as a result of interactions gene-environment	4
Phenotype solely as the expression of genes	1 (TB3)
Examples of the influence of environment	
Two different examples	3 (TB1, TB3, TB4)
One example	2 (TB2, TB5)
Questions and activities requiring the application of the phenotype notion	1 (TB2)
References to determinism and race	2 (TB2, TB3)

- *Definition of phenotype*: We examined first whether textbook authors defined phenotype as the result of interactions between genes and environment or just as gene expression. Second, we examined whether authors discussed the idea that some phenotypic traits are detectable but not visible. Four textbooks defined phenotype as the result of an interaction between genotype and environment and one as solely gene expression: "The genes contained in one individual for a specific character constitutes his/her genotype (AA; Aa or aa) and the expression of that genotype is called phenotype" (TB3). This narrow definition, not consistent with the reference knowledge, revealed determinism. All five textbooks defined phenotype as the "set of visible characters in any individual" (TB2), without mentioning that some characters, like blood types, are detectable but not visible.
- *Examples of the influence of environment on gene expression*: Little space was devoted in the textbooks to explanations or examples of the influence of the environment on gene expression. Three presented two examples, and the other two just one. In all, only four different examples were found: Human height, muscles, hair color in animals, and obesity. Two textbooks (TB1, TB2) presented height as an instance of influence of environmental factors, in particular of nutrition: "Human height is an inherited trait, parents that are tall usually have children that are tall too, but alimentation does definitely influence this trait" (TB2). The development of muscles is explained in two textbooks (TB1, TB3) as a consequence of training.
- *Questions and activities requiring the application of the phenotype notion*: There were only eight questions or activities related to genetics in the five textbooks. Five were problems of Mendelian genetics requiring the use of the Punnett square and did not demand applying the model in different contexts. As Stewart (1983) pointed out, solving this type of problem does not necessarily require an adequate understanding of genetics' content knowledge. We found only one question in TB2 that required applying the notion of the influence of the environment in the genotype expression: *Can two individual with different phenotype exhibit the same genotype?* The others were related to the transmission of human traits from parents to offspring.
- *References to determinism and the "race" issue*: The analysis revealed that while all textbooks addressed some of the social implications related with biotechnology and genetic engineering (transgenic organisms, cloning, DNA tests, etc.), only two (TB2 and TB3) mentioned races and racism. However, neither text explicitly addressed biological determinism and the underlying misunderstanding of the model of gene expression. TB2 mentioned the genetic similarity of humans, using it to justify the lack of a scientific base for the notion of human "races": "Human beings are very similar from each other. 99.9% of the genetic data is common to every person, therefore it does not exist a genetic base for the notion of race."

TB3 discussed three issues in a section entitled "Diversity and racism": First, the old idea of "races" as categories to classify human beings. Second, it emphasized that all individuals are different from each other, highlighting it in bold type: "There are not two people totally identical although neither completely different in everything." Third, it claimed that racism is a social but not a scientific problem. Although raising the issue of races is interesting, the text did not explicitly discuss that races, as hierarchical categories, do not exist in a biological sense or make connections to biological determinism.

In summary, the analysis of these five textbooks indicated that all but one provided an adequate definition of phenotype. It is worth noting that in all books, there was little space devoted to explanations or examples of the influence of the environment: a small number of examples per book and only one question about the explanation of the influence of environment in gene expression. For Toulmin (1972) we only understand the scientific meaning of words and notions when we learn to apply them. If students are not required to transfer the model of gene expression to different contexts we cannot know whether they understand it. Concerning biological determinism, two textbooks mentioned the question of races but did not relate it to the model of gene expression.

Analysis of the state-approved curriculum for tenth grade biology (MEC, 2007) revealed little attention to the notion of phenotype. The evaluation criteria emphasized the student ability to distinguish among primary genetics constructs and to solve Mendelian genetics problems. Concerning the social implications of genetics, the curriculum highlighted the capacity of students to analyze critically "the benefits and risks related to modern biotechnology (genetic therapy and transgenic food)." Therefore, it is not surprising that textbooks did not address biological determinism.

The consequences of this analysis for the didactical transposition were that it can be assumed that if teachers rely on textbooks as a primary resource for teaching, as happens in most cases, they will likely find adequate definitions of phenotype but very little help in terms of activities that support student application of this notion. So the teaching sequence needed to include these types of tasks.

Results of the Pilot Study

The pilot study was developed with a group of ninth graders with the purpose of examining the use of evidence and the students' positions toward Watson's claim about genetic differences in intelligence between blacks and whites (Puig & Jiménez-Aleixandre, 2010a). Grade 9 was chosen as it is the last year when science is compulsory for all students and therefore is the highest level of science for about half of the Spanish population. The purpose was to better understand how the general public would critically analyze a determinist claim. The results revealed that students experienced difficulties in recognizing evidence, as for instance the influence of training or other environmental factors. The students demonstrated problems understanding the influence of the environment in gene expression.

How did we use the results of the pilot study to go from the reference knowledge to the teaching sequence? First, the results highlighted the need for devoting more time to developing the model of gene expression, including a detailed explanation of the phenotype notion and examples of the influence of the environment on gene expression in different contexts. Second, the results suggested the need for learning activities that engage students in modeling the relationships between genotype and phenotype and in using evidence. Third, they revealed a need to modify the items in the Watson task that proved the most difficult for the students to interpret.

Time Constraints

The third empirical element that influenced the transposition was limitations in the time available in classrooms. Our initial draft consisted of six to eight sessions to be delivered over 2–3 weeks but after negotiations with the teachers, it was apparent that they were only ready to devote four or five sessions to this issue. As Jiménez-Aleixandre and Sanmartí Puig (1995) pointed out, the reduction of one-third in the number of science hours in the Spanish curriculum in the last decades was not accompanied by a parallel reduction in expected content coverage. Teachers felt that it was difficult to cover all the topics and to attain all the objectives in so little time. Therefore, it was necessary to adapt the tasks to the number of sessions available for each teacher.

The Knowledge to Be Taught: The Designed Sequence

Taking into account the theoretical and empirical elements and the time limitations, the first decision was to devote session 1 to an explanation of genetics concepts and the remaining four sessions to different tasks, two adapted from previously developed materials (tasks 1 and 4), and two designed by us (tasks 2 and 3). Table 12.3 summarizes the tasks and concepts in the full sequence with an indication of the tasks developed in each case. The development of session 1 is discussed in the next section.

- *Task 1 "Take two people"*: The first task, designed for the second session, was adapted from previously developed materials (Dixon, 1982). It provided opportunities for student groups to create models of inheritance. The objectives of the task were to help students: (a) visualize relationships between phenotype and genotype, and (b) reveal the role of chance in the formation and combination of gametes.
- *Task 2 "Athletics"*: The second task required students to establish relationships between eight pieces of information and three different explanations about the

Table 12.3 Tasks and activities in the teaching sequence

Session/case study	Genetics concepts	Type of task/activity	Content and source
1 Case 2, 3	Gene, alleles, dominant, recessive, phenotype, genotype, heterozygote, homozygote	Lecture	Teachers' explanation of genetic notions
2 Case 1, 2, 3	The same as in session 1	Students' perform a modeling task	Task 1: take two people; adapted from Dixon, 1982
3 Case 2,3	Phenotype-genotype relationships, gene-environment interactions	Students are asked to select and use evidence in the context of choosing a causal explanation	Task 2: athletics achievements of black sprinters
4 Pilot study Case 1, 2, 3	Phenotype- genotype relationships, determinist claim	Students are asked to use evidence to evaluate a scientific claim	Task 3: Watson claim about blacks being less intelligent than whites
5 Case 1, 2, 3[a]	Phenotype- genotype relationships	Students are involved in a modeling activity	Task 4: the doughnuts analogy; adapted from Johnson, 1991

[a]In case study 3, the contents of sessions 4 and 5 were collapsed into a longer session

causes of the outstanding achievements of black sprinters. The prompt highlighted the fact that black athletes have placed in the top three positions at each Olympics and World Championship sprint competitions since 1987. The objectives of this activity were to (a) identify evidence for a given claim and connect evidence and claim through justifications for argument building; and (b) apply the model of gene expression in a real-life context. It had originally been designed for at least two sessions, so we had to simplify it in order to meet the teacher's needs in terms of timing.

- *Task 3 "Evaluation of Watson's claim"*: The third task was tested in the pilot study and modified accordingly (Puig & Jiménez-Aleixandre, 2010a). A new item was introduced with data showing the increase in height of Galician men in the last 70 years. Human height is one of the most frequent phenotype examples in textbooks, but some authors (Diehl & Donnelly, 2008) explain it as a consequence of evolution. In another study with eleventh graders (Puig & Jiménez-Aleixandre, 2010b), we found that a 17% of the students considered this increase as an evidence of evolution.
- *Task 4 The doughnuts analogy*: This modeling activity was adapted from Johnson (1991). The activity used something familiar to students (in our case doughnuts, a traditional Spanish sweet) to explain the unfamiliar (phenotype and genotype). The goal was to help students construct meanings for genetics' concepts and to visualize the influence of environment on the phenotype, as doughnuts made with the same recipe could look quite different.

Two classroom cycles comprising three case studies were implemented in order to test and improve the sequence.

From Knowledge to Be Taught to Taught Knowledge

The second step of the didactical transposition consists of how the knowledge to be taught is transformed into the taught knowledge. Figure 12.2 summarizes the different elements that shape this second step, influencing the different forms of the taught knowledge.

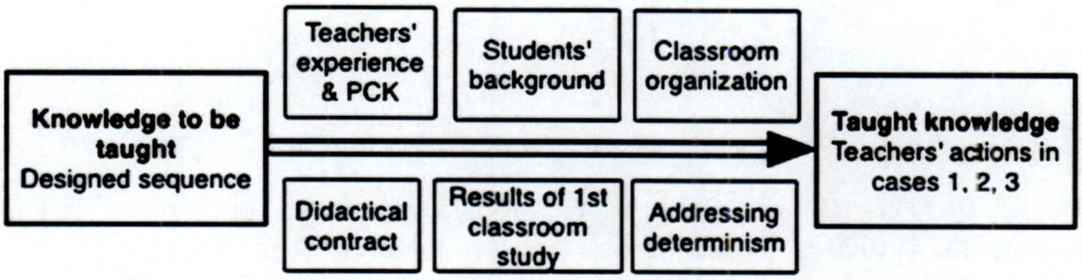

Fig. 12.2 Second step in the didactical transposition: from knowledge to be taught to taught knowledge

Teachers' Experience and Pedagogical Content Knowledge

As discussed in the methods section, the three teachers had different levels of teaching experience and expertise. We framed teacher expertise in terms of Pedagogical Content Knowledge (PCK; Shulman, 1986) or subject matter knowledge for teaching, which includes an understanding of typical student learning difficulties and a repertoire of teaching strategies. The analysis focuses on two teachers Mr. Val (case 2) and Mr. Quiroga (case 3). (Both names are pseudonyms.) Although Mr. Val had more teaching experience, Mr. Quiroga had more sophisticated PCK. Both teachers used TB1.

Students' Background and Previous Experience

Several differences existed among the students in terms of socioeconomic backgrounds and cultural capital (Bourdieu & Passeron, 1970), a notion explaining how social inequalities are reproduced through differences in cultural or symbolic capital (language, accent, dress), influencing academic opportunities and success of students from different backgrounds. There were qualitative differences among the three schools in this dimension, as the cultural capital of students in school A (case study 1) is higher than those from schools B and C (case study 2 and 3). There were also differences in students' experiences working in small groups and engagement in inquiry: the students in case 3 had this experience, while the others did not. In other words, their roles (as perceived or expected by the students) were different. This had consequences for the development of the sequence, as discussed below.

Classroom Organization

The learning environments designed by each teacher differed in their organization (which overlaps with the didactical contract) as seen in the analysis of the teachers' actions. This became especially apparent in task activities that called for cooperative learning and work in small groups.

Didactical Contract

The didactical contract constitutes the reciprocal expectations between teacher and students, the system of norms jointly created in the class. It includes the (usually implicit) roles of students and teachers and classroom management. Although the sequence's design assumed that the students had the responsibility for building models (tasks 1, 4) and arguments (tasks 2, 3), it was also framed in an approach where the teachers supported these practices. The teachers' scaffolding presented a number of differences, discussed below.

Results of the First Classroom Study

The evaluation of case study 1, in two tenth grade classes, helped to identify design problems. An instance was students' difficulties to use evidence, which lead to modifications in the tasks. Therefore, a new version of the sequence (TS2) was developed, to be used in cases 2 and 3. The teacher in case 1 did not consider it necessary to spend one session explaining the genetics concepts including phenotype. Instead, before task 1, "Take two people," he reminded students of the main concepts. Some results of this case study, and consequences for the design were:

- Structure of the sessions: the students could not complete the modeling task "Take two people" because part of the time was used to review genetics' notions. This confirmed our initial design and the need for two sessions, one for explaining the concepts and one for the task.
- Task comprehension: students had some difficulties understanding what they were asked, particularly in one item about black sprinters in the Watson task. The task was split in two, and a new one ("athletics," session 3) was designed, initially for two sessions, including a range of data to illustrate the role of both genes and environment in athletics achievements.
- Use of evidence: as in the pilot study, students struggled to identify the model of genotype expression in the items featured within the Watson task. For instance, some students interpreted human height as an instance of evolution due to mutations.

In summary, the main modifications from TS1 to TS2 were: devoting the first session to instruction about genetics concepts and splitting Watson's task in two, one about athletics, and the other a revised version of the evaluation of Watson's claim, asking students which evidence would be needed to support or rebut it.

Addressing Determinism: Social Implications of Gene-Environment Interactions

One of the three goals underlying the teaching sequence was the development of a critical stance toward biological determinism. Although this dimension was embedded in the design, in particular in tasks 2 and 3, Mr. Val and Mr. Quiroga addressed it differently. Only Mr. Quiroga dealt with it explicitly, first in his presentation in session 1, then in the classroom debate after task 3. Time constraints influenced the two steps in the didactical transposition. In the second step, negotiation with the teachers resulted in four sessions for Mr. Quiroga and five for Mr. Val.

Taught Knowledge: Teachers' Actions

The second objective of the study is the examination of two teachers' actions during the unit about gene expression: in terms of the didactical transposition, the taught knowledge, which is associated with a particular classroom (Tiberghien et al., 2009). This section analyzes the didactical actions focusing on teachers in cases 2 and 3 who taught the same version of the sequence.

First, we examine the teachers' actions and their modes of interaction with students in the first session, during which the teachers directed instruction. Second, their guidance of the students during the tasks in other sessions is analyzed. Third, we examine how they dealt with gene-environment interactions and its social implications. We outline the session development and then analyze the teachers' actions through a revised version of Mortimer and Scott's (2003) scheme. It should be noted that this is not an evaluative analysis; we consider both of these teachers as professionals. The purpose of the analysis is to examine how the same teaching sequence comes to be taught in different ways, thus becoming two different types of taught knowledge.

The Development of Session 1 in Two Classrooms

The development of this 50 min session can be divided into five episodes for Mr. Val and six for Mr. Quiroga, according to the content and the type of discursive moves. The episodes followed a pattern of the consecutive discursive moves of textualization (Mortimer, 2000): *description* as a first approach to a system, object or phenomenon, in this case its definition; *explanation*, which establishes relations between entities and concepts, importing a model to give sense to a specific phenomenon; and *generalization* involving explanations independent from a specific context. To these we add *application*, when the notion is transferred to a new context. The episodes and discursive moves are summarized in Table 12.4.

Episode 1 Introducing the lesson: Definitions. In the case of Mr. Val the episode was primarily a teacher's lecture, with students listening. The predominant discursive move was description or definition. He addressed a question to one student: *Felisa: Would one person be homozygote or heterozygote for all his or her genes? What do you think?* The student answered "No"; but instead of waiting for her justification, he offered an extended explanation.

Mr. Quiroga began by connecting genetics with evolution. He introduced the lesson showing the students a picture of Mendel and a few of Mendel's peas brought from a visit to his laboratory. Then he wrote on the electronic board six terms: genotype, phenotype, allele, homozygote, heterozygote, dominant, recessive. He emphasized the importance of understanding rather than memorizing as they would need to use these ideas to explain different phenomena. Students worked in pairs for 5 min on the definition of the six terms.

Table 12.4 The taught knowledge in session 1: summary of episodes and discursive moves

	Mr. Val	Mr. Quiroga
Episode 1	*Introducing the lesson: Definitions* Teacher defined 14 genetics terms. He asked one yes/ no question to one student.	*Introducing the lesson: Definitions* Teacher contextualized the unit, wrote on the board seven genetics terms and asked the students to work in pairs on their definition.
Episode 2	*Explanation about phenotype* Teacher explained the concept of phenotype with six examples. He addressed two questions to two students, plus one yes/no question, Q3. Students listened; one answered "no" to Q 3.	*Explanation about genotype* Extended dialog (16 turns) among teacher and students about the definition of genotype, with seven questions posed to students. Six students participated in the dialog.
Episode 3	*Genes-environment interaction: Explanation, generalization, and application* Teacher explained the influence of environment in genotype expression, providing six examples. He posed two yes/ no questions, and asked for more examples about gene expression. Students offered three examples.	*Gene-environment interactions: Explanation, generalization, and application* Extended dialog (73 turns) among teacher and students about whether genotype is always manifested in the phenotype. Teacher used two analogies: building drafts and music scores. He asked 30 questions and introduced a metacognitive reflection. Twelve students participated and offered examples about how environment can modify gene expression
Episode 4	*Other genetics concepts* Teacher explained the relationship among dominant/ recessive alleles and phenotype, providing three examples. He asked two questions. Students answered the questions.	*Other genetics concepts* Three teacher-student dialogs of 12, 11, and 14 turns about alleles, homozygote/ heterozygote, and dominant/recessive. Teacher asked 15 questions. Six students replied to the questions.
Episode 5	*Wrapping up and checking questions* Researcher solicited instances of phenotype. Four students answered; they provided the same examples used by the teacher.	*Wrapping up and checking questions* Researcher solicited examples of phenotype. Teacher asked three questions. Eleven students answered the questions.
Episode 6	–	*Social implications of gene-environment interactions* Teacher explicitly addressed determinist views. He asked nine questions about whether traits and behavior depend from genes or from environment. Extended dialog (30 turns) teacher - students.

Episode 2 Explanations about genotype/phenotype: Mr. Val began by asking one student to define phenotype, and followed the same pattern of explaining without waiting for her answer. Mr. Val provided three examples of phenotypic traits, observable (eye color), behavior (mice), and biochemical (lactose intolerance). He explained why phenotypes do not always exhibit in identical ways in parents and offspring. He posed a question about whether phenotype is inherited, to which the students answered "No." Mr. Quiroga initiated a dialog with the students about these terms beginning with genotype after the students finished working on the definitions. He picked up their responses asking about where genes are located and the origin of the genes in our body.

Episode 3 Explanation, generalization, and application about gene-environment interactions: This was the longest episode in both classes, taking about half of the session's time. As gene-environment interactions are the focus of this study, the way each teacher dealt with them are subjected to a detailed analysis in the section about students' processes of construction.

Episode 4 Other genetics concepts: Mr. Val explained the relationship among dominant/ recessive alleles and phenotype, providing three examples. There was a brief dialog initiated by the teacher asking questions related with the concepts of homozygote and heterozygote. Mr. Quiroga and the students reviewed the same concepts, in a Socratic dialog.

Episode 5 Wrapping up and checking questions: In both cases, the researcher asked students about the differences between genotype and phenotype and solicited instances of phenotypic traits. The students of Mr. Val offered the same examples used by the teacher. The students from Mr. Quiroga offered other examples.

Episode 6 Social implications of gene-environment interactions: Mr. Quiroga explicitly addressed determinist views about behavior. In an extended dialog (30 turns) the teacher probed students' understanding with questions about whether traits and behavior depend on genes or come from the environment. There was substantial teacher feedback regarding student answers.

Teachers' Guidance of Students in Tasks in Sessions 2, 3, 4, and 5

There were some general differences related to the didactical contracts in both classrooms. For instance, in Mr. Quiroga's class it was apparent that students were used to working in small groups and were expected to express their opinions and to participate in the debates, whereas these activities were not apparent in Mr. Val's class. Mr. Val's students had difficulties in understanding the purpose of some of the tasks. They needed more support and the researcher had to step in more times as compared to Mr. Quiroga's class. A second difference was that in Mr. Quiroga's class all the students were required to carry out inquiry projects in small groups about their own questions (Jiménez-Aleixandre & Fernández, 2010). There had been much explicit talk in the classroom about working as scientists do, and as a result they were familiar with notions related to scientific work.

Task 1 Take two people: The approaches to introduce this activity were noticeably different. Mr. Val explained what students needed to do, but he did not establish links to the concepts from the first session. Students did not complete the task within the allotted class time, so there was no opportunity for debriefing. Mr. Quiroga began by relating the task to some concepts from the previous session, highlighting genotype-phenotype relationships, and gene-environment interactions.

Task 2 Athletics: Mr. Val read the handout aloud, without clarifying the meaning of each question. He assisted two of the small groups that specifically requested clarification of question 2. Mr. Val highlighted the importance of listening to all views and reaching consensus before writing the report.

Mr. Quiroga began by relating the task to the notion of phenotype and some of the concepts from session 1. He and the researcher cooperated in clarifying question 2. Mr. Quiroga framed the task in scientific practices, making explicit the similarities between this task and scientists' practices, in particular the process of decision making through the use of evidence. Twice during the session Mr. Quiroga initiated a whole class discussion. At the end of the session, he recontextualized the task, asking students about the concepts important for carrying it out.

Task 3 Evaluation of Watson's claim: Mr. Val's students answered individually during session 4 and held a debate in session 5. Mr. Quiroga asked students to briefly discuss the task in small groups and then to write their conclusions. Then he initiated a whole-class debate about the influence of environment on intellectual achievements. He introduced a metacognitive reflection: "We are discussing a definition that is in textbooks and sometimes is not understood: the phenotype is the result of the interactions genotype-environment."

Task 4 The doughnuts analogy: This task was reduced to a part of the last session in both cases. The students brought doughnuts that they had baked. The main difference was that at the end of the session Mr. Quiroga asked students again about the meaning of phenotype and solicited more examples; Mr. Val did not.

How Teachers Dealt with Biological Determinism

The teaching of gene-environment interactions and the process of students' construction of explanations about it, that is teacher-students joint productions, are analyzed in the section about students. It should be noted that the ways both teachers addressed biological determinism were very different. Mr. Val highlighted the influence of environment in the phenotype, providing many examples. However, he never explicitly addressed biological determinism.

Mr. Quiroga explicitly addressed biological determinism in episode 6 of the first session. He mentioned social views about the genetic bases for alcoholism or aggressive behavior. He probed students about this issue, but at this stage he refrained from making explicit his own views about gene-environment interactions. In task 3 he initiated a discussion about intelligence and its basis. Mr. Quiroga also made references to the influence of the environment in gene expression in all the other sessions. He followed up this issue along the sequence, making explicit the continuity of the different tasks and sessions.

Characterization of Didactical Contracts and Their Relevance for SSI-Based Education

For the purpose of characterizing the didactical contracts created in the classroom, an adaptation of Mortimer and Scott's (2003) scheme was used. The scheme addressed two dimensions: the content knowledge including not only genetics but also scientific practices and the communicative approach. It is summarized in Table 12.5.

Table 12.5 Summary of the analysis of the classrooms joint productions in session 1

	Mr. Val	Mr. Quiroga
Content knowledge		
Genetics		
Conceptual load	More concepts (14)	Fewer concepts (7)
Progression of knowledge: how are concepts introduced	All concepts defined and explained by the teacher	Concepts developed through interactions
Genetics knowledge context	Mainly scientific	Combination of scientific and everyday contexts
Use of analogies	–	2 (building drafts, music score)
Use of examples	Teacher provided 17 examples	Teacher asked students for examples
Scientific practices (sessions 2 to 5)	– Students' lack of experience in scientific practices – No references to the role of evidence	– Students' previous experience in inquiry – References to the role of evidence
Communicative approach		
Questions posed to students	$N=12$	$N=65$
Rhetorical	6	7
Application and extension	4	20
Evaluation	2	38
How the teacher took into account students' answers	Teacher answered the questions for students; he did not change his discourse	Teacher developed students' answers; he changed his discourse in some cases
Interaction patterns	Less dialogic, less interactive	More dialogic, more interactive
Teacher's turns	31	86
Students' turns	19	84

Content Knowledge

As the reference knowledge, the content knowledge is composed of two strands: genetics and scientific practices. The first strand explores (a) the conceptual load, (b) how did genetics knowledge progress in session 1, (c) in which context, and (d) the use of analogies and examples. The second strand examines how the development of scientific practices proceeded in other sessions.

As seen in Table 12.5, there were substantial differences between both classrooms. Mr. Val explained more genetics concepts, all of them introduced by him. In contrast, Mr. Quiroga addressed fewer concepts, developing them in interaction with students, so the progression of knowledge took place through social discourse. Concerning the context of the explanations and the use of analogies and examples, Mr. Val focused on the scientific meaning of the genetics notions offering many examples. Mr. Quiroga connected these notions with real-life situations and asked students for examples. He used two analogies in order to clarify the model of gene expression and the relationships between genotype and phenotype.

The development of the scientific practice of using evidence is analyzed as a joint production of teacher and students (Tiberghien et al., 2009). Work on the tasks was influenced by students' previous experiences in the classroom. Mr. Quiroga's students had experience working in small groups, carrying out inquiry projects requiring them to collect and evaluate evidence, and participate in debates. In contrast, Mr. Val's students did not have these experiences; therefore, the level of support required in the two contexts was different. Learning scientific practices combines elements of practice and metaknowledge. Mr. Quiroga explicitly framed the tasks in scientific practices and the use of evidence, for instance, in task 3.

Communicative Approach

The analysis, focusing on session 1, attended to three interrelated aspects, (a) the type and number of teacher questions; (b) how the teachers account for students' answers; and (c) the interaction patterns. There were differences in the number and type of questions: Mr. Quiroga asked five times more questions than Mr. Val, and most of them were either evaluation or application and extension questions. Half of the questions posed by Mr. Val were rhetorical or confirmation questions that did not require elaboration from the students.

There were even greater differences in how the teachers reacted to students' answers: Mr. Val gave very little time for students' responses; on five occasions he answered his own questions, and in the others he did not develop on students' answers, solicit extended explanations, or explore their meaning. Mr. Quiroga developed students' answers, taking some time to discuss them. While Mr. Val's discourse did not experience any change due to interactions with students,

Mr. Quiroga modified his explanations when he detected problems, as for instance when the students were not able to give an example of genes not being expressed in the phenotype. He then presented a second analogy, a music score, which was more successful.

Concerning the interaction patterns, Mr. Quiroga interacted more with students as seen in the number of students' turns, 84, while Mr. Val did not provide many chances for students to participate: there were 19 students' turns in session 1. A second aspect in the interaction patterns is whether the approach is dialogic or not. According to Mortimer and Scott (2003) who draw from Bakhtin, what makes talk functionally dialogic is the fact that more than one point of view is represented and ideas are developed, even when this talk is produced by an individual. It is in this sense that Mr. Quiroga's approach was more dialogic because he took into account students' answers to carry out the lesson. An example is presented in the following excerpt that was taken from episode 3 after the introduction of the music score analogy and the question about what influences gene expression. Mr. Quiroga presented the example of a plant that could reach 8 m, but not in an inadequate environment:

Mr. Quiroga:	If I would plant it there: What would happen?
Students:	It would not grow.
Mr. Quiroga:	It would never grow up to eight meters. Why not?
Students:	Because of the environment.
Mr. Quiroga:	Because the environment does not allow it [...] Look, give me another example about plants, animals or people.
Cristina:	Someone who is prepared to be muscular, but for instance is born in an underdeveloped country.
Mr. Quiroga:	When there is not enough to eat...
Cristina:	... is not going to develop the musculature.
Mr. Quiroga:	A good example. The environment, in this case not enough food to eat, is preventing the development of the amount of proteins that was planned in his or her genes. Good. Another example? One that is not about humans.

In this instance, the teacher built on students' answers to develop the target concepts. We can summarize the classroom discourse patterns in this first session, by saying that in the case of Mr. Val it consisted of detailed explanations interrupted by short question-answer exchanges. The teacher retained the responsibility of the progress of knowledge; he was more concerned about students' understanding of the meaning of genetics notions than about engaging them in its construction. In the case of Mr. Quiroga, there was little time devoted to the teacher lecturing and students listening. The students participated in the definition of each concept and discussed its meaning with the teacher. The teacher modified his discourse when he perceived problems in understanding or applying it, as seen with the first analogy. The progression of knowledge took place through interactive processes between teacher and students. The task of knowledge construction was shared with the students.

Didactical Contracts

As a consequence of these differences, the didactical contracts were very different in the two classes. The didactical contract is the set of teacher behaviors that are expected by students and vice-versa. As Sensevy (2007) points out, this notion positions educational actions as essentially communicative because the actions are grounded in students' interpretation of the situation.

We interpret that, in Mr. Val's case, the students expected the teacher to explain without being interrupted, and sometimes they did not even attempt to answer his questions. They seemed to perceive their own role as one of reproducing the notions and examples presented rather than one of producing new examples or applying the concepts. For instance, when asked about phenotype examples, they offered the same ones presented by the teacher. In Mr. Quiroga's case, the students expected the teacher to interact with them. They answered all the questions posed, sometimes with errors or advancing inadequate examples. The students did not exhibit apprehension in the expression of ideas even when those ideas were contradictory to their teacher's comments. When asked about examples, they knew that they were expected to produce new ones and to apply the concepts, as the teacher explicitly said, in a reflection reproduced above, and they attempted to do so.

The analysis of the development of the tasks in sessions 2, 3, 4, and 5 shows differences in the systems of norms in each classroom: Mr. Quiroga's students perceived their roles to include work in small groups, discussions of different pieces of evidence, and participation in debate. By contrast, for Mr. Val's students these types of tasks seemed not to be part of what they were expected to do, so they asked for clarification more times and required more guidance.

The Relevance of Different Didactical Contracts for SSI-Based Education

In this section we have examined how different the taught knowledge, the instructional approaches, and the didactical contracts can be in two classrooms working with the same teaching sequence and using the same textbook. The question now is: What is the relevance of these differences for SSI-based education? In particular: Are these differences significant for the goals of supporting students in (a) building arguments about the influence of environment in gene expression, and (b) the development of a critical stance toward biological determinism?

In order to answer it in full, it is necessary to analyze how students used the knowledge in making sense of the tasks, which is the focus of the next section. However, drawing from what has been discussed so far, we suggest that the differences have significance for teaching socio-scientific issues in three interrelated dimensions:

1. *Students' autonomy and empowerment*: Goals of SSI-based education include taking an active role to solve controversies, or developing an independent opinion.

As Tiberghien (2008) points out, when introducing SSI in the classroom, we are concerned both with scientific literacy and with citizenship education. In this sense, we interpret that the didactical contract created in Mr. Quiroga's classroom promoted development of students' autonomy to a higher degree. The students' expectations about their own role were to participate in the discourse, to advance their own opinions, and to express their ideas. Developing an opinion and the capacity to participate in society requires having the opportunity to make public positions, to discuss them with others, and to evaluate the evidence supporting them. The students in Mr. Val's classroom had few opportunities to do so.

2. *Students' construction of meanings about the topics discussed*: SSI have social relevance but are also scientific in nature. In order to criticize determinist views, students need to understand causal explanations about phenotypic traits, the influence of the environment in gene expression, and its implications for human performances. This may require conceptual change, modification or refinement of their previous ideas. For this purpose, it is necessary that students' ideas be elicited and that, when made public, the teacher takes them into account. In this dimension, Mr. Quiroga's approach was more aligned with strategies seeking to promote students' construction of their own meanings. He developed ideas suggested in students' answers and changed his discourse, when needed, to address them. In his classroom, work in small groups provided a context for developing and applying ideas. In Mr. Val's classroom the discourse was dominated by the teacher.
3. *Students' acknowledgment of the existence of two views with different social implications*: Working with SSI in the classroom involves, in many cases, supporting students in the acknowledgment of the existence of two positions on a dilemma, two different courses of action, and two views about one issue. In the case explored in this paper, biological determinism and gene-environment interaction are opposing views with divergent social implications. For students, it may not seem contradictory to learn that environment influences gene expression while retaining a view explaining all traits and performances as a consequence of genotype. We suggest that, in order to support this acknowledgement, the opposing views should be explicitly addressed in the classroom. In our case, this means talking explicitly about biological determinism. Only Mr. Quiroga did so.

In summary, we interpret that Mr. Quiroga's didactical contract was more adequate for the specific goals related to teaching about one issue with social implications.

Students' Construction of Explanations Acknowledging Gene-Environment Interaction

The third objective of our study was to examine difficulties students had with the construction of explanations acknowledging gene-environment interactions. We sought to document the meanings that students gave to phenotype and to the model of interaction taught in the sequence and the difficulties encountered in the application of this model to real life contexts. For this purpose, we examine here (a) students' difficulties evidenced during the introduction of the influence of environment on gene expression in the first session; (b) students' application of this knowledge to explanations about human performances, like athletics in task 2; (c) the written results of a retest administered 5 months after. We focus on the students of cases 2 (small groups 2F, 2G, 2H, 2I) taught by Mr. Val and 3 (small groups 3A, 3B, 3C, 3D, 3E) taught by Mr. Quiroga.

Discursive Moves and Obstacles in Gene-Environment Interactions in Session 1

The analysis of teacher-students interactions in the context of the introduction of the influence of environment on gene expression (episode 3 in session 1) showed quantitative and qualitative differences. In Mr. Val's case, the input from students was minimal: six turns of speech out of 17. In Mr. Quiroga's case, about half (36) of the 73 speech turns corresponded to the students.

In the case of Mr. Val's classroom, episode 3 can be divided into three teacher discursive moves following the pattern definition – explanation – application. First, he defined genotype emphasizing heritability and summarized the differences with phenotypes and provided an example (Drosophila). The second move, explanation, began with Mr. Val showing students two photographs of a well-known Spanish politician before and after a period of physical training and asked students whether his genes had changed. After they answered "no," he stated that *His genotype did not change, but his appearance, his phenotype, did change*. Mr. Val posed a second question: *Could the phenotype change without changes in the genotype?* This confirmation question was answered with a simple "yes," and the teacher extended the notion himself, explaining the influence of environment in the expression of genes and offered six examples, two of which are presented here: the lack of vitamin D as a causal mechanism for rickets, and the effects of temperature of incubation on the sex of crocodile hatchlings. In the third move, Mr. Val asked students about other cases of environmental influence, and four students provided examples including obesity and tanning.

Mr. Quiroga and his students engaged in an extended dialog, which can be divided into four discursive moves, which revealed students' difficulties, summarized in Fig. 12.3. The teacher began the first move, definition, by asking students about phenotype. One of the students suggested that phenotype is "what is manifested." Mr. Quiroga followed up with this idea and extended the discussion by asking students about differences between genotype and phenotype. A second student proposed eye color as an example of phenotype "because it is what comes out outside, in our physic." We interpret these statements (and others during task 2) as representative of understandings of genotype–phenotype relationships as making visible or external ("coming out outside") what is invisible and internal ("in the cells"). This is associated with a notion of the genes as entities whose expression is fixed, not modulated.

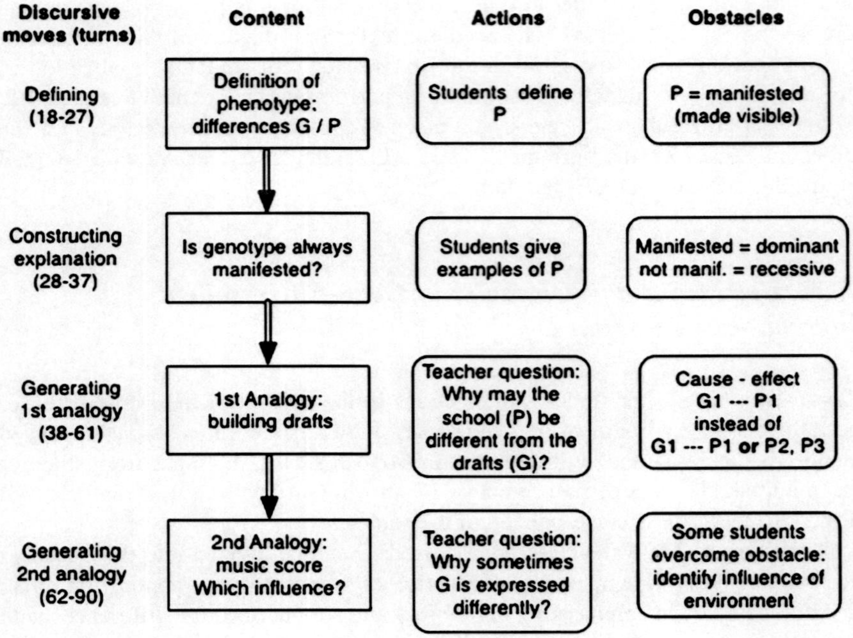

Fig. 12.3 Discursive moves and obstacles about the model of gene-environment interactions in Mr. Quiroga's classroom (*G* Genotype, *P* Phenotype)

The second move, intended to support construction of an explanation about environmental influences, began when the teacher asked if the genotype is always manifested or expressed in the phenotype. Although the students answered "no," their examples (eye color and recessive diseases) it was clear that the only mechanism they could postulate for the nonexpression of a gene was through a recessive allele. They framed the teacher's question in terms of dominant (expressed) versus recessive (not expressed) alleles. Although Mr. Quiroga clarified that the genotype would be constituted by both alleles, they continued giving this type of examples. We interpret that the students gave to "manifested" or "exhibited" the meaning of "dominant" because it corresponds to Mendelian genetics notions.

When Mr. Quiroga saw that the students did not understand his question, he began a third move, using his first analogy: the coincidence or lack of coincidence between the building drafts of a school and the actual school. He asked about possible reasons for the discrepancies, encouraging students to establish correspondence between source and target: *So, which one is genotype and which one phenotype?* He solicited examples from biology, to which students offered mutation, and again the presence of two alleles (blue and brown) in a genome. Extending

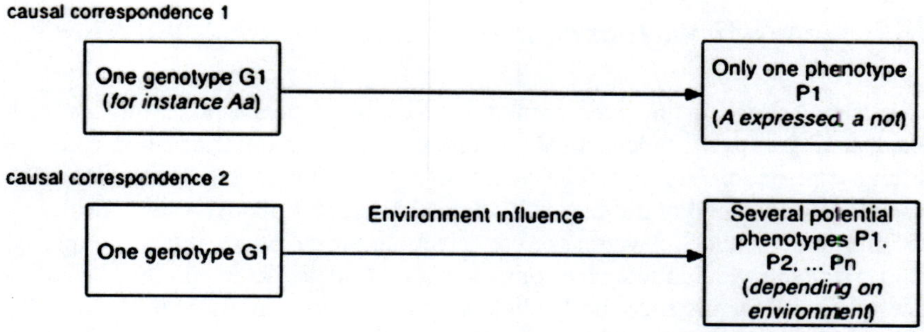

Fig. 12.4 Two models of causal correspondence between genotype (*G*) and phenotype (*P*)

his analogy, he said that in a building draft there could be a room painted brown over a previous blue color: *What I am saying is something that was in the draft but then is not coming out*. We interpret these difficulties to apply the concept of phenotype to a simple, straightforward causal correspondence where one cause produces one given outcome (1), as opposed to the notion of gene-environment interactions, which is more complex in that one causal factor may produce multiple outcomes (2) (see Fig. 12.4).

The awareness of these difficulties led to a fourth move, introduced by a metacognitive reflection by the teacher: *See, this is an instance about how you can clearly define a concept, but you don't understand it. Because until you are able to give me an example, it means that you have not internalized the concept. It is not the same to memorize it than to understand it.* Mr. Quiroga then presented a second analogy, a music score. He asked whether the sound would be the same if a music score was played by different people. This second analogy was more successful, and then he arrived at the episode's central question: *What is this (factor), which has influence and is not in the genes?* After the students answered "environment," Mr. Quiroga offered one example of how plants would not reach its regular height in inadequate conditions and asked students for other examples. Interestingly, the first suggested by Cristina was: *Someone who is prepared to be muscular, but for instance is born in an underdeveloped country*. Some turns of this episode were reproduced in the previous section.

From the analysis of this session, three interrelated problems emerged: (a) an understanding of a genotype being expressed as simply "coming out"; (b) a confusion between "expressed" and "dominant"; and (c) a simple causal correspondence, one cause – only one possible effect. Perhaps Mr. Val's students had similar problems, but there were not enough opportunities for them to become apparent. The analysis of the subsequent sessions, particularly when students were discussing the tasks in small groups, shed light on how in some cases the students overcame these problems, while in others the difficulties persisted.

Explanations About Human Performances in Task 2 "Athletics"

As discussed above, in the description of the teaching sequence, the "Athletics" task asked small groups of students to align eight different sets of information with one of three explanations for the achievements of black sprinters: (A) genes; (B) influence of factors as food or training; (C) a combination of A and B. After considering the information, students were asked to select one of the options. It was expected (a) that students would recognize some pieces of evidence as supporting the influence of genes, for instance the R allele of the ACTN3 gene in chromosome 11, related to fast twitch fibers; (b) that they would recognize other pieces as supporting the influence of environment, such as a table showing race and place of origin for recent medal winners. We distributed the explanations of the nine small groups (noting students with diverging positions) into three categories or stages in the acknowledgment of gene-environment interactions.

Category 1. Genes Are Solely Responsible for the Performances

Two groups, 2H and 2I did not progress beyond this stage in task 2. Group H2 was composed of five students. Although two of them, Hilario and Henrique, mentioned environmental influences at the beginning of the exchange, Hector and Hugo dominated the debate:

Hugo: Is this one [option A, genes], I believe, because it says that all of them [Olympic medalists] are blacks, but not all were born in the same country. They may be from the US, from England.

Hector: It doesn't matter, as they have the genes of blacks…

Hugo: Blacks are better than whites because it doesn't matter where they were born; they always will arrive first.

Hilario: I believe that being black does help, but it also depends on where do you train because it is not the same to train in Africa than in the US.

Hector: Training sometimes does not [influence], because it may be the case that Usain Bolt runs without training and he wins as well.

Researcher: And why it is so?

Hector: (…) It is because of the genes, Usain Bolt's genes.

Group 2I was composed of four students, with two of them, Iolanda and Irma, dominating the debate and the other two speaking very little. The idea of environmental influences was mentioned, but the group ultimately settled on the genes-only explanation.

Irma: It is A, the genes.

Iolanda: And they are all born in the region of… out of Africa, in the US.

Researcher: And, what does that mean for you?

Iolanda: That they are not from Africa!

Isabel: That… it has to do also with climate.

Iolanda: That it has nothing to do with climate! It has to do with genes!
Alba: Why are all of them black and why does not win a white?
Irma: Because they have it in their genes! And whites do have other genes!

Hugo and Iolanda interpreted the information of the table about Olympic medalists, not as a conspicuous absence of African-trained athletes (although the teacher highlighted this absence at the beginning of the task), but as evidence for the preeminence of genes over environment, "it doesn't matter where they were born." Hector in group 2H as well as Irma and Iolanda attributed the performances only to genes. We suggest that these problems are related to three issues: (a) to social representations about deep genetic differences between blacks and whites; (b) to a confirmation bias, in which individuals attend only to aspects of available information that supports a particular hypothesis; (c) to a greater difficulty in perceiving something missing (black sprinters from Africa) than something added (Jiménez-Aleixandre & Pereiro, 2002); (d) to difficulties in making sense of information presented, for instance, some students did not identify the country where an athlete was raised/trained as an environmental factor.

Category 2. Genes and Environment Influence Separately, but Genes are More Relevant

Two groups, 3A and 3C recognized that environment has an influence, but they treated genes and environment as separate factors with genes dominating over environment. One excerpt from group 3C, composed of five students, illustrates this type of explanation.

Roi: I think that it is A [genes]. Because a white may train too, whites eat too, man. Then it has to be because of genes [...]
Bernal: But on this ground [the information], then, I don't know...
Roi: It means that on equal terms, blacks are better.
Rosendo: I think that it is A because on equal terms black people are very superior to white people, and this is due to their genes, because food and environment could be made equal both for blacks and whites, so the only remaining difference would be the genes.
Bernal: Yes, I agree.
Roi: It has to do with genes.
Cristina: And, how do you know that the food is the same?
Roi: Look, if they are going to compete, they would strive to eat as best as they can.

It seems that for these students environmental circumstances, food or training, may be modulated, modified, and "made equal." On the other hand they neither accept a modulation in gene expression, nor the existence of unequal environmental circumstances that may modify performances. For them genes and environment seem to have different status.

Category 3. Combined Influence of Genes and Environment

Five groups, 3B, 3D, 3E, 2F, and 2G, acknowledged the combined influence of genes and environment in their explanation, although they arrived at this conclusion through different processes. Groups 3B and 3E derived their explanation from data while groups 3D, 2F, and 2G chose an explanation before discussing the significance of each piece of information. This is demonstrated in the excerpt below from group 3E:

Roman: It has to be both [genes and environment], because here [Olympic medalists] it talks about their nations, US, US, Jamaica… And that is environment! Because of the place where they are raised. Besides, it is also genetics, because their skin color is black, black, black. And they are all trained in English-speaking countries, except the Jamaican one and that is environment.

Ernesto: I believe that it is both. Yes, because even if you are very good, you need to eat and to train.

Estrela: Besides, they all are in locations where everybody can work and to develop that…

Rexina: Look, I think that it is [option] C.

In this group, students accept the role of genes, and they also identify the relevance of the environment particularly in terms of diet and training, using the data from the table as evidence to support their choice.

In summary, not all small groups progressed to the stage of acknowledging the influences of both genes and environment in their explanations. In two groups, both from Mr. Val's class, the predominant discourse was an attribution to genetics as the only determinant on athletic performance. Two other groups, from Mr. Quiroga's, class recognized the role of environment, but saw it as subordinate to genes. Five groups, three taught by Mr. Quiroga, and two by Mr. Val, arrived at a combined explanation, although they did not express it in terms of interaction. The small number of groups prevents any generalization; however, the fact that both groups that selected the gene-only explanation came from the same class may support a link between this outcome and the teaching approach. In this class, students had few opportunities to develop their ideas and there was no explicit discussion of determinist views.

The analysis of the oral discussions about task 3, "Watson" which asked students to evaluate a claim about differences in intelligence among races showed that almost all groups maintained positions coherent with their explanations in task 2. All explanations and student statements acknowledged the role of environment in intelligence, except those from group 2H, which claimed that intelligence depended only on genotype. When asked to justify that claim, Henrique said, "You are born with a capacity," and Helena added "You may do as much as you can, but you cannot develop it more than what you have." It seems that Henrique changed her position, which, in task 2, acknowledged the influence of environment. This may be related to the group leadership by Hugo and Hector who strongly supported the genetics-only option.

Table 12.6 Categories in the answer to the retest

Categories	Case 2 (N=15)	Case 3 (N=31)
3. Gene-environment interaction	1	3
2. Influence of environment	14	27
1. Genes only	–	1

Written Results of a Retlest

Five months after the completion of the unit, at the end of the school term, a question about identical twins raised in different countries was included in a final written test. The item asked whether at 16 both twins would have similar or different height and muscles and would develop similar or different reading and mathematical skills (see Table 12.6).

All students but one in both classes acknowledged the influence of environment in physical traits, as well as in intellectual achievements. However, only four of them integrated both genotype and environment in their responses, for instance:

> Braulio (3D): No [it would not be the same], because departing from the same genotype it would not be equally developed in favorable conditions than in precarious circumstances. They would not be identical because their phenotype is different; in B [raised in a developed nation] it could be developed optimally and in A [raised in an underdeveloped nation] would be limited.

The majority of students acknowledged the influence of environment, but did not mention genes or genotype. We interpret that they took for granted the role of genes, as the question was about identical twins. There was only one student, Camilo (3A), who had supported in all the tasks the notion that only genotype mattered, and continued supporting it:

> Camilo: The physical traits would be the same for A and B, because environmental conditions cannot alter their genes.

In summary, it seems that all students but one took into account environment in their explanations. This may suggest that the teaching sequence was successful in both classes in overcoming initial explanations that in some cases were close to determinism. However, we also need to consider that the academic context, a final examination, may prompt some students to give the answer that they believed that the teacher expected.

Discussion and Implications

Most topics in genetics are difficult for students, but some of them have received greater attention in the literature. Our work seeks to add to this knowledge about genetics teaching and learning an examination of the challenges posed by the model of gene expression, in particular, the interactions between genotype and environment, resulting in the phenotype traits. This model has social implications as biological determinism ignores environmental influences presenting human traits and performances as solely determined by genes, a view lending support to political agendas that challenge notions of equity.

This chapter presents a developmental study in five classrooms, framed in Chevallard's (1991) notion of didactical transposition, or transformation from reference to taught knowledge. First we discussed the process of transformation of the reference knowledge, including genetics and scientific practices, into the knowledge to be taught. This transformation draws from goals related to science and to citizenship education, and from an analysis of textbooks, revealing little attention to the application of the model. Second, we analyzed how this designed sequence was transformed in two different types of taught knowledge in two classrooms. Third, we examined the process of construction by students of explanations taking into account gene-environment interaction.

The analysis of the taught knowledge is characterized through the notion of didactical contract (Brousseau, 1998), students' expectations about the teacher and teacher's expectations about the students. The examination of several dimensions, collapsed in content knowledge and communicative approach, showed substantial differences between the didactical contracts in both classrooms. In the case of Mr. Val's class, students expected the teacher to be in charge and lecture, and they perceived their own role as one of reproduction. The students requested more guidance when they were required to engage in modeling and using evidence because they usually were not expected to collaborate in group work. The expectations of the teacher seemed to be that the students would listen to him and respond "yes" or "no" to his questions. On the other hand, he is a good professional, holding students' attention all the time, and providing a structured explanation.

The case of Mr. Quiroga was different, both in content knowledge, as he was guiding the concept development trough interactions and in the communicative approach which was more dialogic and interactive than Mr. Val. A relevant dimension is how he took into account students' answers, changing his discourse in reaction to students' problems with the notion of phenotype. He expected students to participate and the students' expectations were to be engaged in all the textualization moves: they defined concepts, explained their meaning, and tried to apply them, although not without difficulties. They seemed to feel comfortable openly expressing ideas even when those ideas contradicted the teacher because there were no negative academic consequences when they said something wrong.

An important question is the relevance of the two different teaching approaches for the goals of SSI-based education. We suggest that didactical contracts like the one Mr. Quiroga created in his classroom may support (a) students' autonomy and the development of independent opinions; (b) students' acknowledgment of the existence of two views with different social implications; and (c) the construction of appropriate meanings for the conceptual topic underlying the SSI. While the two first dimensions are common to all or most SSI, our topic, biological determinism, may be different in requiring mastering the model of gene expression. However, we believe that most SSI also require a deep understanding of the concepts involved and, for this purpose, engaging students in active participation is an important feature of teaching.

The analysis of discourse in both classrooms enabled identification of student difficulties. These difficulties included meaning construction for phenotype and application of this concept to real life contexts. The discursive moves used by Mr. Quiroga and his students illustrated both the students' difficulties for constructing appropriate meaning about one notion (that genotype can be sometimes expressed in different ways due to environmental factors) and the changes introduced by the teacher in his discourse for overcoming them. We interpret that these difficulties, and the students' confusion between "not manifested genotype" and "recessive alleles," are related to the fact that the linear causal correspondence, G1–P1, is more intuitive than the idea of a complex correspondence where one genotype may yield several potential phenotypes, G1– P1 or P2, or… Pn. This may be related to epistemological assumptions about causal mechanisms and to difficulties in accepting uncertainty. The findings suggest that this complex question needs to be taught through application in different settings and that a simple lecture is not sufficient.

These epistemological assumptions may be the reason why the first analogy (building drafts) was not successful, as usually there is only one building corresponding to one draft. The second analogy, music score, helped students to understand and apply the notion, probably because they saw that a score could be played differently by different people. This may suggest features for successful analogies when teaching this issue.

The analysis of small group discussions during completion of tasks revealed variance in the application of the notion of interplay among genes and environment to a real context. Five groups acknowledged that both genes and environment had an influence on performances, but for two other groups genes would dominate over environmental influence. The remaining two groups (as well as some individual students, disagreeing with their group) accepted only the effect of genes, a position that closely approximates biological determinism. We suggest that these difficulties may be related to social representations that assume deep genetic differences between blacks and whites, as well as to a high status of genetics in the social imagery. Genetics is equated to "scientific"; therefore, the influence of genes is not to be doubted.

The initial design of the teaching sequence contained more time and tasks related to the use of evidence. It is a complex scientific practice that scarcely could be developed in a few sessions. The students' difficulties with the Athletics task points to the need to devote more sessions to engage students in selecting and using evidence. Educational authorities, in Spain and elsewhere, need to be aware that introducing recommendations about argumentation and use of evidence in the curriculum is not enough when teachers have too much content to be covered.

A final remark concerns the potential utility of didactical transposition and didactical contract to frame analysis of teaching sequences and teachers' actions. It is important to make public the goals and decisions underlying particular tasks and activities. To reflect about how reference knowledge is transformed in a teaching sequence may help teachers to decide which issues need more time and emphasis. An instance could be the scientific practices of modeling and using evidence, which make part of the reference knowledge in this sequence. Didactical contracts conceive educational actions as essentially communicative, and in this sense are aligned with current attention to communication in the classrooms and to the development of the scientific practice of knowledge communication.

Acknowledgments Work supported by the Spanish Ministry of Science and Innovation, code EDU2009–13890-C02–01. Blanca Puig's work supported by a University of Santiago de Compostela predoctoral grant, partly funded by S-TEAM (Science Teacher Education Advanced Methods), code SIS-CT-2009–234870, project financed by the European Union, 7th Framework Program. The authors thank Troy Sadler for his helpful suggestions to the first draft.

References

Berland, L. K., & Reiser, B. J. (2009). Making sense of argumentation and explanations. *Science Education, 93*, 26–55.

Bourdieu, P., & Passeron, J.-C. (1970). *La reproduction. Eléments pour une théorie du système d'enseignement*. Paris: Les Éditions de Minuit. Translated as: Reproduction in education, society and culture. London: Sage, 1977.

Brousseau, G. (1998). *Théorie des situations didactiques*. Grenoble, France: La Pensée Sauvage.

Chevallard, Y. (1991). *La transposition didactique [Didactical transposition]* (2nd ed.). Grenoble, France: La Pensée Sauvage.

Cuvier, G. (1817). Extrait d'observations faites sur le cadavre d'une femme connue à Paris et à Londres sous le nom de Vénus Hottentotte. *Mémoires du Muséum d'Histoire naturelle, 3*, 259–274.

Diehl, D., & Donnelly, M. P. (2008). *Inventors and impostors. How history forgot the true heroes of invention and discovery*. Richmond: Crimson.

Dixon, R. (1982). Take two people. A genetics teaching kit. *Journal of Biological Education, 16*(4), 229–230.

Duncan, R. G., & Reiser, B. J. (2007). Reasoning across ontologically distinct levels: Students' understandings of molecular genetics. *Journal of Research in Science Teaching, 44*(7), 938–959.

Duncan, R. G., Rogat, A. D., & Yarden, A. (2009). A learning progression for deepening students' understanding of modern genetics across the 5th–10th grades. *Journal of Research in Science Teaching, 46*(6), 655–674.

Gelbart, H., & Yarden, A. (2006). Learning genetics through an authentic research simulation in bioinformatics. *Journal of Biological Education, 40*(3), 107–112.

Gould, J. (1981). *The mismeasure of man*. New York: W. W. Norton.

Herrnstein, R. J., & Murray, C. (1994). *The Bell curve. Intelligence and class structure in American life*. New York: The Free Press.

Jensen, A. (1969). How much can we boost IQ and scholastic achievement? *Harvard Educational Review, 33*, 1–123.

Jiménez-Aleixandre, M. P. (2008). Designing argumentation learning environments. In S. Erduran & M. P. Jiménez-Aleixandre (Eds.), *Argumentation in science education: Perspectives from classroom-based research* (pp. 91–115). Dordrecht, the Netherlands: Springer.

Jiménez-Aleixandre, M. P., & Federico-Agraso, M. (2009). Justification and persuasion about cloning: Arguments in Hwang's paper and journalistic reported versions. *Research in Science Education, 39*(3), 331–347. doi:10.1007/s11165-008-9113-x.

Jiménez-Aleixandre, M. P., & Fernández, L. (2010, March 21–24). *What are authentic practices? Analysis of students-generated projects in secondary*. Paper presented at the NARST annual meeting, Philadelphia.

Jiménez-Aleixandre, M. P., & Pereiro, C. (2002). Knowledge producers or knowledge consumers? Argumentation and decision making about environmental management. *International Journal of Science Education, 24*(11), 1171–1190.

Jiménez-Aleixandre, M. P., & Puig, B. (2010). Argumentation, evidence evaluation and critical thinking. In B. Fraser, K. G. Tobin, & Mc Robbie (Eds.), *Second international handbook of science education*. Dordrecht, the Netherlands: Springer. In press.

Jiménez-Aleixandre, M. P., & Puig, B. (2011). *The role of justifications in integrating evidence in arguments: making sense of gene expression*. Paper presented at the ESERA Conference, France: Lyon.

Jiménez-Aleixandre, M. P., & Sanmartí Puig, N. (1995). The development of a new curriculum for secondary school in Spain: Opportunities for change. *International Journal of Science Education, 17*(4), 425–439.

Johnson, S. (1991). Food for thought. The cookie analogy. Center for Biology Education. Madison: University of Wisconsin. Downloaded April, 2010, from http www.wisc.edu/cbe/assets/docs/pdf/FoodForThought/cookie-analogy.pdf

Kaplan, C., & Llomovatte, S. (2009). Revisión del debate acerca de la desigualdad educativa en la sociología de la educación: la reemergencia del determinismo biológico. In S. Llomovatte & C. Kaplan (Eds.), *Desigualdad educativa: la naturaleza como pretexto* (pp. 9–21). Buenos Aires, Argentina: CEP.

Knippels, M. C. (2002). Coping with the abstract and complex nature of genetics in biology education: The Yo-yo learning and teaching strategy (PhD dissertation, University of Utrecht, Utrecht, the Netherlands).

Knippels, M. C. P. J., Waarloo, A. J., & Boersma, K Th. (2005). Design criteria for learning and teaching genetics. *Journal of Biological Education, 39*(3), 109–112.

Kolstø, S. D. (2006). Science students' critical examination of scientific information related to socioscientific issues. *Science Education, 90*, 632–655.

Levy, R. S., Selles, S. E., & Ferreira, M. S. (2008). Examining the ambiguities of the human race concept in biology textbooks: Tensions between knowledge and values expressed in school knowledge. In M. Hamman et al. (Eds.), *Biology in context: Learning and teaching for the twenty –first century* (pp. 338–346). London: University of London.

Lewis, J., & Kattmann, U. (2004). Traits, genes, particles and information: Re-visiting students' understandings f genetics. *International Journal of Science Education, 26*(2), 195–206.

Lewis, J., & Wood-Robinson, C. (2000). Genes, chromosomes, cell division and inheritance– do students see any relationship? *International Journal of Science Education, 22*(2), 177–195.

Lewontin, R. C. (1991). *Biology as ideology. The doctrine of DNA*. New York: Harper Collins.

Ministerio de Educación y Ciencia, Spain (MEC). (2007). *Real Decreto 1631/2006 por el que se establecen las enseñanzas mínimas correspondientes a la Educación Secundaria Obligatoria* (pp. 677–773). Boletín Oficial del Estado 5/01/2007.

Mortimer, E. F. (2000). Microgenetic analysis and the dynamic of explanations in science classroom. *Proceedings of the III Conference for Sociocultural Research*, Campinas, Brazil [Cd-Rom].

Mortimer, E., & Scott, P. (2003). *Meaning making in secondary science classrooms*. Buckingham, UK: Open University Press.

Moscovici, S. (1961–1976). *La psychanalyse, son image et son public* (2nd ed.). Paris: PUF.

Puig, B., & Jiménez-Aleixandre, M. P. (2010a). What do 9th grade students consider as evidence for or against claims about genetic differences in intelligence between black and white "races"? In M. Hammann, A. J. Waarlo, & K Th Boersma (Eds.), *The nature of research in biological education* (pp. 137–151). Utrecht, the Netherlands: University of Utrecht.

Puig, B., & Jiménez-Aleixandre, M. P. (2010b). *Students understanding about evidence for evolution*. Paper presented at the ERIDOB conference, Braga, Portugal.

Sensevy, G. (2007). Des categories pour décrire et comprendre l'action didactique. In G. Sensevy & A. Mercier (Eds.), *Agir ensemble: Élements de théorisation de l'action conjointe du professeur et des élevès* (pp. 13–49). Rennes, France: Presses Universitaires de Rennes.

Shulman, L. S. (1986). Those who understand: Knowledge growth in teaching. *Educational Researcher, 15*(2), 4–14.

Stewart, J. (1983). Student problem solving in high school genetics. *Science Education, 67*, 523–540.

Tiberghien, A. (2008). Foreword. In S. Erduran & M. P. Jiménez-Aleixandre (Eds.), *Argumentation in science education: Perspectives from classroom-based research* (pp. ix–xv). Dordrecht, the Netherlands: Springer.

Tiberghien, A., Vince, J., & Gaidioz, P. (2009). Design-based research: Case of a teaching sequence on mechanics. *International Journal of Science Education, 31*(17), 2275–2314.

Toulmin, S. (1972). *Human understanding: Vol. 1. The collective use and development of concepts*. Princeton, NJ: Princeton University Press.

Tsui, C. Y., & Treagust, D. F. (2007). Understanding genetics: Analysis of secondary students' conceptual status. *Journal of Research in Science Teaching, 44*(2), 205–235.

Venville, G., & Donovan, J. (2005). Searching for clarity to teach the complexity of the gene concept. *Teaching Science, 51*(3), 20–22.

Chapter 13
Metalogue: Design and Enactment of SSI Curriculum: Critical Theory, Difficult Content, and Didactic Transposition

Timothy Barko, Shirley Simon, María Pilar Jiménez-Aleixandre, and Troy D. Sadler

Critical Theory

Sadler: I am very interested in the author's application of critical theory to the teaching and learning of SSI. My work tends not to be guided by this perspective; not because I do not see value in critical theory and pedagogy but rather because a critical perspective raises certain issues with which I struggle to deal. I hope the authors can discuss ways that they deal with or conceptualize some of these issues. I want to frame this discussion by highlighting two quotes from the chapter:

> We think that understanding the model of expression is a necessary but not sufficient condition to evaluate determinist claims and that students need to develop critical thinking. In our characterization of critical thinking (Jiménez-Aleixandre & Puig, in press), there is a component related to social emancipation and the capacity to develop your own opinion as opposed sometimes to the mainstream ideas of a community or society. (p. 5)
>
> Therefore, this sequence has three goals for students, two related to science education and the third to citizenship: 1) to be able to apply the model of gene expression to real life contexts; 2) to develop the competency of using evidence and building arguments; and 3) to be able to develop a critical stance toward biological determinism. (p. 9)

I think the goal of helping learners to engage in reasoning related to important, socially relevant issues and to develop their own opinions is at the heart of SSI based education. But what if these efforts result in development of nonemancipatory thinking? The authors presented cases in which biological determinism has been used to reify conservative political ideologies. Despite the intention of curriculum designers and teachers, is it not possible for students learning through SSI to gravitate toward

T. Barko (✉) and T.D. Sadler
School of Teaching and Learning, University of Florida, Gainesville, FL, USA
e-mail: tim.barko@ufl.edu

S. Simon
Institute of Education, University of London, London, UK

M.P. Jiménez-Aleixandre
University of Santiago de Compostela, Santiago de Compostela, Galicia, Spain

T.D. Sadler (ed.), *Socio-scientific Issues in the Classroom: Teaching, Learning and Research*, Contemporary Trends and Issues in Science Education 39,
DOI 10.1007/978-94-007-1159-4_13, © Springer Science+Business Media B.V. 2011

conservative ideologies rather than more liberal ideas? The fact that James Watson adopts a biological determinist view is particularly problematic for SSI approaches that prioritize science content knowledge and maintains a goal of promoting critical perspectives. This clearly represents a case which illustrates that knowledge of science is not enough to guarantee the adoption of progressive perspectives. Should SSI-based curriculum and instruction explicitly promote critical and progressive perspectives? If so, do such approaches run the risk of teaching a particular ideology rather than supporting learner development of their own ideas?

Barko: Knowledge of science is most definitely not enough to guarantee the development of progressive perspectives. I think one of the problems is we assume, influenced by enlightenment positivism, that science is a form of knowledge generation that is essentially neutral and amoral. Science creates knowledge but makes no claim on the moral efficacy of that knowledge. Science creates nuclear bombs but provides no insight as to their use and distribution. It makes no statement as to whether we should have created them to begin with. To quote Richard Dawkins, "[b]e warned that if you wish, as I do, to build a society in which individuals cooperate generously and unselfishly towards a common good, you can expect little help from biological nature (Dawkins, 2009, p. 3)." The story of science, to Dawkins, oftentimes is not the story of generosity and altruism. He even states it is through education and not scientific practice per se where we should learn the ethics associated with the common good. So science should not be burdened with politics. But I am willing to respectfully disagree with Dawkins. I believe science cannot be disassociated with politics. As long as humans practice science, science will be partially political. What becomes problematic is when we assume it is not value-laden and take the stance that scientists often do, where they say "we'll do the science" and then point to the philosophers, educators, and policy makers and continue "you do the ethics." In order to approach science with a critical theory perspective we must assume science is "burdened" by politics, but in so doing, we must also acknowledge the inherent political assumptions of critical theory, that in doing science from a critical theory perspective we too are using science to support specific political agendas, that the finger too, can be pointed back at us.

Jiménez-Aleixandre: We are still struggling to integrate critical theory as a part of our approach to teaching and learning. As Troy points out, there are always two risks: the first is that bringing these issues into the classroom could result in the development of positions supporting inequality. The second is teaching a particular ideology rather than supporting students' development of their own independent thinking. Our approach to these risks is to reveal or uncover positions that are not grounded in science but rather in prejudice. In the case of Watson's position, we would argue that the claim advanced is based on racism. Of course there are students who may share these determinist views, but our data in these five classrooms do not suggest that they develop them as a result of explicitly discussing biological determinism. Based on the data presented in this case study, we suggest that the teaching sequence helped make implicit views explicit, develop understanding of the consequences of these positions, and encourage students to reflect on the coherence of their positions and scientific data. The expectation is that reflection about determinist explanations (contrasted with other views based in gene-environment interaction) may support students in developing a more sophisticated view.

In terms of the second risk, teaching a particular ideology, Simonneaux (2008) offers an interesting discussion of this issue. We think that there is an important distinction in a teacher making explicit her or his position in cases that deal with equally respectable positions (for example, different options for waste management) and in cases when one of the positions violates human rights or constitutional values. In the case of the two teachers in this study, we do not interpret that they made explicit their position against racism while teaching, but certainly Mr. Quiroga addressed biological determinism in such a way as to make it clear implicitly.

Gene Expression

Barko: As a former graduate student in Zoology, studying bioinformatics and molecular systematics, gene expression was one of the more difficult concepts that I dealt with on a regular basis. It was difficult to learn and difficult to teach and explain. Part of this problem is our understanding of gene expression has been in a constant state of transformation since Mendel's work was rediscovered by Hugo DeVeries and the early twentieth century genetic pioneers. Our notions of gene expression are still in a state of flux. I remember originally learning that one gene equaled one trait. This was always confusing because "traits" never made sense to me, particularly when concepts such as genotype and phenotype were introduced. Which one was the "trait," the genotype or the phenotype? Later, in my education, the relation became one gene codes for one protein. Then it became one gene codes for one peptide. Now we are not even sure the one gene one polypeptide correspondence holds. A recent article in the journal Heredity (Johnson & Tricker, 2010) discussed the role of epigenetics in regulating gene expression. We are finding that not all genetic variance comes from the DNA sequences that code for genes (or polypeptides). Changes in chromosome structure can alter gene expression without actually altering the DNA sequence of any one gene. The classic example is methylation where simple methyl groups are added to specific regions of the chromosome to regulate expression of certain genes. These methylation patterns are in fact heritable and are often known as maternal effects. The embryo inherits certain noncoded proteins and RNA from the mother and this directly affects the expression of genes. The child may have a different genotype from the mother, but because of the maternal effects, the child displays the same phenotype as the mother. That is, it is not always a gene that codes for a trait. Johnson and Tricker make an argument that these epigenetic traits are responsible for much of the plasticity we see in the human genome. So, gene expression is anything but clear cut. When we begin to talk about something as simple as the expression of eye color we must take into account not just our genetics but also our environment and even the epigenetic factors that predict gene-by-environment influences. Expression is a highly complex and plastic mechanism. My intention is not to give an object lesson in gene expression, but to further emphasize the difficulty of the subject matter.

This is a topic that is not always understood on the first pass or even second and third. I have been at it most of my adult life and still struggle with it. In fact, with the pace of modern genomic and genetic sciences, it is most likely that any notion of gene expression will have to be renegotiated. It is this very complexity that makes Puig and Jimenez-Aleixandre's article so compelling. The SSI perspective begins to look at how gene expression has intersected with social and political ideologies. We begin to see what influence this intersection has on how we view our own potential, whether we see ourselves as plastic and free or fatalistic and determined. Different environments can lead to different gene expressions from the same DNA sequences. Actual geneticists still struggle with the models of gene expression; no wonder it is such a difficult subject to teach.

Jiménez-Aleixandre: Certainly, as Tim points out, gene expression is an exceedingly difficult topic. That also happens with other genetics topics that are part of the curricula. The problem is that curriculum designers, textbook writers, and sometimes teachers assume that they will be learned after a short lecture illustrated with one or a few examples. Research allows us to glimpse students' difficulties and to think hard about how to face them.

Barko: These concepts rarely ever stick on the first pass and often a "one size fits all" mentality can and will obscure the ongoing processes of learning. When we look at something like SSI, we begin to see a process that "builds to suit." This more flexible process is more likely to engage students and create opportunities for them to access ideas in multiple ways. It promotes more critical reflection and less "memorize and repeat" formulas for understanding.

Didactic Transposition

Simon: In this chapter, I was particularly interested in the processes of design and enactment of the learning tasks and the authors' use of didactic transposition which they define as:

> the process of transformation of knowledge from one community, scientists (reference knowledge), to another, classrooms (taught knowledge)… there are two steps in the transposition: (1) from the reference knowledge to the knowledge to be taught and (2) from the knowledge to be taught to the taught knowledge. The knowledge to be taught consists of official curricula, textbooks, and other resources and the taught knowledge is related to the way a teacher enacts it in a particular classroom. (p. 4)

The authors' use of the notion of didactical contract allows for a dynamic view of the relationship between teacher, students, and knowledge in that it takes into account not only the norms in the relationship that are established and lasting but also the changes in the relationship as new knowledge is introduced. Much of our understanding of the realization of didactical transposition in the study arises from the comparison between the cases and the didactical contract between each teacher and his students. Given the central role of the teacher in the didactical contract, the characteristics of the teachers are clearly influential on the nature of didactic transposition in terms of "taught knowledge"; that is, how each teacher adapts the teaching sequence. For their analysis, the authors chose two experienced teachers but each was situated in a different environment. It would be interesting to explore and compare these teachers' personal beliefs and values, and how these might influence their enactment of the teaching sequence (Clarke & Hollingsworth, 2002).

In the comparison of the two cases, the authors draw on Mortimer and Scott's (2003) communicative approach, showing the more dialogic nature of Mr Quiroga's practice. The differences are clearly demonstrated in the examples of classroom discourse. I found the authors' link between this analysis and the notion of didactical contract potentially useful as it opens up questions about students' expectations when teachers adopt different communicative approaches. Such findings could be used to inform teachers on how their students respond to their practice and what kinds of expectations become embedded as norms. Such feedback can provide a stimulus for changing teachers' views of what is important in their practice such as changes in how they view the consequences of their teaching. It is possible that these kinds of teacher changes can lead to more sustained pedagogical development.

The design of the teaching sequence is 'top down,' but clearly informed by theoretical considerations of the content, context, and the nature of knowledge transformation. However, our work, like this study, has shown the intricacies of how teachers interpret activity designs (Simon & Richardson, 2009). An analysis of the steps needed in the interpretation of designs, and a knowledge of teachers' underlying goals and values, can inform us of ways in which to guide and enhance the experiences provided for students through the enactment process.

Jiménez-Aleixandre: We think, with Shirley, that the didactical contract approach helps us in capturing the dynamic development of classrooms. We are sharing the results (in Galician) with the teachers, and certainly hope that this feedback may inform their professional development.

References

Clarke, D., & Hollingsworth, H. (2002). Elaborating a model of teacher professional growth. *Teaching and Teacher Education, 18*, 947–967.

Dawkins, R. (2009). *The selfish gene* (30th Anniversary ed.). Oxford, UK: Oxford University Press.

Jiménez-Aleixandre, M. P., & Puig, B. (in press). Argumentation, evidence evaluation and critical thinking. In B. Fraser, K. Tobin, & C. McRobbie (Eds), Second International Handbook of Science Education. Dordrecht, the Netherlands: Springer.

Johnson, L. J., & Tricker, P. J. (2010). Epigenomic plasticity within populations: Its evolutionary significance and potential. *Heredity, 105*, 113–121.

Mortimer, E., & Scott, P. (2003). *Meaning making in secondary science classrooms*. Buckingham, UK: Open University Press.

Simonneaux, L. (2008). Argumentation in socioscientific contexts. In S. Erduran & M. P. Jimènez-Aleixandre (Eds.), *Argumentation in science education: Perspectives from classroom-based research* (pp. 179–199). Dordrecht, The Netherlands: Springer.

Simon, S., & Richardson, K. (2009). Argumentation in school science: Breaking the tradition of authoritative exposition through a pedagogy that promotes discussion and reasoning. *Argumentation, 23*(4), 469–493.

Chapter 14
Learning Nature of Science Through Socioscientific Issues

Siu Ling Wong, Zhihong Wan, and Maurice Man Wai Cheng

School Science Curriculum Reform in Hong Kong

A series of reforms in science education in Hong Kong started at the turn of the twenty-first century. In line with international trends, science education in Hong Kong has undergone considerable changes in the last decade since the implementation of the revised junior secondary science curriculum (grades 7–9) (Curriculum Development Council [CDC], 1998). The new curriculum encourages teachers to conduct scientific investigations in their classes, advocates scientific investigation as a desired means of learning scientific knowledge, and highlights the development of inquiry practices and generic skills such as collaboration and communication. Most importantly, it was the first local science curriculum that embraced understanding of nature of science (NOS), for example, being "able to appreciate and understand the evolutionary nature of scientific knowledge" (CDC, 1998, p. 3) was stated as one of its broad curriculum aims. In the first topic, "What is science?", teachers are expected to discuss with students some features about science, for example, its scope and limitations, some typical features about scientific investigations, for example, fair testing, control of variables, predictions, hypothesis, inferences, and conclusions. Such an emphasis on NOS was further supported in the revised secondary 4 and 5 (grade 10 and 11) physics, chemistry, and biology curricula (CDC, 2002). Scientific investigation continued to be an important component while the scope of NOS was slightly extended to include recognition of the usefulness and limitations of science as well as the interactions between science, technology, and society (STS).

In preparation for the implementation of a new curriculum structure (from a 7-year secondary education system to a 6-year one) in September 2009, a new set

S.L. Wong (✉) and M.M.W. Cheng
Faculty of Education, University of Hong Kong, Pokfulam Road, Hong Kong
e-mail: aslwong@hku.hk

Z. Wan
Department of Science and Environmental Studies, Hong Kong Institute of Education, Hong Kong

T.D. Sadler (ed.), *Socio-scientific Issues in the Classroom: Teaching, Learning and Research*, Contemporary Trends and Issues in Science Education 39,
DOI 10.1007/978-94-007-1159-4_14, © Springer Science+Business Media B.V. 2011

of Curriculum and Assessment Guides was devised for senior secondary level science subjects (CDC-HKEAA, 2007). We note a further leap forward along the direction of earlier curriculum reforms in the curriculum and assessment guides. The importance of promoting students' understanding of NOS is more explicitly spelt out. To put greater emphasis on environmental issues, students' appreciation of STS is extended to STSE, where "E" stands for environment. For example, in the physics curriculum, students are expected to "appreciate and understand the nature of science in physics-related contexts," "develop skills for making scientific inquiries," "be aware of the social, ethical, economic, environmental and technological implications of physics, and develop an attitude of responsible citizenship," and "make informed decisions and judgments on physics-related issues" (CDC-HKEAA, 2007, p. 4). There is a clear intention to develop students' awareness and understanding of issues associated with the interconnections among science, technology, society, and the environment. A separate subsection entitled *STSE connections* is embedded in each science topic of the Curriculum and Assessment Guides. It suggests examples of issues that teachers could make use of in developing students' awareness and understanding of STSE connections. For example, in the topic mechanics, one of the issues suggested is a dilemma of choosing between convenience and environmental protection in modern transportation.

Preparing Teachers for the New Curriculum Goals

Based on the literature and our local context, we identified a host of problems and challenges we had to tackle in encouraging and training teachers to teach NOS. In particular, we have been cognizant of the disappointing conclusions that were consistently reached by various studies. That is, both students and science teachers have inadequate understanding of NOS (Lederman, 1992) and STSE (Rubba & Harkness, 1993). There is however emerging empirical evidence that can inform efforts to improve NOS and STSE understandings. Explicit and reflective approaches in teaching NOS can support learner development of sophisticated NOS ideas (Abd-El-Khalick & Lederman, 2000; Khishfe & Abd-El-Khalick, 2002). Teachers with good understanding in NOS still face many constraints including concerns for student abilities and motivation (Abd-El-Khalick, Bell, & Lederman, 1998; Brickhouse & Bodner, 1992), lack of pedagogical skills in teaching NOS (Schwartz & Lederman, 2002), and lack of teaching resources (Bianchini, Johnston, Oram, & Cavazos, 2003) particularly those in local contexts and language (Tsai, 2001). Effective NOS teaching also depends on teachers' belief in the importance of teaching NOS (Lederman, 1999; Tobin & McRobbie, 1997) and their conception of appropriate learning goals and teaching role (Bartholomew, Osborne, & Ratcliffe, 2004).

In addition to the problems identified above, Hong Kong teachers very rarely have experienced learning of NOS or STSE during their own schooling. Noting teachers' inadequate NOS understanding and hence their lack of appreciation and less than effective use of the NOS instructional materials, we then restructured our teacher training programs to allow more time on these aspects. For example, in early 2000s, we proposed the use of science stories, such as on the discovery of penicillin, the development of cowpox, Newton's proposition of Law of Universal Gravitation, and the treatment of stomach ulcers (Tao, 2002), as a medium through which NOS could be introduced to students. However, due to the lack of both understanding of NOS and experience in learning and teaching NOS, many teachers made use of them only for arousing students' interest. Hence, the availability of teaching resources would not by itself result in teachers' learning of NOS and an effective teaching of NOS. Unless teachers had the ability to understand and appreciate the rationales behind the design of the instructional materials, it was likely that they would overlook or miss the targeted learning objectives (learning NOS features) and gravitate toward the parts which were more appealing to them (dramatic stories which promote students' interest). Such a situation was reflected in the comment made by a junior science teacher who had been telling the interesting science stories to his students. He came to realize his oversight of not having made good use of the stories for teaching NOS after he attended our NOS training workshop:

> I found the story on stomach ulcers very interesting....Marshall tested his hypothesis by trialling out himself....Students all enjoyed the story... I only realised now that there are deeper meanings behind the story and other important learning outcomes to be achieved through it and other stories.

We also reckoned that there were some inadequacies of these relatively "old" stories. Teachers and students expressed that though these stories aroused their interests, they happened quite a while ago. Those who did not have the historical and cultural backgrounds of the scientific discoveries and inventions would fail to develop an in-depth understanding of, and hence appreciate, the thought processes of the scientists related to what they encountered at their time.

In summer 2003, when the crisis due to the Severe Acute Respiratory Syndrome (SARS) in Hong Kong was coming to an end, we saw a golden opportunity to turn the crisis into a set of meaningful instructional resources which might help address the issues raised above. The SARS incident was a unique experience that everyone in Hong Kong had lived through and the memories of which would stay for years to come. At the beginning of the outbreak, the causative agent was not known, the pattern of spread was not identified, mortality and morbidity seemed soaring, yet an effective treatment regimen was uncertain. It attracted the attention of the whole world as scientists worked indefatigably to understand the biology of the disease, develop new diagnostic tests, and design new treatments. Extensive media coverage kept people up-to-date on the latest development of scientific knowledge generated from the scientific inquiry about the disease. We believed that the incident would have much to reveal about NOS.

As anticipated we identified many interesting aspects of NOS based on our interviews with key scientists who played an active role in the SARS research, analysis of media reports, documentaries, and other literature published during and after the SARS epidemic. The SARS incident illustrated vividly some NOS features advocated in the school science curriculum. They included the tentative nature of scientific knowledge, theory-laden observation and interpretation, multiplicity of approaches adopted in scientific inquiry, the inter-relationship between science and technology, and the nexus of science, politics, social, and cultural practices. The incident also provided some insights into a number of NOS features less emphasized in the school curriculum. These features included the need to combine and coordinate expertise in a number of scientific fields, the intense competition between research groups (suspended during the SARS crisis), the significance of affective issues relating to intellectual honesty, the courage to challenge authority, and the pressure of funding issues on the conduct of research.

The details on how we made use of the news reports and documentaries on SARS, together with episodes from the scientists' interviews to explicitly teach the prominent features of NOS can be found in (Wong, Kwan, Hodson, & Yung, 2009). Since January 2005, we have been using the SARS story in the training of hundreds of preservice and in-service science teachers about NOS. The feedback has been encouraging. The SARS story was particularly successful in promoting teachers' understanding of NOS in terms of (1) the realization of inseparable links between science and the social, cultural, and political environment, (2) deeper understanding of how science and technology impact on each other, and (3) a richer appreciation of the processes of authentic scientific inquiry and the humanistic character of scientists. The effectiveness was mainly attributed to immediacy, relevance, and familiarity of the SARS story which made the abstract tangible. Teachers' personal experience of this unique piece of "history" of science and the powerful affective impact of the interviews with scientists also contributed to the favorable learning outcomes (Wong, Hodson, Kwan, & Yung, 2008).

The encouraging results of using the SARS story with teachers prompted us to produce NOS instructional materials that should also be grounded in the local contexts (and language) for school students. At the same time, we were fully aware that high quality curriculum materials do not automatically result in student learning. Such materials have to be mediated by teachers with necessary content knowledge, pedagogical skills, beliefs, and intention of meeting the curriculum goals (Bartholomew et al., 2004; Lumpe, Haney, & Czerniak 1998; Schwartz & Lederman, 2002). Thus, in September 2005, we embarked on a 2-year project which aimed to produce local NOS/STSE curriculum resources while preparing more teachers for NOS/STSE teaching. We also envisioned that the project was timely in view of the new science curricula to be implemented in 2009. Having made reference to the reminder by Hodson (2006) that,

> [a]ction research offers the most likely route to such far reaching curriculum changes, committed teachers work together in mutually supportive groups to address the theoretical and practical issues surrounding the implementation of a learning about science curriculum and to develop suitable and effective curriculum resources. Curriculum materials need to have a "street credibility" that can only be gained when they are developed *by* teachers, *for* teachers. (p. 305)

we deliberately involved teachers at the beginning stages of the design process of instructional materials. More than 50 senior secondary science teachers worked together with the university team members to develop 12 sets[1] of teaching resources which integrate NOS knowledge with subject knowledge of the new senior secondary biology, chemistry, and physics curricula. Efforts were made to include as many local examples as possible. The topics included development of the Disneyland in Hong Kong, consumption of shark fins, an abridged version of the SARS story, etc. In doing so, teachers would be more ready to make use of the materials in their own classrooms. This was important for our project as we wanted to collect data to refine the teaching materials as well as provide opportunities for the teachers to learn how to teach NOS with the support of the university team members. In brief, the teacher development project comprised the following key components:

1. A 6-h NOS session conducted by the first author using SARS and other contexts in illustrating various aspects of NOS.
2. Workshops on refinement of NOS instructional materials drafted by the university team members.
3. Prelesson discussion with the university team member(s) on the main ideas and fine adjustment of the instructional materials to be tried out in the teacher's own classroom.
4. Teaching the lesson(s) using the instructional materials.
5. Reviewing the lesson video and reflecting on the trial lesson(s).
6. Meeting between the university team member(s) and the teacher to discuss critical incidents in the videotaped lesson (followed by written feedback from university team members) in preparation for a sharing session with other in-service teachers joining this project (teacher members).
7. Sharing of the trial-run experience with other in-service teachers of the project and receiving their comments.
8. Dissemination seminar: Over 500 teachers from other schools in Hong Kong attended. Sharing of teaching materials and trial-run experiences with a wide audience.

Due to the limited financial and human resources, seven of the teacher participants went through all the above training components. The rest participated only in components (1), (2), (7), and (8). The development of the seven teachers in the understanding of NOS, the pedagogical skills in teaching NOS, and the intention to teach about NOS were probed at various stages of the project. These teachers found learning through a contextual approach, as what they did in the NOS workshop, very effective in promoting, enriching, and consolidating their NOS understanding. They also treasured both the training materials and some activities in the teaching resources that they tried out in their classroom as they could modify or had modified them to incorporate teaching of NOS in their teaching of subject knowledge teaching. The actual classroom implementation and detailed review and discussion of the lesson video "worked best" and more importantly were "treasured most" by

[1]12 sets of teaching resources can be accessed through the website: http://learningscience.edu.hku.hk

the teachers in enhancing their pedagogical skills in teaching NOS and intention to teach NOS. The review and analysis of the lesson videos facilitated teachers' reflection on which areas needed improvement. It also allowed them to appreciate and value the teaching of NOS when they saw students' ability and interest in learning NOS. The proof of workability of both teaching and learning NOS had prompted teachers who were reserved about teaching NOS to follow suit. Details of their development, their concerns, and their learning outcomes were reported in Wong, Yung and Cheng (2010).

While encouraged by the promising results in the project, we have been mindful that a lack of collegial support may lead to a decline of teachers' commitment in achieving curriculum change. Thus in September 2008, we initiated another 2-year professional development project. It aimed to cultivate a mutually supportive environment where teachers would collaborate and develop their pedagogical content knowledge of NOS. A key feature of the new project was the formation of study groups among teachers of the same subject discipline from different schools (subject-based approach) and science teachers of the same school (school-based approach). We believe that putting like-minded teachers together in a study group is more likely to sustain and even enhance their commitment to teach NOS. While retaining the components that teachers who participated in the previous project valued and treasured most (such as the SARS case study, the detailed review and discussion on their lesson videos with the university team members and their peers, etc.), we also encouraged teachers not to simply modify and adapt the available teaching resources, but to proactively design their own instructional materials. This is indeed a goal that we wish teachers could ultimately achieve. Thus at the outset of the current project, we explained our intention to the teachers by a Chinese proverb, "Give a man a fish and you feed him for a day. Teach a man to fish and you feed him for a lifetime." (In Chinese, it says 授人以魚,三餐之需;授人以漁,終生之用。).

Two physics teachers, Wayne and Kyle, who were from the same school, participated in the school-based approach of the project (with three other colleagues teaching other science subjects). Inspired by the contextual approach in which they learned various NOS aspects through a series of socio-scientific issues (SSIs) in the SARS incident, they designed and implemented their own teaching materials in which NOS and SSIs were drawn upon. In the following sections, we expound on their learning and teaching experience in the current project by covering in detail three important components that they found most helpful in their development, namely (1) contextual learning of NOS through SSIs in the SARS crisis and other related contexts, (2) viewing and responding to videos of exemplary practice of NOS teaching; (3) their own practice of designing and implementing a NOS lesson. At appropriate places, we include views and comments expressed by Wayne and Kyle during our interaction with them in the following occasions: brainstorming sessions on how to incorporate NOS ideas in their physics lessons, pre- and post-lesson discussions, video workshops with the team of science teachers of their school in which we shared and reflected on video episodes of how the teachers instilled NOS in their respective subjects, and an exit interview near the end of the project.

We then compare the instructional materials and the teaching strategies of the NOS lesson collaboratively planned by Wayne and Kyle, and discuss the extent to which the SSI and NOS features demonstrated in the materials and the strategies are mediated by their learning experience.

Learning NOS Through SSI During the SARS Crisis

A detailed review of 17 experimental studies of context-based or STS-oriented approaches at secondary school level by Bennett, Lubben and Hogarth (2007) indicated that such approaches had positive impacts on students' understanding of scientific concepts and attitudes to science and school science, especially when attention focuses on recent scientific research and innovations. Though none of these studies focused specifically on NOS understanding, we successfully demonstrated that the use of the SARS context with a number of remarkable SSIs was effective in enhancing NOS understanding (Wong et al., 2008). This provided support to the compelling arguments that critical consideration of SSIs offers an ideal forum for students to deploy and develop their NOS understanding (Sadler, 2004; Sadler & Zeidler, 2005; Zeidler, Sadler, Simmons, & Howe, 2005), especially when the issues are controversial (Kolstoe, 2000; Oulton, Dillon, & Grace, 2004) or have strong personal relevance (Ryder, 2002). Khishfe and Lederman (2006) remind us that although locating NOS teaching within controversial topics (in their case, discussion of global warming) can be effective, the essential requirement, as in internship and apprenticeship approaches, is that NOS teaching is both explicit and supportive of critical reflection. The case study of SARS contextualized in many examples of SSIs was intensely personal for our learners and had many controversial aspects. Moreover, the learning materials were designed to draw explicit attention to important NOS items and to encourage reflection.

To explicate the learning experience of Wayne and Kyle through these SSIs, we describe in detail four key SSIs during the SARS crisis, namely (1) the hunt for the causative agent of SARS (2) the tragic outbreak at Amoy Gardens and the subsequent quarantine of all residents in the severely infected block, (3) the disclosure of Dr. Jiang on the discrepancy of the number SARS cases announced by the mainland Chinese government, and (4) the ban on eating of civet cats to prevent reoccurrence of SARS infection. These SSIs were most cited by preservice and in-service teachers taking the teacher training programs as very effective in prompting their deeper understanding of how science and technology impact on each other, and their realization of the inseparable connection between science and society, in particular the influence of scientific findings about SARS on the actions of different bodies in society during and after the epidemic (Wong et al., 2008). The detailed historical development of each of the four SSIs with the NOS elements exemplified, and the learning and teaching activities included in the multimedia instructional materials are given in Tables 14.1–14.4.

Table 14.1 Details of hunting for the cause of SARS

Time	Historical development related to the SSI	Multimedia items/teaching and learning activities	Embedded elements of NOS
Mid-Mar 2003	Dr. Klaus Stohr, a scientist with the WHO, initiated the establishment of an international collaborative network among 11 laboratories, located in 10 countries.	A 60-s video showing Dr. Stohr making several phone calls and inviting different labs to form an international research consortium through telephone conferencing – thus enabling rapid sharing of data and information about SARS. Learners are asked to identify and describe the nature of collaboration shown in the video. They are also asked if they can see beyond collaboration (we were looking for students to identify competition, as reflected in the hidden message of the video narrative).	Technology can change the ways in which science is conducted and the speed with which it is completed. Without recent advances in telecommunications and availability of the Internet, the large-scale collaboration could not have been established so rapidly. Scientists are competitive by nature; they compete with each other, themselves, time, resources, etc.
		A 40-s video edited from the interviews with Dr. LP and Dr. RC, in which they talk about how the pace of their research was accelerated by "competing" against time, resources, their own self-esteem, and social demands.	Scientists are not detached from society and they often work strenuously to meet social demands.
18 Mar 2003	CUHK first announced they had found evidence that paramyxovirus was the causative agent of SARS. Laboratories in Germany, Canada, and Singapore immediately announced that they had also found evidence of paramyxovirus in samples collected from patients.	Learners are shown a picture of a landscape with trees and rocks on the seaside. The picture also depicts a praying mother and son as an "embedded view" which is not immediately obvious to most observers. [It is anticipated that when the first learner sees the embedded view, more learners will start seeing it. Learners will then be explained about the possible subjectivity in observation which can be affected by one's knowledge base.]	Subjectivity of scientists and theory-laden observation. Scientists' observations are influenced by their knowledge and the theoretical framework they employ – i.e., their observations may be affected by what they expect to see or prepare themselves to observe.

22 Mar 2003	HKU found evidence suggesting that the causative agent was coronavirus. A laboratory in Rotterdam and CDC in Atlanta quickly announced that they had also found evidence in favor of coronavirus.	Learners are then asked to identify and describe examples of theory-laden observations from an 80-s video showing the progression from the agreement among different laboratories that a paramyxovirus was the causative agent of SARS to the agreement that a coronavirus was the real causative agent of SARS. The video includes excerpts from the press conferences of both announcements and the characteristics of both types of virus.	Subjectivity of scientists and theory-laden observation. Scientists' observations are influenced by their knowledge and the theoretical framework they employ – i.e., their observations may be affected by what they expect to see or prepare themselves to observe. Uncertainties and controversies in science.
		Learners are explained that coronavirus was initially ignored as it is well known for causing mild common cold only. [It is expected that some learners may even be able to identify that the virologists who identified the viruses from their characteristic appearance were basing their views on their own expert knowledge, whereas a lay person would be unable to recognize the patches on the microscopic slides as viruses.]	Tentative nature of scientific knowledge (linked to the skepticism and open-mindedness of scientists.)

Table 14.2 Details of the tragic outbreak at Amoy Gardens and the subsequent quarantine of all residents of block E

Time	Historical development related to the SSI	Multimedia items/teaching and learning activities	Embedded elements of NOS
28–31 Mar 2003	A residential building, Block E in Amoy Gardens, Hong Kong, was found to have an alarmingly high number of cases of SARS infection.	An animated graph showing the accumulated total number of infected cases in Amoy Gardens, rising from 7 to 185 within 4 days.	
31 Mar–1 Apr 2003	The Hong Kong Department of Health imposed quarantine on Block E of Amoy Gardens. An unprecedented order from the Hong Kong government to move all residents of Block E to isolation camps to allow a thorough investigation of the mysterious severe outbreak.	A 20-s video showing government health workers and residents of Amoy Gardens, all in masks, being sent to isolation camps in a rural area. It also shows the cross-disciplinary investigation team entering the building to find clues to the cause of the outbreak. Learners are prompted to consider the diverse sentiment of people living inside versus those living outside Amoy Gardens.	Science and political decisions are interrelated.
Early- to mid-Apr	Scientists identified the presence of SARS-coronavirus in rats, cockroaches around the residential area, sewage from the drainage system of the building.	Pictures showing the pattern of infection.	Modeling is a typical process in scientific research to provide explanation to fit the observations.
	The pattern of infection (most of the infected residents lived in higher floors of Flat 7 and 8 which are connected to the same drainage system) suggested the causes of the outbreak was related to the drainage system. The model proposed by the local scientists explaining the infection was only accepted by the scientists of the WHO when they saw the over-packed conditions of the buildings.	Pictures illustrating the proposed explanation by scientists: The bathroom floor drains with dried-up U-traps opened a pathway for small droplets containing coronavirus into the bathroom. The exhaust fan in the bathroom then sucked the contaminated droplets to the light well where they were carried to higher levels by the warm humid air.	Uncertainties and controversies in science.

(continued)

Table 14.2 (continued)

Time	Historical development related to the SSI	Multimedia items/teaching and learning activities	Embedded elements of NOS
Now	Ensure proper functioning of the U-traps by keeping it filled.		Social practice shaped by scientific findings (taking evidence-based decisions and actions for home hygiene).

Table 14.3 Disclosure of Dr. Jiang on the discrepancy of the number SARS cases announced by the mainland Chinese government

Time	Historical development related to the SSI	Multimedia items/teaching and learning activities	Embedded elements of NOS
Since late Mar 2003	In Beijing, Dr. Jiang disclosed the truth to the media and drew international attention to the severe outbreak of SARS in mainland China. He was put under house arrest.	Learners are asked to give examples of similar stories from the history of science when political or social and cultural factors have affected the progress of science [e.g. house arrest of Galileo]. 60-s interview excerpts of Dr. AC and Prof. GZ expressing their views on the Chinese government's decision to suppress information about the spread of SARS.	Political, social, and cultural factors influence the development of science.

Hunt for the Causative Agent of SARS

Wayne indicated the hunt for the causative agent of SARS (Table 14.1) was most remarkable. He found many NOS aspects illustrated in this SSI rather new to him and even beyond his expectation. These included ideas about the theory-laden nature of observation and the related subjectivity of scientists.

> Yes, the theory-laden one was really good. In the competition between CUHK [Chinese University of Hong Kong] and HKU [University of Hong Kong]…to be the first research team to identify the cause of SARS…when the professor [in CUHK] announced that he found it [paramyxovirus], other expert virologists quickly agreed with him…It's a big surprise to me that these renowned scientists also made mistake…they're also affected by their prior knowledge and belief…You know, SARS, it's a big thing…it's large-scale…but they're still heavily affected by their background knowledge. [Coronavirus has been well known for causing mild common cold. The scientists only realized later that SARS-coronavirus can become deadly after gene shuffling due to cross-species transmission.]

Table 14.4 Details of the ban of eating of civet cats to prevent re-occurrence of SARS infection

Time	Historical development related to the SSI	Multimedia items/teaching and learning activities	Embedded elements of NOS
23 May 2003	Researchers from Hong Kong and the CDC and Prevention of Shenzhen, mainland China, successfully isolated the coronavirus causing SARS from civet cats.	Photos of civet cats in cages in restaurants and newspaper headlines about research in Hong Kong and Shenzhen CDC revealing that many civet cats carried a coronavirus that was 99.8% genetically identical to the human coronavirus.	Scientific claims are based on logical deduction.
	They also found that some animal traders involved in slaughtering the animals had antibodies of the virus carried by civets.	Researchers commented that civets might have been infected from yet another unknown animal source in the markets, where many different animals are caged in close proximity. Such market practices could provide a venue for the spread of the virus.	
	The unique dietary habits in Guangzhou might have caused the disease.		
Since late May 2003	The sale of civet cats in mainland China and other exotic animals was banned when evidence identified civet cats as the origin of SARS.	A 18-s video of a scientist who commented on the livelihood of many people working in businesses related to the sale of civet cats and exotic animals was greatly impacted by this political decision.	The findings of science affect political decisions, which in turn affect social and cultural practice.
Oct 2005	After SARS, newer evidence reported by scientists pointed to bats as the natural reservoir of SARS-coronavirus.	Learners were asked if they were aware of the most updated research findings in the search of the origin of SARS virus.	Not uncommon to have political decisions based on or inadequate scientific findings.

Wayne also frankly admitted that he was not quite aware of the competition between research teams as he described above or even collaboration among scientists before he experienced the SARS instructional materials, as he put it:

> The collaboration of those scientists was also memorable…the science we teach now was done in old days…we just know somehow the scientist got it…we don't even exactly know how he got it…In SARS, we saw the collaboration of many scientists.

Apparently Wayne was recalling the video clip from the program, *SARS: The True Story*, produced by the British Broadcasting Company (Higgins and Learoyd, 2003) which described how Dr. Klaus Stohr, a scientist with the World Health Organization (WHO), initiated an international collaborative network, utilizing telephone conferencing, among 11 laboratories spread across 10 countries. The interview of Dr. Stohr and the narrator of the program gave a sense of urgency at that moment in time.

> We knew that it was a race against time, so we had to find very quickly the pathogen, the causative agent for this disease... These scientists are competitive by nature but we called up 11 laboratories in 10 countries... (Dr. Klaus Stohr)
>
> 24 hours later, all the labs had agreed to forgo their rivalries and collaborate. For the first time in history, the full force of the world's scientific might was united. (Jack Fortune – Narrator)

The SARS crisis also provided Wayne with an excellent example of how society impacts science in the STS connections:

> This disease...very life-threatening...there was a huge societal demand...there was an immediate need to do research, to find out what caused it and how to cure it. I recalled in STS, society and science should be interrelated...but does science affect society or society affects science?...This one (SARS epidemic) is a good example of society driving the development of science...so many scientists thus worked together to tackle the problem.

The inclusion of an interview with Dr. RC conducted by a local television company gave Wayne a strong impression of the urgent societal need for scientists to find out more about SARS disease. It was filmed during the most serious and critical moment of the SARS crisis, when many doctors in her hospital were being infected. She was almost in tears as she shared her frustration and helplessness.

> After the two colleagues passed away, some of our colleagues are in a life threatening situation. We are now receiving phone calls if our research can give immediate help. We need to work on urgent requests...hoping that we can help stop or reduce the rate of infection... We have become so stressed out...We also wish earnestly that we can lessen the harm and have been asking ourselves "how come our research still can't be any help?"

The feelings expressed by Dr. RC reinforced the need for scientists to compete against time and highlighted the pressure on them to meet social demands. Her distress and expressions of concern demonstrated that scientists were not isolated from the social environment in which they worked.

Tragic Outbreak at Amoy Gardens

In the tragic outbreak at Amoy Gardens (Table 14.2), Wayne and Kyle saw a clear impact of scientific findings on political decisions. In response to the finding of the medical statisticians identifying an alarmingly high infection rate and peculiar infection patterns of a particular residential building, the Hong Kong government imposed an unprecedented order of quarantine. All residents of the building were moved to a rural campsite to permit a thorough and unimpeded examination of the building. In relation to the issue of whether the government should isolate the residential building and its residents, they appreciated that a decision could arouse very diverse sentiments among people of different roles.

Disclosure of Dr. Jiang About the Infection Situation in Beijing

In the disclosure of actual number of infected cases of SARS by Dr. Jiang Yanyong (Table 14.3), they saw the opposite relationship between science and political decisions. There were, sadly, some occasions when scientific research on SARS was impeded by political decisions. In mainland China, in a misguided effort to prevent widespread panic, strenuous efforts were made to conceal the true extent of the epidemic. When Dr. Jiang exposed the cover-up in a letter to a Beijing television station and *Time magazine*, he was put under house arrest. The comment made by Dr. AC, a Hong Kong medical researcher who played an active role in fighting the disease, reflected how political decisions could hugely exert negative impact on the advance of science:

> In Mainland China, when the government said there were no cases of SARS at the beginning of the epidemic…how could the scientists in China do research on samples collected from the SARS patients there if there were no cases?…Certainly scientific research is oftentimes controlled by the will of the government.

Professor GZ, a mainland Chinese molecular biologist who played a key role in tracing the molecular evolution of the SARS-coronavirus during the course of the SARS epidemic in China, said that he could see "both sides of the coin" with regard to the delay in announcing details of the SARS infection in China. In an interview conducted 18 months after the SARS crisis was over, he commented:

> We have learnt from SARS that if the government wants to keep this kind of thing secret, it's not good… The government didn't want to release information in the early stages because they were afraid that the media was going to make the situation worse…
>
> Now the WHO doesn't think SARS is a very severe disease. Even if SARS comes back, the WHO will treat it very calmly…Media may sometimes be dangerous in causing overreaction among the people… and that probably would influence the research and also the medical treatment…It's really not necessary to panic, but the media didn't know that… I think we have to educate the government, the media and the people…

Clearly, Professor GZ held the view that irrational panic could be suppressed by educating the general public and promoting a better understanding of science and nature of science.

Ban of Sale of Civet Cats

In the ban on the sale of civet cats, Wayne and Kyle further consolidated the impact of science on political decisions and social practice. They also felt the helplessness on the immediate impact on livelihood of a particular group of people in the society due to the findings of scientific research as commented by Dr. RC about the data leading to the ban on eating of civet cats. She put particular emphasis on the social impact as a result of research findings.

The data [including early SARS patients were mostly restaurant workers who handle wild animals and serve exotic food like civet cats; civet cats carried a coronavirus almost identical to human coronavirus] directly affected the business of merchants selling exotic animals. These people may not even have learnt science at all in their life, but their lives have been heavily affected by science.

Her feeling of helplessness and sympathy for the merchants further deepened when she learnt that later compelling data pointed to bats instead of civet cats as the natural reservoir of SARS-coronavirus when the epidemic was over. Such an episode prompted Wayne and Kyle to recognize the intrinsic problems with the uncertainty of science and that decisions could only be made based on data available which at the time may turn out to be incorrect.

Exposure to Good Practice of Teaching NOS

Through the SARS story, as another component of the professional development project, Wayne and Kyle also reviewed and shared their thoughts on videos of exemplary NOS teaching integrated in the teaching of science. These exemplary lessons were conducted by teachers who adapted some instructional materials developed in our previous project (2005–2007). Thus teacher participants of the current project, including Wayne and Kyle, have been exposed to good examples of different ways of teaching NOS, for example, situating the scientific concepts being taught in a historical context, making use of teachers' demonstrations or students' inquiry experience to highlight NOS, and capturing critical incidents to elaborate on certain NOS aspects.

As demonstrated in the study by Wong, Yung, Cheng, Lam, and Hodson (2006), exposure to exemplary science teaching prior to the commencement of formal preservice teacher training course is effective in getting prospective teachers "ready to 'think like a teacher' and to begin to be cognizant of the complex ways in which the actions of teachers impact on their students" (p. 17). Wong and her colleagues provide evidence that videos could reinforce and develop prospective teachers' conceptions of good science teaching in one or more of the following ways: (1) recognizing exemplary practitioners in the videos as role models who can inspire them to formulate personal goals directed toward these practices; (2) broadening their awareness of alternative teaching methods and approaches not experienced in their own learning; (3) broadening their awareness of different classroom situations; (4) providing proof of existence of good practices; and (5) prompting them to reflect on their existing conceptions of good science teaching.

Teaching NOS was fairly new to Wayne and Kyle. There are many similarities in enculturating preservice teachers into good science teaching and in enculturating teachers into teaching NOS. For instance, Wayne and Kyle were very impressed by a physics teacher who demonstrated diverse teaching strategies in teaching

NOS while covering the topic 'wave nature of light'. The lessons included the story-telling of how Fresnel tried to overcome the challenge by Poisson on his argument on the wave nature of light, the reconstruction of Fresnel's demonstration of diffraction of light to simulate how Fresnel provided empirical evidence in convincing Poisson of his argument that wave nature of light, etc. The lesson series broadened Wayne and Kyle's awareness of alternative teaching methods and approaches and provided them proof of existence of good NOS teaching. In another instance, after watching an early career physics teacher trying a different way (from the standard lesson sequence in local physics textbooks) of introducing the concept of momentum and its relevant NOS, Kyle commented with a pleasant surprise.

> …less than one year of teaching experience? …he was so good…and his students got the ideas [both the concept of momentum and NOS aspects]!…Students also practised the skills in extracting data from the graphs [generated by the data-logging software] and data handling skills.

Apparently his comments reflected an initial doubt about students' ability in discovering the conserving property of the total sum of the products of mass and velocity of each colliding object, (i.e. total momentum) before and after different types of collision, and students' appreciation of pattern- or rule-seeking as a typical and important activity of scientific inquiry.

Making Use of a Timely SSI to Teach SSI

With the encouraging proofs of existence of successful teaching of NOS by other physics teachers, Wayne and Kyle decided to make their first collaborative attempt in teaching the interconnection of science, technology, and society. In August 2009, Kyle and Wayne had a meeting with the first author, Alice, to share their preliminary idea about making use of the recent controversial decision of FINA[2] that the use of shark skin swimsuits[3] would be banned in international swimming competitions after 1 January, 2010. Their intent to cover STS connections was guided by an apparent focus of the new physics curriculum guide. They felt that this SSI was a very suitable context to be integrated in the topic of mechanics. It was also timely and relevant to the interest of their students.

After Wayne and Kyle had introduced the fundamental physical quantities related to the motion of an object (time, distance, displacement, speed, velocity,

[2]The international governing body of swimming, diving, water polo, synchronized swimming, and open water swimming. FINA stands for Fédération Internationale de Natation, meaning International Swimming Federation.

[3]Swimsuits made of technologically advanced fabrics biomimetically designed with a surface that mimics the shark denticles to reduce drag resistance through the water.

acceleration), they decided to conduct the lesson on the *Ban of Shark Skin Swimsuits* which was dedicated to promote students' understanding of the STS connections. It was discernable that the choice of the context and their design of the instructional materials bore considerable similarities to the SSIs that they learnt in SARS. For example, the context chosen was immediate, contemporary, and familiar to students. The topic itself and the decision of the ban were controversial. It was a good example to illustrate the impact of science and technology on social practice (in sports). The contents of the 50-min lesson are shown in the first column of Table 14.5. The second and third columns are the teaching and learning activities and the embedded NOS aspects, respectively.

Table 14.5 Details of the lesson on STS connections using the ban of the use of shark skin swimsuits in swimming competitions

Lesson sequence	Multimedia items/teaching and learning activities	Embedded elements of NOS
Introduced the purpose of the lesson: STS connections.		
Quick revision of quantities in describing motion of an object (t, s, v, a).	Asked students (Ss) to recall how to calculate velocity and acceleration.	Society demands for more accurate timing devices
Highlight the need of accurate timing device for accurate measurement of these quantities.	Teacher (T) highlighted their relationship with time and hence need accurate measurement of time for calculation.	Technological advances may bring advantages to societal events (e.g. sports competitions)
	T showed photo of a digital stop watch and an electronic timer which avoids errors due to reaction time.	
	T elaborated improved technology can generally benefit sports competition, and timing device is one of them.	
Use of a highly-attended 110 m hurdle race to illustrate why accurate timing device (or technology in general) is important to the fairness in sports competitions.	T capitalized on the first return of Liu Xiang on the field (the best Chinese 110 m hurdler) in an international hurdle match in Shanghai held just 2 days ago.	Highlighted once more accurate device is needed for fairer judgment.
	Showed a video clip in which Liu Xiang and Terrence Trammell made the same time record of 13.15 by digital timer (difference <1/100 s).	Linked the need of technologies to fairness of sports competitions.
	T then explained the use of ultra high-speed multiple photo shots in differentiating the position.	

(continued)

Table 14.5 (continued)

Lesson sequence	Multimedia items/teaching and learning activities	Embedded elements of NOS
Drew Ss awareness to the use of other technologies beyond accurate timing. Prompted students to reflect on the issues of fairness brought by the use of technology in sports.	Invited Ss to give examples of the use of technologies in sports for other purposes in sports competitions. Ss gave appropriate examples including: IR gun used in measuring distance of short put, sports shoes with air cushion, electronic scoring machine. T showed a series of video clips of technologies used within competitions and (e.g. set-running aid, shark skin swimming suit, man-made legs for running competition in Olympics for the handicapped), during training (swimming pool with resistive flow, video replay for identification of imperfect action for focused improvement) T raised a question to Ss: If an athlete or his country is so poor that these kinds of technologies could not be obtained, will it be fair?	Various technologies improve the performance of sports competitions and change the practice both during competition and during training which results in much improved world records.
Focused on shark skin swimsuits – the technological product which just got banned.	T showed Ss the video of the recent swimming competition in July 2009 in Rome in which Michael Phelps of USA lost to Paul Biedermann of Germany in the 200 m-free style due to a considerable progress of the technology used in shark skin swimsuits.	Technological advances which result in a jump beyond social expectation can lead to controversy of its use.
Highlighted the impact of the shark skin swimsuits on the world records in swimming competitions.	Showed a table of the world records in swimming competitions to highlight that almost all events were broken in 2009 and then contrasted this situation with track events which were much less frequent.	

(continued)

Table 14.5 (continued)

Lesson sequence	Multimedia items/teaching and learning activities	Embedded elements of NOS
Decision-making with provision of arguments.	Role play – Should shark skin swimsuits be banned? Ss discussed in groups whether they agreed with the ban of shark skin swimsuits. Each group played a different role (spectators, athletes, scientists developed the swimsuits, sponsors of swimsuits) and noted down their views.	
Presentation of views from different groups.	Representative of each group presented their views. T made encouraging remarks on the some thoughtful arguments. T also encouraged Ss of different groups to comment on each other's views	
Concluding remarks on the STS interconnection.	T made use a figure showing the interconnection of STS.	It is common to have different views by different people on an SSI.
	T further explained that if a link in the figure is broken due to societal decision (in this case – ban of shark skin swimsuits), newer technology will not be produced and the science related to that technology may be inhibited, which may indirectly affect the birth of other useful products. In this case, it could be new types of materials for the use on boats to make it move faster.	Societal decision can (negatively) affect the advances in science/technology.

Immediate Reflection and Modification of Lesson Implementation

Wayne and Kyle conducted their own class on the same day. Wayne's lesson with class 4F took place in the morning while that of Kyle with class 4E took place in the afternoon. Kyle and Alice joined Wayne's morning lesson. Wayne and Kyle had put great effort in locating relevant videos, historical world records of swimming and track events, and other informative data for various lesson activities. During the lesson, videos like the return of Liu Xiang back on the track, which obviously captivated students' interest and attention, Kyle and Alice felt the lesson was too loaded in the first half that the time left for student discussion in the role play activity and their subsequent reporting was limited (with less than 15 min). Opportunities for quality student interaction and arguments were scarce. Kyle was very sensitive and reflective in identifying components with good engagement of students, for example, the 110 m-hurdle race with Liu Xiang. As an observer of Wayne's lesson, Kyle acted as a critical friend for Wayne, identifying several lesson components that could be run more efficiently by (1) rephrasing some questions to ensure clarity and good understanding by students, (2) showing only selective parts of some videos while retaining its original purposes, and (3) rearranging some videos to make better linkage from one component to the next. Kyle's immediate postlesson reflection on the codesigned lesson plan turned out to be very successful in addressing some inadequacies identified in Wayne's implementation.

Learning Outcomes of NOS Teaching

Throughout Kyle's lesson, the students were actively engaged and they contributed relevant and interesting ideas. With the modifications, the pace of the lesson flow was swifter as compared with Wayne's class in the morning. The time allocation for the role play and subsequent reporting was about 23 min which allowed more well-thought-out views in supporting the students' stances for each role. The following excerpts taken from student presentations reveal the stance and arguments presented by a group of representing scientists who developed the swimsuits and a group of sponsors of the swimsuits.

Views of Scientists

We do not agree to ban the use of the shark skin swimsuit. Our intention of designing shark skin swimsuit is to enhance the swimming speed of athletes. It is great that our product (shark skin swimsuit) is effective. However, the Olympic Council banned it…Improvement is made through competitions. If the Council does not allow athletes to use it, how can our invention spread all over the world? As scientists, we want to improve the quality of our technology. We design such kinds of products for humans. We hope that every country would support us. The use of our product by participants proves our success. Therefore, we absolutely support the use of shark skin swimsuit in sports competitions.

Views of Sponsors

We do not agree. The swimming sport is not to compete for the slowest, but for the fastest. We can give the shark skin swimsuits to all the swimming athletes, so that it can be fair and all the athletes can be faster. If we ban the shark skin swimsuit just because it is different from the traditional one, should all athletes be required to use the same swimsuit?

Right after Kyle's lesson and before the postlesson discussion with Kyle and Wayne, Alice chatted with a girl of Kyle's class to see if she captured the key message of the attempt of the teachers in teaching NOS with a specific focus on the STS connections. Alice delightfully shared the student's responses to her questions with Wayne and Kyle.

(A – Alice; K – Kyle)

A: I asked her "... if you were to chat with your family about this lesson, what would be the take-home message of this lesson that you would share with them?" What message do you think she remembers best?

K: The message that she remembered best?

A: Yes...impressed her most...you tried convey quite a few messages in this lesson.

K: Surely I hope she can remember my summary made at the end of the lesson

A: Yeah...

K: ...however, I am afraid she won't remember this main point.

A: OK, what did you wish she could get from your conclusion?

K: Um ... Now we have banned the use of shark skin swimsuits, which would affect the development of science and technology. It will also influence our society. There will be fewer technology products.

A: Um...You will be very happy. She actually mentioned this message...Yes, she did.

K: That's great!

A: Yes. She did mention that.

K: Wow, at least I know one student got this message, right? This is so pleasing!

A: Yes. She actually pondered for a while before answering me. She didn't simply recall what you just mentioned... She thought seriously and told me that ... what she remembered best was that "now, the use of shark skin swimsuits is banned...however, shark skin swimsuits may be applied to athletic games some days in the future". You know, probably she was prompted by your joke you made at the end of the lesson [would bring along a shark skin swimsuit for the running in the School Sports Day tomorrow]. She said further, "The ban may affect the development in other things...In athletic games, there are also many technological products, but they are not banned."

K: She could even get this point?

A: Yes!
K: She did very well!
A: Yes, she did very well indeed.

What else could be more rewarding for teachers to know that their effort in preparing a lesson has turned nicely into students' learning? To us, as teacher educators, what else could be more rewarding when hearing teachers' excited acclaims on their students' learning? We shall conduct more interviews with the teacher participants and their students toward the end of this professional development project to understand more about the relationship between the teaching NOS and students' learning outcomes.

Implications for Professional Development

The data collected from the teachers and students seems to suggest that their learning of NOS through the use of timely SSIs has been promising. Teachers who experienced the learning more than 18 months ago through the SSIs in the SARS incident still possess very good NOS understanding and still recall the various contexts in which they learnt relevant NOS. Students' discussion during the role play and their subsequent presentations also indicated that understanding of certain NOS features has been achieved, in particular the STS connections which were highlighted in the lesson on the *Ban of Shark Skin Swimsuits*.

We see several similarities of our teacher training lesson in promoting teacher NOS understanding and Wayne and Kyle's lesson in promoting students NOS understanding (in particular STS connections). First, both lessons used timely contexts that were relevant to learners. SARS has captured much attention of all people in Hong Kong where it was most hard-hit by the epidemic. Thus it has been highly relevant to teachers in Hong Kong and its timeliness would last for years due to the unforgettable experience and the scare of a similar crisis due to another outbreak, like the recent swine flu. For Wayne and Kyle's lesson, sports have been mostly liked by teenage students. Wayne and Kyle deliberately included examples of a diverse range of sports. Kyle reminded us that the effectiveness would be enhanced if the teachers also find the context relevant. He said, "I love sports very much and enjoy looking up information about sports…feel excited and at ease when talking about them to students." They also capitalized on the timely context of the ban of shark skin swimsuits that happened a couple of months before their lessons. The use of the return of Liu Xiang (favorite sports star of many Hong Kong students) to the 110 m-hurdle competition 2 days before their lessons was both timely and relevant to the students. Students were obviously attracted to the many video clips of sports competitions and sports training used by them in the lesson. Similarly, the SARS teacher training materials also made heavy use of videos of scientists who fought against SARS and documentaries produced by reputable media about the inside story of SARS. Teachers felt excited to hear these inside stories. These videos were immensely helpful in the reconstruction of the relevant SSIs to give learners a sense of authenticity about the events shown. This is the second common feature of the SARS training materials and Kyle and Wayne's instructional materials that motivated the learners. Third, both lessons engaged learners in critically reflecting on dilemmas, issues, and problems faced in decision making. Instead of being a purely rational and logical exercise, such decision making mandated considerations of social demands, sentiments of public, incomplete scientific findings and possible consequences to different stakeholders, etc.

Although we have reported the unique features of the SARS instructional materials that may account for its effectiveness in promoting teachers' understanding of NOS (Wong et al., 2008), we have not shared these features explicitly with the teacher participants who joined our professional development project. Our oversight was due to our belief that another similar incident as SARS would be unlikely, which embodied so many NOS features in one case. It was interesting to find that Kyle and Wayne somehow translated some of the unique features of the SARS training session to their teaching of NOS. Kyle and Wayne's lesson demonstrated that the same features could still be nicely applied to a simple SSI embedded with just a few NOS aspects. Indeed Kyle said that he felt fairly at ease in running the role play activity as he had experience using one of the sets of teaching resources, LASIK, developed in our previous project from 2005 to 2007. In the LASIK package, there was an activity in which students were asked to advise, with justification, people of different backgrounds, careers, and eye problems whether they are suitable for taking laser-assisted in situ keratomileusis (LASIK) surgery. Kyle further commented that if he had not used the LASIK package, it would have been quite daunting for him to design a lesson teaching STSE on his own.

> I used the LASIK package before...all information are there...I didn't need to prepare anything extra, only implemented the activities...After the implementation, I felt relatively easily to design a new one.... If I have to design one right from the beginning, it would be hard. But if there is a sample, like LASIK...also on STS...I can follow it like a model.

The learning of Wayne and Kyle seems to suggest that teachers did learn from exemplars, or through the modeling of exemplars. They were able to make use of some features of exemplar SSIs, and to apply them in designing novel SSIs that fitted the needs of their students. The capability of identifying key features of exemplar materials and transferring such features to personalized teaching resources seems to be essential for teachers' development. Developing such capability seems to be a way for our future teacher training programs. We envisage that such modeling is potentially applicable to other approaches of teaching NOS, for example, doing inquiry, history of science, etc.

References

Abd-El-Khalick, F., Bell, R. L., & Lederman, N. G. (1998). The nature of science and instructional practice: Making the unnatural natural. *Science & Education, 82*, 417–437.

Abd-El-Khalick, F., & Lederman, N. G. (2000). Improving science teachers' conceptions of the nature of science: A critical review of the literature. *International Journal of Science Education, 22*, 665–701.

Bartholomew, H., Osborne, J., & Ratcliffe, M. (2004). Teaching students "ideas-about-science": Five dimensions of effective practice. *Science & Education, 88*, 655–682.

Bennett, J., Lubben, F., & Hogarth, S. (2007). Bringing science to life: A synthesis of the research evidence on the effects of context-based and STS approaches to science teaching. *Science & Education, 91*(3), 347–370.

Bianchini, J. A., Johnston, C. C., Oram, S. Y., & Cavazos, L. M. (2003). Learning to teach science in contemporary and equitable ways: The successes and struggles of first-year science teachers. *Science & Education, 87*, 419–443.

Brickhouse, N. W., & Bodner, G. M. (1992). The beginning science teacher: Classroom narratives of convictions and constraints. *Journal of Research in Science Teaching, 29*, 471–485.

CDC. (1998). *Science syllabus for secondary 1–3*. Hong Kong: Curriculum Development Council.

CDC. (2002). Physics/Chemistry/Biology curriculum guide (Secondary 4–5). Hong Kong: Curriculum Development Council. Retrieved August 15, 2008, from http://www.edb.gov.hk/index.aspx?nodeID=2824&langno.

CDC-HKEAA. (2007). *Physics/Chemistry/Biology/Integrated science curriculum guide and assessment guide (Secondary 4–6)*. Hong Kong: Curriculum Development Council and Hong Kong Examinations and Assessment Authority.

Higgins, J., & Learoyd, S. (Producer). (2003, May 29). *SARS: The true story* [Television broadcast]. North Sydney, N.S.W.: BBC Worldwide.

Hodson, D. (2006). Why we should prioritize learning about science. *Canadian Journal of Science, Mathematics and Technology Education, 6*, 293–311.

Khishfe, R., & Abd-El-Khalick, F. (2002). The influence of explicit/reflective versus implicit inquiry-oriented instruction on sixth graders' views of nature of science. *Journal of Research in Science Teaching, 39*, 551–578.

Khishfe, R., & Lederman, N. G. (2006). Teaching nature of science within a controversial topic: Integrated versus nonintegrated. *Journal of Research in Science Teaching, 43*, 395–418.

Kolstoe, S. D. (2000). Consensus projects: Teaching science for citizenship. *International Journal of Science Education, 22*(6), 645–664.

Lederman, N. G. (1992). Students' and teachers' conceptions of the nature of science: A review of the research. *Journal of Research in Science Teaching, 29*, 331–359.

Lederman, N. G. (1999). Teachers' understanding of the nature of science and classroom practice: Factors that facilitate or impede the relationship. *Journal of Research in Science Teaching, 36*, 916–929.

Lumpe, A. T., Haney, J. J., & Czerniak, C. M. (1998). Science teacher beliefs and intentions to implement science-technology-society (STS) in the classroom. *Journal of Science Teacher Education, 9*, 1–24.

Oulton, C., Dillon, J., & Grace, M. M. (2004). Reconceptualizing the teaching of controversial issues. *International Journal of Science Education, 26*(4), 411–423.

Rubba, P. A., & Harkness, W. L. (1993). Examination of preservice and in-service secondary science teachers' beliefs about science-technology-society interactions. *Science & Education, 77*, 407–431.

Ryder, J. (2002). School science education for citizenship: Strategies for teaching about the epistemology of science. *Journal of Curriculum Studies, 34*(6), 637–658.

Sadler, T. D. (2004). Informal reasoning regarding socioscientific issues: A critical review of research. *Journal of Research in Science Teaching, 41*(5), 513–536.

Sadler, T. D., & Zeidler, D. L. (2005). Patterns of informal reasoning in the context of socioscientific decision making. *Journal of Research in Science Teaching, 42*(1), 112–138.

Schwartz, R. S., & Lederman, N. G. (2002). "It's the nature of the beast": The influence of knowledge and intentions on learning and teaching nature of science. *Journal of Research in Science Teaching, 39*, 205–236.

Tao, P. K. (2002). A study of students' focal awareness when studying science stories designed for fostering understanding of the nature of science. *Research in Science Education, 32*, 97–120.

Tobin, K., & McRobbie, C. J. (1997). Beliefs about the nature of science and the enacted science curriculum. *Science & Education, 6*, 355–371.

Tsai, C. C. (2001). A science teacher's reflections and knowledge growth about STS instruction after actual implementation. *Science & Education, 86*, 23–41.

Wong, S. L., Hodson, D., Kwan, J., & Yung, B. H. W. (2008). Turning crisis into opportunity: Enhancing student teachers' understanding of the nature of science and scientific inquiry through a case study of the scientific research in Severe Acute Respiratory Syndrome. *International Journal of Science Education, 30*, 1417–1439.

Wong, S. L., Kwan, J., Hodson, D., & Yung, B. H. W. (2009). Turning crisis into opportunity: Nature of science and scientific inquiry as illustrated in the scientific research on Severe Acute Respiratory Syndrome. *Science & Education, 18*, 95–118.

Wong, S. L., Yung, B. H. W., & Cheng, M. W. (2010). A blow to a decade of effort on promoting teaching of nature of science. In Y.-J. Lee (Ed.), *The world of science education: Handbook of research in Asia* (pp. 259–276). Rotterdam, the Netherlands: Sense Publishers.

Wong, S. L., Yung, B. H. W., Cheng, M. W., Lam, K. L., & Hodson, D. (2006). Setting the stage for developing pre-service teachers' conceptions of good science teaching: The role of classroom video. *International Journal of Science Education., 28*(1), 1–24.

Zeidler, D. L., Sadler, T. D., Simmons, M. L., & Howes, E. V. (2005). Beyond STS: A research-based framework for socioscientific issues education. *Science & Education, 89*(3), 357–377.

Chapter 15
Metalogue: Preconditions and Resources for Productive Socio-scientific Issues Teaching and Learning

Siu Ling Wong, Dana L. Zeidler, and Michelle L. Klosterman

The Use of Mass Media

Klosterman: It seems natural to include news broadcasts, clips from documentary films, and other media sources to introduce teachers and students to socio-scientific issues (SSI) and to highlight current features of nature of science (NOS). Like the SSI used as contexts in this chapter, media is timely, captures student attention (and therefore qualifies as being "relevant"), and can highlight the different perspectives of individuals concerned with SSI. As someone interested in classroom use of media and how science is represented in the media, I would like to know more about how and why the media clips were selected. In the Severe Acute Respiratory Syndrome (SARS) example, it was clear that multiple perspectives were represented through the clips and the accuracy of the information presented was considered. Was the goal to present an overall picture of the issue? Did the teacher use media for a similar purpose? Did any of the other teachers you observed incorporate media from multiple perspectives within one lesson? What impact, if any, do you think the type or content of media might have on student's decision making around the issues?

Wong: We produced the SARS instructional package intent on making use of the unforgettable SARS story to demonstrate a rich list of NOS elements. In our choice of media clips, we perused all accessible documentaries, news records, interview data of scientists, etc., to represent the historical development of the epidemic and the associated rapid scientific developments. However, I cannot claim that we presented an overall picture of all issues seen in the SARS example. We mainly focused

S.L. Wong (✉)
Faculty of Education, University of Hong Kong, Pokfulam Road, Hong Kong
e-mail: aslwong@hku.hk

D.L. Zeidler
College of Education, University of South Florida, Tampa, FL, USA

M.L. Klosterman
Department of Education, Wake Forest University, Winston-Salem, NC, USA

T.D. Sadler (ed.), *Socio-scientific Issues in the Classroom: Teaching, Learning and Research*, Contemporary Trends and Issues in Science Education 39,
DOI 10.1007/978-94-007-1159-4_15, © Springer Science+Business Media B.V. 2011

on elaborating the details of a series of SSI in the historical development of the epidemics which were closely linked to the series of scientific findings related to the SARS disease. There were other issues on which we did not elaborate (e.g., the delay of government warning about the seriousness of the disease due to the potential economic impacts on tourism-related industries, the suspension of all schools due to parental concerns, and challenges faced by hospitals related to resource allocation). However, for each SSI in our package, we incorporated media clips, pictures, and scientists' interviews, which presented the situation from multiple perspectives.

In the shark skin swimsuits unit developed by Kyle and Wayne, they made use of video clips to help students see a broader view about fairness of sports competitions from different perspectives as seen in Table 14.5. I believe that such information would invite students to be mindful of seeing fairness not limiting to what was used in the sports grounds. However, due to the nature of the role play activity, students were asked to consider the issue from the perspectives of a certain role (spectators, athletes, scientists developed the swimsuits, sponsors of swimsuits). As the key goal of the current projects is to develop teachers' pedagogical content knowledge in teaching NOS in their teaching of science, most teachers used the media to present the relevant historical background of the scientific development (also a key feature of the SARS Story) related to the topics of science that they taught.

Modeling Teacher Practices

Klosterman: I think this chapter does an excellent job of highlighting the importance of modeling for the development of teacher practice. The authors highlight the potential benefits of a three-step sequence to teacher training. The teachers, Wayne and Kyle, were able to effectively address elements of NOS using SSI contexts after first engaging in professional development around NOS and SARS, followed by viewing and discussing models of success, and then developing their own units. Progressively removing levels of support (scaffolding) has proved very successful in a variety of teaching contexts. However, as I was moving through the chapter, I wondered how Wayne and Kyle were so successful at developing their own units after learning about NOS through the SARS unit and viewing exemplars of teaching practice. The transformation from seeing to doing is a big leap. At the end of the chapter the authors mention that Kyle tried to use another previously developed package – laser-assisted in situ keratomileusis (LASIK) – before developing his own unit with Wayne. I do not think we should underestimate the potential impact that this "trial" run had on Kyle's success with implementing the shark skin swimsuit unit. As the researchers, did you notice if any of Kyle's experiences using the LASIK package contributed to the planning discussions for the shark swimsuit unit? And did Wayne have any similar practice using one of the previously developed units prior to designing the shark swimsuit unit?

Wong: As you rightly pointed out, Kyle shared with us during his reflection on his own learning of teaching NOS that the use of LASIK package had paved way to his own design of NOS teaching. The LASIK package along with other packages produced in an earlier project completed in summer 2007 was indeed intended to provide science teachers with some exemplars of teaching resources as references for teaching NOS.

Like Kyle, Wayne also used the LASIK package; however, I must admit that during the planning discussions on the development of the shark skin swimsuit unit, I was not aware that both of them had read and used the LASIK package. I did not become aware of these experiences until reading their reflective statements later in the process. Kyle suggested that without the reflection activities, he might not have come to realize the favorable impact of the use of the LASIK package on his own development of NOS teaching resources. His views reflected that exemplars of instructional materials provide an invaluable intermediate step to planning a more independent unit. More exposure to teaching ideas through sharing and use of well-designed resources will enhance teachers' pedagogical content knowledge in teaching NOS and SSI. They will be more sensitive to possible contexts and materials for turning them into instructional units.

SSI as a Context for Instruction

Dana: I was very interested in this chapter inasmuch as I recognize the opportunity and importance of using SSI to provide both a context for engaging students in conceptual understanding of content and for providing a framework for epistemological understanding of NOS. The tables the authors provide clearly illustrate the conceptual links between SSI and NOS. As I explored the chapter, three issues related to the connections between SSI and NOS emerged for me.

First, the authors suggest that because of a lack of conceptual understanding as to the robust epistemological nature offered by historical NOS stories, and not fully understanding the rationale behind the design of (NOS) instructional materials, teachers (and consequently their students) were not able to develop an appreciation of NOS in social contexts. While some teachers found the stories interesting, this begs the question as to whether a minimum threshold of epistemological sophistication is needed before any new curriculum or approach (e.g., NOS, SSI, STSE, Inquiry, Collaborative Learning) can be effective.

Wong: Historical stories of science and scientists are commonly found in many science textbooks in both the West and East. Yet there have been many studies reporting on teachers' and students' lack of adequate understanding of NOS. Such findings indicate that an appreciation of the embedded NOS aspects does not come naturally. Indeed, most of the NOS elements are the theorized understandings about science crystallized from years of academic studies about science. It is rather difficult for science teachers and students to figure out these ideas by themselves through just listening to the historical stories of science. Similar to the learning of scientific concepts, we are skeptical about an extreme discovery approach. We rely on a more guided approach that uses targeted activities to help teachers and students appreciate NOS in history of science or linkages between SSI and NOS.

Interestingly, but disappointingly, we found that some myths about science and distorted images of scientists were reinforced by some science stories if we did not guide or make explicit connections to help teachers see the relevant NOS features embedded in the stories. For example, when we told the story about the treatment of stomach ulcers to our preservice and in-service teachers, many of them were most attracted to the episode in which Dr. Marshall tested his hypothesis by being a clinical trial subject himself. Some comments from teachers like "See…only scientists would be so odd and crazy" reflected a reinforcement of the image of weird scientists who are detached from the world and different from normal people. Some focused on 'incidental discovery' rather than appreciating scientists' perseverance in collecting empirical evidence and the courage required in challenging long-standing beliefs when they noted clues to the cause of stomach ulcers through careful observation and attention to details. Such outcomes are not unsurprising as 'observation and data interpretation are theory-laden' – an important aspect of NOS! When teachers do not have an adequate epistemological understanding or minimum threshold of epistemological sophistication, it is very easy for them to miss the intended targets. Thus any curriculum reform with new teaching approaches will likely fail if teachers have not acquired the expected level of understanding.

After a series of projects in promoting teachers' pedagogical content knowledge (PCK) in teaching science, technology, society, and environment (STSE), NOS, SSI, and scientific inquiry, in the past decade, we also know that it is most important for teachers themselves to value such ways of teaching or approaches. In one of our ongoing projects, when teachers were encouraged to design their own teaching units, we noted different teachers have their own preference of contexts when they infused NOS aspects in their lessons. Many favored the use of history of science, some preferred doing scientific inquiry, and others tended to place NOS teaching in SSI. We are conducting interviews with these teachers to probe their perception of the value of teaching NOS. Our preliminary data suggest that there were strong linkages between their choice of contexts/instructional activities and their perceived values of teaching NOS. We also find some teachers' perceived values are influenced strongly by the rationales put forward in the new curriculum guides while some are more influenced by available instructional resources. We hope to report the full findings of these follow-up interviews in an independent article.

Zeidler: Second, you cite Hodson's (2006) claim that in order for curriculum materials (NOS, STSE, SSI, etc.) to have "street credibility" – it needs to be "developed by teachers, for teachers." I think much of this is true for most professional development settings. Owning horses, I also know that the most nutritious food is of little consequence if it is also not palatable. If the horses won't eat it, it matters not how good it is for them! While I am not equating students to livestock, our research has shown that for SSI to be effective, students must find the ideas contained therein personally relevant and meaningful. Therefore, I would like to suggest that we be sensitive to providing the opportunity and conditions necessary for students to raise their own questions and develop their own units of study within the goals of the curriculum. Given certain parameters and guidance, they can often do this quite well – and move a little further down the trail.

Wong: I like the suggestion of providing students opportunity and necessary conditions to raise their own questions and develop their own units as much as I like your analogy of feeding horses. Your suggestion reminds me of a few lessons I observed some years ago in Wayne and Kyle's school by their vice-principal, Larry. Larry is an experienced physics teacher who was one of the recipients of the Award for Teaching Excellence organized by the Hong Kong Education Bureau. He has a practice of letting his students take up the teaching of selected physics topics in turn. I was in one of these lessons when a group of students were explaining how lenses are used for correction of eye defects. Apparently this was a topic highly relevant to students (over 80% of Hong Kong students by the age of 16 suffer from nearsightedness and many of their grandparents also rely on reading glasses). I was impressed by Larry's patience and 'tolerance' in keeping quiet when the students-in-charge of the lesson got the concepts wrong about presbyopia (lack of accommodation of near objects upon advancing age). His tolerance was paid off after the 'incorrect' explanation went on for about 5–10 min when a few of their fellow students started to raise questions based on observations of the glasses of their grandparents. The inaccurate concepts were corrected through active interaction, negotiation of conflicts, and provision of evidence in support of one's arguments. I was convinced then that when students got interested in a topic, self- and peer-learning could be more fun and effective. Although the lessons I observed in the school were mainly on subject knowledge, I can imagine when students are encouraged to go beyond subject knowledge to integrate related NOS/STSE/SSI, their enthusiasm in preparing materials for teaching and learning of the topics will be even greater.

Zeidler: The third issue I will raise here deals with the framing of SSI. You properly suggest that SSI are controversial and can provide a context for epistemological understanding of NOS. However, I also think it important that SSI contain some feature of ethical tension – to create some degree of moral dissonance. This is important in terms of generating interest, resolving conflict, challenging presuppositions of evidence and norms, creating character, advancing developmental reasoning, and the like. I can see the potential in using the SARS scenario of where this may exist (e.g., Tragic Outbreak at Amoy Gardens), but am left wanting to know more about how this potential was leveraged and tapped? This is a key element of SSI, as I envision it.

Klosterman: I would like to extend Dana's third comment and ask: What do we consider SSI as? This article raises the issue that STS issues and SSI are related, but to what extent? As those interested in SSI-based instruction and outcomes, I think we need to be careful that we clearly define the differences between SSI, STS, and even problem-based learning (PBL) scenarios.

Wong: We agree with your views that many SSI can provide good contexts to induce moral dissonance which challenges preconceptions and norms, encourages reasoning and balance of pros and cons, inculcate values for character-building. Your questions prompted us to consider if the Tragic Outbreak at Amoy Gardens, which aroused intense ethical tension among different stakeholders (regarding the unprecedented government order to quarantine the residents of the seriously infected block to a rural camp site), could serve as a good context to create moral dissonance and subsequently achieve a number of invaluable learning outcomes. Upon reflection, our team considered the tragic outbreak might not be as effective when compared with other SSI that deal with situations that are still unfolding. For the Amoy Gardens incident, all stakeholders including those who strongly disagreed with the government's order initially could see that the government order turned out to be effective in terms of halting the mysterious spread even if it was not the best decision. Due to the known outcomes, this issue did not generate the ethical tension Dana references.

Dana: These responses were quite illuminating and provided much insight into this important area. It seems the authors agree, in principle, with the conceptual notion of a "Threshold Model" of epistemological understanding that drives subsequent socioscientific reasoning, NOS understanding, and the like. I look forward to your further work on how teachers' perceived values are influenced either wittingly or unwittingly by knowledge of instructional contexts and pedagogy.

Your were truly fortunate to have someone like Larry work with your students and be able to take a "back seat" to the ideas that students were generating in class. Your anecdotal observations of increased student interest and participation are consistent with our observations of student engagement with a robust SSI approach. In our case, a perceptive teacher, like Larry, was able to honor the students' ability to propose their own arguments and subtlety guide them when necessary. It is sometimes difficult to turn over the reigns to students in pursuit of their own understanding but the dividends can pay off in terms of engagement and authentic learning.

I also appreciated the author's nuanced interpretation of how ethical tensions – an important part of SSI, may be ameliorated by the known outcomes of historical events. This is something I will personally give more thought to and it has important implications. It would seem that moral dissonance – hence ethical tensions – central to SSI, is more gripping when the outcomes are ambiguous, uncertain, probabilistic in nature, and where "experts" have fundamental disagreements – all the proper precursors for an effective SSI.

Reference

Hodson, D. (2006). Why we should prioritize learning about science. *Canadian Journal of Science, Mathematics and Technology Education, 6*, 293–311.

Chapter 16
Enacting a Socioscientific Issues Classroom: Transformative Transformations

Dana L. Zeidler, Scott M. Applebaum, and Troy D. Sadler

> The first semester, I was constantly frustrated because the students seemed to be incapable of understanding Nature of science (NOS), the relevance of Socioscientific Issues (SSI) in their daily lives, and the value of evidence-based argumentation. Actually, my first impression was that this was an exceptionally unintelligent group…though I often doubted my teaching ability, as the problem was global throughout six classes. (High School Teacher's Reflections of First Half of First Quarter)

Sociomoral discourse, argumentation, and debate are necessary elements in a socioscientific issues-centered classroom. While these factors are fundamental in realizing a socioscientific issues (SSI) curriculum, related pedagogical factors, such as a commitment to inquiry, enacting opportunities for the cultivation of character, and conceptualizing the role of the nature of science (NOS) are consistent with progressive views of science teaching and scientific literacy (Sadler & Zeidler, 2009; Zeidler & Sadler, 2010). Further, classroom research has demonstrated that a fully enacted SSI approach to science education becomes a transformative process for participating students and their teacher. Successful transformation occurs when the teacher-centered approach shifts to a student-centered classroom and the science curriculum becomes issues-driven. Further, the results of this shift may be said to be transformative when students' discovery of scientific concepts emerges out of socioscientific issues.

Introduction of novel pedagogy is often met with resistance from experienced teachers, as well as students who have become comfortable with classroom and instructional expectations. The unique dynamics of SSI-based instruction requires establishment of new relationships among teachers, students, and researchers, a series of transitions likely to impinge on established classroom social norms that can

D.L. Zeidler (✉)
College of Education, University of South Florida, Tampa, FL, USA
e-mail: Zeidler@usf.edu

S.M. Applebaum
College of Education, Palm Harbor University High School, Palm Harbor, FL, USA

T.D. Sadler
School of Teaching and Learning, University of Florida, Gainesville, FL, USA

T.D. Sadler (ed.), *Socio-scientific Issues in the Classroom: Teaching, Learning and Research*, Contemporary Trends and Issues in Science Education 39,
DOI 10.1007/978-94-007-1159-4_16,

be both subtle and overwhelming. The social norms consist, in part, of implicit and explicit expectations established about the roles of the teacher and the students in that classroom. Frequently, we discover that such norms are firmly established after years of entrenched teacher-centered instruction. Historically, the education initiatives and changes to classroom norms have been structurally superficial in nature, where a new pedagogical approach may be implemented, yet students remain dependent on the teacher for instruction regarding the method of learning and which information has value. In contrast, an SSI curriculum provides for fundamental and deep structural changes that reorganize those norms at a core level of social networking and understanding. In the former case, one may think about how to rearrange desks and chairs or initiate a "new teaching strategy" without much thought to the deeper core structures of the classroom. In this case, there would certainly be changes in the classroom and between student-teacher interactions. We could speak about that experience as having undergone a certain type of transformation that was not necessarily transformative, in that the introduction of those new elements merely represented changes in surface features of normative and structural relationships. By contrast, if a teacher and his or her students experience pedagogical changes in ways that alter the *fundamental nature* of social discourse and community, we should understand that the experiences bringing about this deeper shift in epistemic beliefs are different in kind, and may properly be said to be *transformative*.

Many contemporary educative experiences consist only of transforming surface level characteristics of the classroom setting. It has become commonplace to find examples of reform in education that addresses only surface structure issues (e.g., high stakes testing, redistricting, the No Child Left Behind mandates and the like). Comprehending the shift from traditional classroom practice to an SSI framework requires an understanding of the distinction between the pedestrian transformation from new surface structure reform mantras and deep structure transformative practice—the latter represents fundamental normative shifts in core pedagogical expectations on the part of teachers and a sense of empowerment in terms of assuming responsibility for learning on the part of students. Our approach with SSI is an instantiation of progressive education, a concept that necessitates transformative shifts in how we understand science education and science teaching.

The purpose of this chapter is to present a description of a comprehensive research project that used a SSI framework to dramatically alter a high school level science curriculum and implemented the necessary pedagogical practices that transformed the teacher and his students from actors within a very traditional classroom context to participants in a progressive educational setting. It is our claim, that this accomplishment was based upon the transformative nature of a robust SSI approach that facilitated deep structural changes necessary to accept and understand the complexities of developing a progressive science teaching curriculum. We initiated this project to explore issues and challenges associated with the implementation of an SSI-driven curriculum. Specifically, the two primary objectives of this chapter are to: (1) Describe the conceptual design and implementation of a year-long SSI-driven course and (2) Outline a framework for SSI instruction, with suggestions and caveats that emerged from the design-based research associated with this implementation.

Pedagogy and Deep Structure Reality

> My mood as an educator has improved. Once I realized what needed to be accomplished, the projects progressed easily, rapidly, and with a great deal of enjoyment for the class and me. It should be noted that part of the differences was the result of my decision to put the projector, computer and transparencies to rest. The technology was beautiful and well organized, but exceptionally impersonal. When I rolled the white board to the front of the class and drew pictures and wrote down what the class was saying, they became involved in their own education. (Teacher's Reflections at the End of the First Quarter)

As we suggested above, many reform attempts to impact the educative experiences of our children consist of transforming surface level characteristics of the school setting. In understanding the shift from traditional classroom practice to the SSI framework, it is important to note the distinction between transformations that occur in classroom practice and transformative practices—the latter represents deep structural shifts both in teacher pedagogy and students' conceptual understanding of subject matter and reflective thinking. Figure 16.1 illustrates the contrast between two far ends of a continuum of instructional paradigms: traditional methods of instruction on the one hand, and progressive instruction on the other. We view our approach of SSI as an instantiation of progressive education—an approach that necessitates transformative mind-shifts in how we think about science education. The SSI framework we propose, necessitates deep restructuring—and recreating pedagogical reality in science education if we wish to arrive at the outcomes (autonomy, responsibility, etc.) often associated with progressive education.

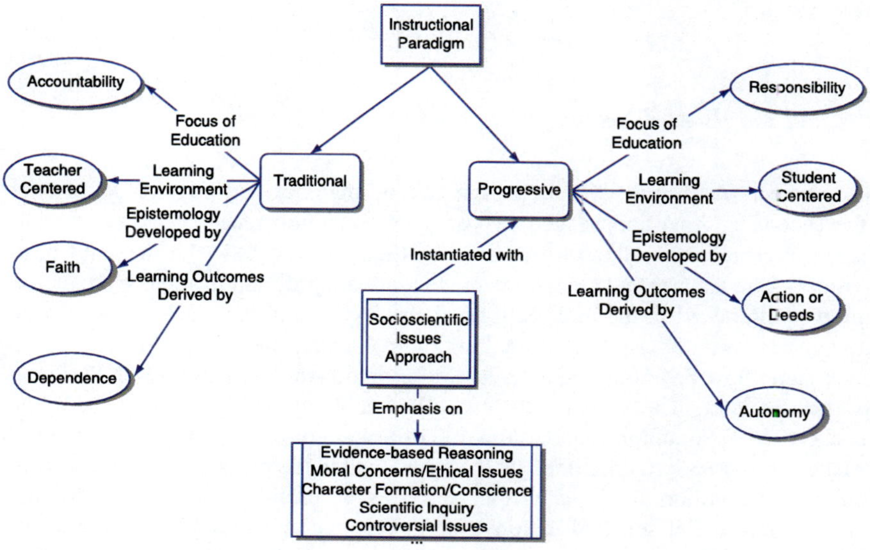

Fig. 16.1 Continuum contrast of instructional paradigms

The progressive venture has its historical roots in the experientialists (e.g., Jean Rousseau (1712–1778); Friedrich Froebel (1782–1852); John Dewey (1859–1952)), who viewed education as an individual (student)-centered social process where epistemic meaning is derived from the collective meanings of shared social experiences (actions and deeds) where autonomous thinking emerges as a natural outcome. This stands in stark contrast to a traditional approach, influenced by the thinking of social behaviorists (e.g., Johann Herbart (1776–1841); Wilhelm Wundt (1832–1920); Edward Thorndike (1874–1949)), which is dominated by a teacher-centric emphasis that focuses on mastery of prescribed bodies of fixed and discrete knowledge. Such an approach tends to produce in students, a dependence on the teacher where, in the extreme form, epistemic meaning is an act of faith. In this case, students are taught in an autocratic fashion, where text and authority produce unreflective students and inflexible knowledge that appears immutable. Hence, whereas the traditional approach develops knowledge and beliefs associated with the justification of that knowledge, through dogma or nonevidential (faith) methods, SSI begins with evidence-based reasoning, and challenges the normative assumptions of the knowledge claims students hold. Emphasis is placed on engaging students in the activity of scientific inquiry, and connecting that inquiry to contextualized social-scientific issues. In prioritizing personal and collective responsibility as an outcome of progressive philosophy, the cultivation of conscience through the formation of character is achieved by students engaging in discourse and making decisions about various moral problems (Zeidler & Sadler, 2008). It is important to emphasize that the final epistemological claims students hold are less important than the means by which they were developed. Under our SSI framework, students ought to be able to provide evidence-based justification for a position and exhibit an openness to reflect on that position in light of new evidence.

Project Goals and Setting

The project was initiated when a high school anatomy and physiology teacher (the second author) approached two science education researchers with established track records related to research and teaching of SSI (the first and third authors). The teacher had recently begun a graduate program in science education and was interested in the intersections of SSI, NOS, and the teaching of science content. He wanted to conduct a longitudinal experiment with his classes to understand how SSI could be leveraged to promote student learning of NOS and science content. The researchers enthusiastically agreed to partner with the teacher. The first author assumed the role of project director. He worked with the teacher on a weekly (sometimes daily) basis, visited the classroom frequently (to monitor curriculum implementation, serve as a resource and mentor for the teacher, and model selected lessons), and coordinated a group of graduate students who collected various forms of data. The third author served in a consulting role for both the teacher and the project director. All three authors collaborated on the design of the year-long curriculum and the development of specific learning activities within the curriculum. The three also maintained frequent communications (via phone, email, and face-to-face visits) to work through implementation challenges and research issues as they arose.

Teacher–Researcher Relationships

Finding a teacher with the fortitude to commit to a longitudinal intervention incorporating SSI, NOS, argumentation, and a dose of reflective judgment into their classroom for one academic year is no doubt a rarity. Such a teacher must be comfortable with his content and teaching abilities and demonstrate an unwavering commitment to inquiry into his own teaching practices and results. Labaree (2003) notes many of the unique challenges of teacher-as-researcher. Both teachers and researchers tend to conceptualize their roles as transformational, but the targets of their efforts are somewhat different. Teachers are personally invested in the lives of their students and work to transform those lives. Researchers typically seek more global effects: they work to improve education systems, curricula, and/or theory by creating better understanding of teaching, learning, and learning environments. Ultimately, these goals are complementary, but in the immediacy of an intervention study, the variable perspectives can create challenges for teachers trying to help their students learn and researchers trying to understand how students learn. Table 16.1 reveals some of the key issues that arose regarding the tension between the teacher and the researchers, and the ultimate resolution of those issues in the context of the current project. Because of the high degree of professionalism and mutual respect between the teacher and researchers, problems were brought to light and discussed with candor and humor. All three authors were very closely involved with not only the research but also curriculum design and implementation.

Because planning SSI units was done in concert with the teacher, we were able to question one another about the meaning and intent of particular investigations without undermining anyone's sense of ownership or professionalism. At times, where there may have existed a disconnect between the researchers' vision of how the various components of SSI were to be implemented relative to design features of the study, and the teacher's concern for immediate observable outcomes, we were able to have spirited discussions and come to a consensual resolution, much like we expected from the students who underwent this SSI curriculum. As the academic year unfolded, as suggested by the subtext of Table 16.1, we experienced transformations that were, indeed, transformative in nature. Perhaps the best way to describe how we fundamentally changed our understanding of working through various issues that arose is to suggest that all of us continually experienced a succession of mini-epiphanies about what constituted critical issues in the context of the research design and in the context of the real needs for the teacher and his students, and were able to negotiate and achieve common understanding and resolution as issues arose.

Table 16.1 Teacher-researcher action research issues and resolutions

Context	Critical issue	Teacher perspective	Researcher perspective	Resolution
The researchers ask the teacher to administer several pretreatment assessments	Appropriateness of the assessment instruments.	Students will not perform well on the assessments. "The nonhonors students have little chance to no ability of understanding the questions nor the type of answer they are expected to provide… They have a long way to go in the area of cognitive ability or skills."	The point of the project is to assess the extent to which a curricular intervention can promote the skills assessed by these instruments. Low scores on the pretreatment assessment are useful data points.	Administer the assessments with protocols informed by the teacher for instances in which students do not understand an item.
At the beginning of the project, the teacher uses a newspaper article to stimulate discussion and writing assignment for all students.	Communication between the teacher and researchers regarding the nature of each treatment group.	A news article of interest and relevance is available, so it should be shared with all students. The "experimental" classes will receive explicit instruction on argumentation later. "Yesterday, on the front page of the St. Pete Times was an article entitled, 'In stem cell debate, truth lies between the lines.' Michael Reagan …was quoted as referring to embryonic stem cell research, 'junk science.' I take issue with his uniformed opinion and had 150 copies prepared for the classes to read and evaluate."	Topics related to socio-scientific issues and/or argumentation should not be specifically addressed with the "comparison" classes. Presenting the article blurs the distinction between the "comparison" and "treatment" classes.	The Teacher and Researchers discuss the nature of each treatment. They take steps to enhance communication among all parties especially with regard to classroom activities.
The researchers produced a classroom activity related to course content,	Development of materials that address issues	The activity uses inaccurate information to support one side of the controversial issue. Students	The extent to which information is deemed inaccurate is heavily influenced by one's	The Teacher and Researchers compromised to

which is an area of the teacher's expertise, and argumentation themes, which is an area of the researchers' expertise. The activity attempted to point out two sides of a controversial issue.	mportant for course content as well as research themes.	would likely make inappropriate conclusions "My concern is that the risk/benefits are not equal and would prejudice most open-minded individuals."	perspective. When considering SSI, individuals are exposed to contradictory information, and the activity provides a real demonstration.	produce a draft acceptable to all parties.
The teacher asked students to explore NOS characteristics in the context of scientific theories. He became very frustrated with the products that were not nearly as well developed as had been expected.	Student learning.	Students should be learning, and it is very discouraging when they do not reveal anticipated gains. "I believed that they would be able to describe the topics using NOS as an historical basis and for the varied current theories that are still tentative and changing with cultural and technological influences…The essays reflect that they still do not get what the f**k NOS is all about and they are unable to understand how science could exist outside of the statements printed in their texts."	Improving student understanding of NOS themes is a seminal goal of the project. Data suggesting that students are not achieving this goal are every bit as important as positive results.	The Researchers are able to stay relatively detached from the situation; whereas, the Teacher maintains a much greater personal investment in the success of his students. Follow-up NOS activities coupled with more explicit instruction produces more encouraging results. In this case, one of the researchers modeled further NOS inquiry for the teacher.

Settings and Context

Since the focus of this chapter is on the pedagogical aspects relevant to the teaching of SSI, we omit certain methodological features that can be found elsewhere (e.g., Zeidler, Sadler, Applebaum, & Callahan, 2009) and concentrate on providing a brief overview of the learning treatment. The study involved a full academic school year that enabled the teacher to observe and monitor growth in students' perspectives of characteristics of science, scientific inquiry, and the relevance of science to daily decision—making through debates, argumentation, class discussions, small group and individual projects.

Participants were from four intact classes of the eleventh and twelfth grade students (typically ages 16–18) from a large suburban public high school in Florida. Two classes were honors and two classes were nonhonors anatomy and physiology sections. Each class had between 29 and 31 students. One honors and one nonhonors class were assigned as a comparison group while the two remaining classes became the treatment group sections. It is important to note that all sections (including those constituting the comparison group) were to have explicit emphasis on NOS constructs. Our rationale to include NOS in all groups stems from our belief that while SSI can elucidate features of NOS (Zeidler, Walker, Ackett, & Simmons, 2002), NOS in and of itself is a central component to teaching all science. Additionally, any changes in students' ability to use SSI-contextualized evidence-based reasoning could better be attributed to the interaction of NOS with SSI, rather than NOS alone. The teacher, who holds terminal degrees, taught all sections to control for variation in teacher attributes. Observations by researchers and extensive weekly journals helped to guide instructional decisions. An overview of the pedagogical approaches used with the two groups (i.e., comparison and treatment) is presented in Table 16.2.

We found Kolstø's (2001) general framework of eight "content transcending" themes quite useful in guiding the scientific dimensions of contextualized SSI instruction. While we have described these themes in previous research (Zeidler et al., 2009), it is important to restate them here: (1) Science-in-the-making and the role of consensus in science; (2) science as one of several social domains; (3) descriptive and normative statements; (4) demands for underpinning evidence; (5) scientific models as context-bound; (6) scientific evidence; (7) suspension of belief; and (8) scrutinizing science-related knowledge claims. These themes provided a template that served as a pedagogical mind-set for the researchers, and especially for the classroom instructor. Therefore, we were mindful that both the SSI modules and the pedagogy incorporate elements of one or more of these themes, as they provided guidance for developing important scientific habits of mind. For example, during discussions, debates, or advancing oral position narratives, the teacher would constantly question students' positions and demand that they provide supporting evidence for their claims. Statements like "I heard that…" or "My friend told me…" were forever banned from the students' vocabulary. He went further to challenge the veracity of the evidence students provided. It was interesting to observe how students eventually began to adopt these criteria

Table 16.2 Pedagogical framework for SSI study

	Comparison group	Treatment group
Approach	Traditional Approach: content topics follow textbook chapter topics.	Socio-scientific Issues Approach: content-related course topics embedded with SSI.
Teaching methods	Lecture, lab, discussion of content-related concepts, worksheets, predesigned lab activities.	Focus on argumentation and discourse, small group activities, role-play, and student research into SSI. Limited lectures and traditional labs.
Nature of science	Explicit activities and connections are made.	Explicit activities and connections are made.
Intended outcomes	Mastery of structure, function, and pathology of anatomical systems; more sophisticated views of NOS.	Improved critical thinking and decision-making particularly in the context of SSI; engagement in scientific discourses; sociomoral development; content mastery; more sophisticated views of NOS.
Classes	2 classes: 1 regular and 1 honors	2 classes: 1 regular and 1 honors

in reviewing their own work and when questioning one another. He provided opportunities for authentic inquiry investigations so students could engage in the activity of science, not as a fixed body of information, but as a process where people construct knowledge through collective understanding and examination of evidence. Avoiding dogma was achieved by encouraging the notion that students may question authority and think about the social perspectives under which knowledge claims were advanced. The work of Keefer (2003), Ratcliffe (1997), and Pedretti (2003) also informed our thinking in terms of ensuring the conditions necessary to focus on argumentation and discourse.

We were further influenced by features related to the Reflective Judgment Model and its use in classrooms such as the use of evidence-based reasoning, consideration of the role of authority, understanding the relationship between the role of knowledge and the status of epistemic beliefs (Baxter Magolda, 1999; Kegan, 1994; King & Baxter Magolda, 1996; King & Kitchener, 1994, 2002, 2004). The congruence between these factors and the type and quality of reasoning and discussion within the SSI framework are synergistic. In practice, the following strategies proved to be useful guideposts:

1. Show respect for students' assumptions, regardless of the developmental stage(s) they exhibit. Their assumptions are genuine, sincere reflections of their ways of making meaning, and are steps in a developmental progression. If students perceive disrespect or lack of emotional support, they may be less willing to engage in challenging discussions or to take the intellectual and personal risks required for development.
2. Discuss controversial, ill-structured issues with students throughout their educational activities, and make available resources that show the factual basis and lines of reasoning for several perspectives.

3. Create many opportunities for students to analyze others' points of view for their evidentiary adequacy and to develop and defend their own points of view about controversial issues.
4. Teach students strategies for systematically gathering data, assessing the relevance of the data, evaluating data sources, and making interpretive judgments based on the available data.
5. Help students explicitly address issues of uncertainty in judgment-making and to examine their assumptions about knowledge and how it is gained (King & Kitchener, 2002, p. 55).

The units chosen for the SSI Project were designed to move students toward deeper understandings of scientific concepts and their application to SSI. The issues were carefully selected in a manner that aligned students' interests with the course content embedded in the SSI, challenge core beliefs, and apply new content knowledge to the appropriate scientific context in a manner that was personally relevant and meaningful. The treatment(s) was intentionally designed to consistently challenge deeply held core values by offering opportunities to confront and defend or reject new information. The curriculum included multiple activities that required participants to evaluate claims, analyze evidence and their sources, come to a decision on a personal position, make moral decisions, and present the information within a group of peers to negotiate a consensus opinion. Each SSI unit required between 3 and 7 days; however, content was reinforced and reiterated on multiple occasions. Topics ranged from organ transplant allocation, the safety of marijuana and fluoridated water, the morality of stem cell research and euthanasia, quality of life issues, fast food consumption, and other contemporary subjects that were socially relevant. The learning opportunities were carefully crafted so as to highlight the idea that scientific knowledge is theory-laden and socially and culturally constructed.

Teacher Transformation in a SSI World

> Most importantly, each section began with a discussion or project of some SSI that they [students] resolved in groups of four, and then presented to the class... the content knowledge was extracted from these discussions, mostly from questions they had to clarify certain issues. They were using NOS references without realizing it and they have become aware of the relevance of science in their daily decision-making .... mostly, they learned that scientific knowledge is evolving and some of the empirical information is distinct for different groups of people. (Teacher's Reflections during the Second Quarter)

Intervention

The SSI project was developed and designed to feature an issues-driven curriculum. Activities and investigations were intended to provide personal experiences that are individually relevant and socially shared, while promoting enduring reasoning and decision-making skills. The goal for each unit was for the teacher to create opportunities for students to discover and acquire scientific content knowledge from an investigation of an SSI context. The motivating goal of the ten SSI units was to create activities that strategically directed students to essential subject matter content and concepts of anatomy and physiology. (Please refer to Appendix 1 to view all ten units). It should be noted that while the subject matter is narrowly defined, the contemporary application and moral implications of each unit could broadly connect to students' daily decision-making.

In preparation for the design of individual units, we developed a design framework to help inform this work. We used the design framework to explicitly highlight common elements to be introduced across all of the SSI based units, including the evolution of subject matter awareness and comprehension through contextual examination of corresponding social issues. This framework is presented in the outline below. We used this framework as a basic sequence for planning and implementing instructional activities but this list does not necessarily prescribe a fixed sequence. This issue highlights an important caveat to the presentation of the outline: following a prescribed sequence of steps is no path to assured success with SSI. It does take a flexible and insightful teacher to take advantage of opportunities when they arise and orchestrate these many components in creative ways to mesh with the moment, context, and students. The outline below provides a template of a typical SSI unit.

Development of an SSI Unit

1. Topic/Subject Matter Introduction
   a. Magazine headlines, articles, and advertisements
   b. YouTube video presentation of controversy associated with subject matter
   c. Photographs
   d. Models
   e. Other media formats
2. Challenging Core Beliefs
   a. Contentious questions that "attacks" commonly held beliefs
   b. Challenging "Common knowledge" of subject matter
   c. Misconceptions
3. Formal Instruction
   a. Anatomy
   b. Physiology
   c. Related science information
4. Group Activity
   a. Development of related, but unconventional topic/subject matter questions
   b. Individual investigation of data and evidence

c. Small group negotiation of evidence
d. Group presentation of consensus understanding

5. Develop Contextual Questions

a. Fundamental science concepts of subject matter
b. Defeating misconceptions
c. Contemporary claims regarding subject matter

6. Class Discussion

a. Evidence reliability of contemporary issues
b. Importance of specific knowledge for informal decision-making

7. Teacher Reiteration of Content/Subject Matter

a. Essential learning of subject matter content
b. Purpose and relevance of specific knowledge
c. Application of content knowledge
d. Negotiating contemporary issues

8. Knowledge and Reasoning Assessments

a. Group presentations
b. Posters
c. Argumentation/debate activities
d. Paper production of selected topics
e. Written tests of subject matter

A contextual example of the development of a particular SSI unit is provided to demonstrate how an SSI lesson plan for learning the digestive system for an anatomy and physiology course can be achieved. Specifically, investigation of popular diet plans and outrageous claims of weight loss from consuming exotic fruit (acai berries), taking diet pills, or wearing patches and creams can introduce specialized subject matter and engage a classroom of high school students to enthusiastically investigate esoteric science concepts. When confronted with Internet and tabloid advertisements that proclaim, "Lose weight without diet or exercise!", "Lose weight permanently! Never diet again!" "Lose weight no matter how much you eat of your favorite foods!" or "Block the absorption of fat, carbs, or calories!" students were challenged to utilize knowledge, reasoning, and argumentation skills obtained in science classrooms to decipher physiological facts from science fiction. Following the template for a typical SSI unit is a detailed example to provide clarity to this unique pedagogy.

1. Topic Introduction: magazine articles, advertisements, and headlines; 5 min YouTube video of subject matter controversy, photographs, models, or media format.

   The development of a representative SSI lesson plan began with subject/topic introduction, engaging students with interesting demonstrations of recent magazine or newspaper headlines, articles, and advertisements. Visual presentations of subject matter controversy, such as YouTube and other Internet, photographs, models and other methods of capturing adolescent attention were used. We have realized that "more is better" when providing sufficient stimulus to encourage all students to become interested and engaged in learning new subject matter. Our introduction of science concepts, using contemporary issues related to the digestive system included photographs of teenage and adult obesity and inquiry about the absence of pictures of an overweight elderly population. Determining student knowledge and "pre-conceptions" of the subject matter was determined by posing non-threatening inquiry, such as how to relate individual diet, and health, and the potential side effects of a fatty diet besides excess weight. The number of students involved in discussions and the conversation volume of the classroom is a reasonable method of determining the quality of the topic introduction.

2. Challenge Core Beliefs with Contentious Questions.

   A fundamental element of negotiating scientific issues is the extraction of content knowledge from the controversial context. In this regard, a focus on the "learning" of anatomy and physiology was adjusted to consider contentious SSI contexts related to the digestive system. Specifically, controversial questions were written on the white board for students to evaluate and reference during their continuing investigation and argumentation. The debatable claims were intended to challenge bias and misconceptions that form the basis for many core beliefs. Typical and customary questions included: "If someone deliberately consumes food they know is both harmful to their health and detrimental their future well-being, is that choice an immoral decision?" or "Should high fat foods be taxed, since their consumption affects health care costs for the general population?" or "Since more people die of heart disease than drug overdose, should fried foods be considered an endangerment to the community and therefore illegal?" The moral implications of the questions challenged students' core beliefs and the varying responses enabled us to develop group activities, based upon "uninformed" responses.

3. Formal Instruction

   We recognized that formal instruction of anatomy and physiology of the digestive system (in this example) was necessary to provide students with a fundamental vocabulary of relevant structures and functions, which enabled better comprehension of information obtained through individual investigation. The anatomical structures of the digestive system and pathway of food movement were demonstrated, with PowerPoint photographs and drawings of related organs, tissues, and cells. Students were reminded of the complementary relationship between structure and function, while tracing the peristaltic passage of food from the mouth to the anus. While it is beyond the scope of this chapter to include all of the instruction regarding mechanical and chemical digestion, peristalsis, absorption, and excretion, it should be noted that students need to be "reminded" that formal science knowledge and application is a necessary adjunct to intelligent argumentation and negotiation of evidence and data reliability and validity. Students discovered that simple food selection requires sophisticated knowledge of the digestive system. However, new knowledge was now anchored in more meaningful contexts so students could create conceptual understandings and better transfer concepts to novel situations.

4. Construct Group Investigations and Presentations

   Because SSI instruction requires student engagement and commitment to active discovery, individual and socially shared group activities must be utilized. Our construction of the activities included multi-tiered projects involving individual investigation, evaluation of validity and reliability of evidence, group negotiation of verifiable data, creation of knowledge displays and presentation materials, and group presentations. Reintroduction and presentation of contentious popular claims related to the digestive system was stressed, to stimulate student interest. Typical group activities included:

   a. Group #1: Why are my biceps small and my butt so big?
   b. Group #2: How can I lose weight without dieting or exercising?
   c. Group #3: Are my bad dietary habits an eating disorder?
   d. Group #4: What are good, better and best diets?
   e. Group #5: Why does older generally mean fatter?

   The rationale for individual investigation and small group negotiation activities is based upon the recognized importance of developing and practicing skills related to evidence evaluation for reliability and accuracy. Science is ultimately an exercise in generating and testing new understandings particularly in light of misconceptions; therefore, group presentations were not intended to be amusing demonstrations, but synthesized instruction of the science related to the claims made and corresponding to the question posed.

5. Develop Contextual Questions Directed Toward Content and Concept Discovery.

   Subsequent to formal instruction, students were reintroduced to SSI, using less controversial, contextual inquiry, directly related to recent scientific instruction. Students are encouraged to investigate formal and practical aspects of the subject, with overtones of personal relevance. The formality of information is intended to promote inquiry of concepts and misconceptions. Contextual questions included:

   - How can you break down your cheeseburger into a molecular size so that 37 trillion body cells can receive nutrients to survive (make ATP)?
   - What is the difference between digestion and indigestion?
   - Why does alcohol and medication rapidly enter the cardiovascular system?
   - How do food molecules enter the cardiovascular system?
   - What are the differences between carbohydrates, protein and fats?
   - What is a calorie and how many do I need?

   The primary intention of questions like these was to encourage students to apply their informal reasoning skills and utilize their newly acquired knowledge to resolve contentious issues, evaluate evidence reliability and make informed decisions.

6. Class Discussion, Argumentation and Debate

An established method for promoting reasoning skills and winnowing concepts and misconceptions is argumentation around debatable themes. Further, a significant component of SSI curriculum is realizing that decision-making is a moral exercise and a reflection of individual character. For high school students, the topic of morality connotes perceptions of personal behavior. For this reason, it was imperative that formal instruction included an opportunity for students to better understand that community living standards and relationships are based upon contemporary understanding of moral attitudes. Providing students with contentious issues that confront moral dilemmas is an important part of the acquisition of formal knowledge. Teachers can find excellent sources of ethical controversy in newspapers, on television, and conversations with their students. SSI activities should encourage students to develop personal understanding and informed opinions based upon reliable evidence. The cumulative purpose and goal of these activities is student maturation, character development and skillful negotiation of ethical dilemmas, as well as formal knowledge of the digestive system.

In a designed digestive system-SSI activity, several topical issues were presented as the topic for argumentation and debate. In the midst of our implementation, the case of Terri Schiavo became headline news in our area, and we used the case as a center-piece of debates and discussions within the digestive system unit. In 1990, Schiavo collapsed and fell into a persistent vegetative state for 15 years. It was hypothesized that the initial condition was precipitated by a severe eating disorder that included frequent use of diet pills. The case gained national prominence when Shaivo's husband petitioned the court for permission to remove her feeding tube. Shaivo's parents strongly opposed this action and a legal and ethical battle ensued that involved the state and national supreme courts and the President of the United States. Ultimately, Schaivo's feeding tube was removed and she died. The patient's initial physical condition, her end of life condition and the pathology of her deteriorating digestive, cardiovascular and nervous systems provided a rich and compelling context for students to discuss physiological functions and connecting science to real-life events. Other topics that could be used in a similar fashion include:

- Diets for Sale: Nutri-System, Jenny Craig, Weight Watchers, etc.
- The Fast Food Highway to Cardiovascular Disease
- Fat Tax for Unhealthy Foods
- Health Care Penalties for Obesity
- Involuntary Camps for Overweight Children

Students explored these topics (and others relevant to the other curricular topics) through investigations of media, Internet resources and interview activities. SSI curriculum provides opportunities for enhancing scientific literacy, with students translating their understanding of SSI subject matter to research and position papers. Historically, students have demonstrated an ability to learn and "store" large quantities of scientific information, yet struggle to apply conceptual understanding to random claims, stated with authority. Internet search engines provide an easy access to vast amounts of information and concepts that may or may not be valid. It is the further purpose of utilizing an SSI program of study that students are encouraged to learn and develop skills that will facilitate intelligent and prudent application of knowledge.

7. Teacher: Final Instruction and Clarification of Concepts

   At the completion of the SSI/subject matter unit, the teacher acted as a moderator to revisit topics and clarify concepts, so students could confirm the understanding that science is innately organized and can be meaningful when understood in relation to the perception of their universe. The unit is best summarized by explaining that knowledge of the digestive system, and organ systems in general, cannot be understood in isolation, but as pieces of an incomplete biologic puzzle. In a perfect classroom setting, teachers will continue to discuss the elements of the section as other subjects are introduced.

8. Knowledge and Reasoning Assessments of Anatomy, Physiology and SSI related to the Digestive System

   Public and private schools rely on number and letter grading systems to measure student ability, and SSI instruction and curriculum encourages practical and conceptual understanding of science in the "real world." In these contexts, measurement of empirical knowledge is standard. Student presentations, arguments, posters and evidence provided rich insights into student understandings—but assessment of these products can be subjective. Written examinations can be arguably objective, but neither traditional or performance assessments provide complete insight into a student's ability to develop comprehensive understandings of subject matter. We made a concerted effort to utilize process and product assessments, which included evaluating quality of evidence used to defend opinions, the depth of understanding demonstrated in investigative papers, and student ability to recognize the inherent value of considering opposing opinions. While individual perception is difficult to assess, student evaluation should include a measure of their awareness that content knowledge and reliable evidence are fundamental building blocks in formulating well-measured stances. The definitive final examination occurred when students were confronted with controversial socioscientific issues that required understanding of empirical information and the skills of informal, moral reasoning.

Points of Consideration

Because SSI instruction often introduces issues with a moral dilemma, conflicting evidence, as well as multiple sources of evidence, teachers are expected to evaluate claims regarding the students' sources of information. Adaptation to this new approach requires the teachers to transform their perception about being a singular source of knowledge and encourages students to make individual decisions, even when personal beliefs are mistaken for scientific concepts. Recent studies have demonstrated that when SSI is presented in science pedagogy, students can handle conflicting evidence by drawing upon past experiences and combining them with new information, to explain actions in a scientific context (Driver, Leach, Millar, & Scott, 1996; Driver, Newton, & Osborne, 2000; Kolstø, 2001, 2006; Sadler, 2004; (Sadler, Klosterman, & Topcu, Chapter 21, this volume); Zeidler & Keefer, 2003; Zeidler & Sadler, 2008; Zeidler, Sadler, Simmons, & Howes, 2005; Zeidler et al., 2009).

Perhaps even more important is recognizing the significance of the transformative nature of classroom climate and culture. Under the SSI canopy, classroom management includes developing a relationship of trust where the students gain confidence in their ability to learn important science concepts. Creating a relationship of trust with students is necessary when using designed activities that investigate personal use of cigarettes, alcohol consumption, recreational drugs, and steroid abuse. The advantage of using this format was the ability to adjust the themes to accommodate both the academic abilities and interests of the students as well as the different science disciplines.

Using issues-based curriculum, teachers are compelled to reveal explicit nature of science connections by demonstrating that scientific knowledge is not absolute but forms as a result of social knowledge construction from argumentation and discourse. These new goals require deep epistemological conceptual shifts so teachers can transform their pedagogical orientation from being purveyors of scientific knowledge to moderators and mediators of a classroom culture that mirrors society in which students are challenged to make informed scientific decisions and exercise moral reasoning.

And Now for Something Completely Different

> They are having fun in class, taking notes without encouragement.... and best of all, their test grades are exceptionally good. Their ability to remember esoteric information has increased because the information makes sense to them. I constantly remind them that the earth is flat and that images are projected into the openings in the front of their eyes, like movies. They laugh because they understand that this is my "make sense epistemology." They know that people (including scientists) create answers and theories to explain phenomena, even when they are still unsure; because they have abandoned the excuse, "it must be magic." (Teacher's Reflections during the Third Quarter)

Outcomes and Discussion

SSI instruction is more than an instructional strategy. It can foster the development of content knowledge and a range of skills and dispositions, such as curiosity, problem solving, communication and collaboration skills, decision-making, and self-directed learning. Instead of presenting a prefabricated lesson plan, teachers present science content through the introduction of open-ended and messy problems. Delivering science content is replaced with argumentation and discourse-based instruction, developing collaborative group, communication, and problem-solving skills. Convincing students that investigating and arguing issues related to real-world problems or simulations of real-world problems that have more than one solution are effective methods of learning content knowledge that can best be accomplished by a teacher with confidence in the method, good presentation skills, understanding of performance-based assessment, and the willingness to transform roles from teacher-centered to student-focused.

We have previously reported on research outcomes associated with advances by students in reflective judgment and reasoning (Zeidler et al., 2009), moral sensitivity (Fowler, Zeidler, & Sadler, 2009), NOS (Walker & Zeidler, 2007) and embedded content knowledge (Sadler & Zeidler, 2005). Here, we wish to present other facets of the research project more directly related to the pedagogical practice of SSI. Therefore, the scope of this chapter is intentionally limited to descriptive observations by the classroom teacher, in conjunction with two researchers, as students' core beliefs about issues related to SSI topics were challenged. Students were constantly challenged to align their core beliefs with evidence supporting and/or opposing various perspectives related to socioscientific issues.

The teacher utilized multiple opportunities for noting daily responses and gradual changes in individual and class understandings of anatomy and physiology. Weekly discussions between the teacher and the two researchers served to identify and clarify observed trends in students' behavior. During the course of the academic year, and at the end of the school year, the teacher and researchers collectively identified and synthesized the main outcomes. Next, we offer descriptive indicators of each outcome with respect to the major factors competing with or facilitating students' epistemological understanding of course content and SSI. Contextual pedagogical factors that impacted the quality of the classroom ecology were also identified and described in a similar manner.

Confronting Core Beliefs

Many deeply held beliefs about the world and scientific issues originate from the mistaken concept that we are separate from the source of knowledge and understanding. Historically, science education has introduced scientific concepts and discoveries attached to the names of famous scientists. A long discussion could be inserted that describes the general home life of adolescents and their belief that they do not possess the necessary knowledge and experience to offer a valuable opinion. Parents and teachers convey the message directly or by subtle commentary that the voice of authority is reserved to a select group, without specifically addressing the knowledge and experience needed to enter the elite assembly. When underlying sources of science information originate from hearsay and secular sources, or is a generalized proclamation handed down from the court of public opinion, students generally adopt the information as core belief because they do not have believable, conflicting evidence. Once embedded, even contradictory data rarely dislodge a core belief.

Throughout our project, students were capable of evaluating and synthesizing data. However, when *SSI provided information that conflicted with their core beliefs,* several interesting patterns emerged. These patterns are summarized in Table 16.3.

Table 16.3 Factors identified when students beliefs were challenged

Major factor	Outcomes	Examples
Core belief persistence and discrepant data	Students often dismissed data (e.g., graphs, charts, and statistics) that were in conflict with their core beliefs or failed to meet the criteria of personal experience.	SSI[a]: Fluoridation of public water supplies. The majority of students believed fluoride was harmful, ignoring substantial evidence that demonstrated 350 million people drank fluoride daily, without side effects or illness. An opposing article provided statements that indicated a possible link to cancers and dental disfigurement. The value of "potential" harm or negative consequences, even unsubstantiated, was more important than well-documented benefits.
Lack of critical reasoning	The perceived value and relevance of information was based upon its fit with personal experience(s).	SSI: Stem cell research. Current media assertions by nonscientist "authorities" (government leaders) proclaimed that stem cell research was comparable to abortion. Without personal experience in areas of fallacious reasoning, the students reverted to fundamental, core beliefs and expressed a genuine fear of possible illegality and religious sin. In contrast, evidence of demonstrating the connection between unhealthy diets, smoking, and heart disease seems only remotely relevant.
Normative reasoning	When students were compelled to defend their opinions to their respective group, the class, and to the teacher, they included their core beliefs and personal experiences in their defenses.	SSI: Animal rights and the use of animals for scientific research. When students were forced to defend a position that was not parallel to their personal belief system, it provided an opportunity to challenge the credibility of their opinion, which had been developed entirely around the love of the family pet and nature-related television programs. Students struggled to develop arguments substantiated with evidence, demonstrating that science requires empirical data.
Reasoning with conflicting data	Students were generally surprised that reliable sources of scientific information at times provide conflicting claims and conclusions.	SSI: Legalization of marijuana. In the activity regarding the safety or harmfulness of marijuana, conflicting evidence regarding the potential harm/benefit of marijuana confused students but encouraged them to evaluate various sources of data and information from "authorities."

[a]SSI refers to the issue which served as a context for the example listed

Our observations of students on a day-to-day basis and over the course of an academic year confirm our position that prevailing cultural perceptions, particularly in the realm of SSI, may be understood as core beliefs. These perceptions reveal themselves as the prevalent beliefs that society (the general population) accepts as true, generally without reflection. Such perceptions form the basis of "socialization," which only involves blind acquiescence to a social norm and does not entail any form of internal evaluation (Zeidler & Sadler, 2008). Many of these shared perceptions form the basis of human awareness. We observed students to be easily influenced by generalized information that was presented as authoritative, with little attention paid to the source of that information. (It is noteworthy that even conspiracy theories that have managed to reach print status inform students' reactions to, and reasoning about, many SSI.) We therefore found it to be extremely important to make the effort to teach students to question whether or not their intuitive (initial) responses were actually true and subsequently had them ascertain, question, examine data from varied sources, and, through active discourse, form judgments about the credibility of information relevant to the subject matter at hand. In this manner, the process of norm acquisition and the formation of judgment in finding the fittingness of conduct to context can be allowed to reflectively develop (Green, 1999).

Confronting Contextual Factors

Contextual factors linked to teaching and learning scientific concepts interact not only with students' learning characteristics, but also with understanding of principles as a group. *When SSI were used as context*, then the content became personally relevant and accessible to students. Table 16.4 summarizes the main outcomes observed as they related to contextual factors.

Students' personal belief systems were, quite often, challenged while at the same time compatible science concepts, when contextualized in a manner that made subject matter personal and relevant, allowed students to frame their understanding of the content in more sophisticated ways. One of the most striking achievements across the class was the development of more mature attitudes toward the formation of consensus resolutions to dilemmas even when individual students' personal beliefs conflicted with the decision of their respective groups. We found that the process of challenging deeply held, personal beliefs and, perhaps their subsequent rejection, is extremely difficult. Indeed, a great deal of anxiety can result in a classroom where personal values are questioned. Thus, it was imperative that we established a learning environment conducive to the safe expression and exploration of ideas and thoughts by individuals and groups. We made constant adjustments to the kinds of contextual factors that would ultimately convey a kind learning environment that valued open inquiry about SSI and independent thinking, one that presented a coherent and consistent experience for the learners, and one that sought to be self-improving through processes of reflection, feedback, and critical inquiry.

Table 16.4 Summary of main contextual outcomes for students using SSI

Major factor	Outcomes	Indicators
Evaluation of evidence and claims	Students' ability to evaluate claims provided by media and other sources was improved when scientific concepts were related to relevant SSI.	Using the students own personal observations and experiences regarding the use and abuse of alcohol, difficult concepts, including the movement of the sodium potassium pump was learned because they "made sense" in the perspective of muscle and nerve failure.
Real-world relevance	Students demonstrated improved understanding of scientific concepts when they were able to attach the concept(s) to relevant SSI.	The case of Terri Schiavo (termination of life support for a brain-dead person) provided the SSI background and an instruction opportunity to discuss the anatomical structure of the brain and the related physiology. Further, students were able to construct meaningful discussions on the various cultural "definitions" of life.
Understanding contextual interrelationships	When presented with contemporary SSI, students were able to transfer conceptual understanding from one context and apply to a new and/or different context.	Examination of the stem cell issue, diseases of the nervous and muscular system, the effect of smoking on respiratory tissue, osteoporosis and contagious diseases such as AIDS and influenza allowed students multiple opportunities to investigate cell structure and the driving principles of homeostasis. Students demonstrated a better understanding of complementarity and the relationship between form and function.
Role playing and Role reversal	Students were able to identify and manipulate key variables (component parts) within a specific context to understand the direct and indirect effect on related concepts.	Students participated in role-playing activities during investigations of SSI, such as organ allocation, animal rights, and the matter of marijuana safety. The random selection of roles allowed students to challenge and defend their beliefs, using evidence they considered reliable. The use of various forms of evidence over time improved their skills in evaluating conflicting information.

Pedagogical Issues: Student-Centered Context

Moving SSI from theory to practice is essential in contemporary classrooms. Science education that includes SSI offer unique opportunities to challenge students' moral reasoning, and in the process, present concepts that seem to make sense because of their relevance and inherent interest. Consistently, the main competition to understanding and coherence are core beliefs, pseudoscience, and lack of personal experience in moral decision-making. The challenge to science teachers is to allow students to discredit their own belief system by having opportunities to be able to formulate new perspectives. Our experiences have allowed us to identify several areas that are potentially problematic for students when engaging in SSI. Student impediments to success included:

- Core beliefs
- Scientific misconceptions
- Lack of personal experiences
- Lack of content knowledge
- Underutilized scientific reasoning skills

In presenting this list, we do not mean to dissuade teachers from attempting an SSI approach. In fact, it is our position that insofar as students have such impediments, teachers have a responsibility to provide them with opportunities that challenge their personal belief systems about the social and natural world. Our experiences in the classroom over an academic year (along with other supporting studies previously cited) have revealed that the SSI approach fostered students' conceptual understanding of course content as well as more sophisticated views of NOS, empathy, and reflective judgment. When science is embedded within current SSI, students become motivated to participate in discussion that presents multiple opportunities for engagement in activities that require understanding of scientific concepts and content. Students demonstrate a greater acceptance and understanding of requisite information when it is connected to a contemporary issue that has personal relevance.

While encouraging students to consider evidence-based alternative arguments is of primary importance, it is equally important that teachers who are interested in using debate or discussion-focused activities also consider the match between their own pedagogical expectations and the theory base guiding the research. For example, an effective teacher engaged in SSI would need to rely on research to better direct classroom debates through various lines of questioning (e.g. epistemological probes, issue-specific probes, role reversal probes, and moral reasoning probes). The importance of exposing students' to discursive activities in the science classroom cannot be overstated if our goal is to increase science literacy. Of course this cannot be accomplished without the development of teacher training programs that focus on the pedagogical techniques necessary to create content-specific and NOS-embedded learning activities that emphasize discourse and debate. This requires that teachers become adept at guiding students in the process of applying their understandings of the nature of science as they decide on and evaluate the worthiness of competing scientific claims. Strategies similar to our SSI approach are valuable in that they allows teachers to reveal and become familiar with epistemological factors of students' reasoning including possible scientific misconceptions, moral reasoning, the ability to interpret and evaluate data, and fallacious reasoning.

Similarly, a teacher looking to the web for SSI fodder recognizes that Internet and issues-based learning activities can also be an invaluable resource in terms of exposing students to diverse perspectives on current scientific reports and claims. Again, current research can suggest important ideas to inform practice. With scaffolded learning interfaces (e.g. Walker & Zeidler, 2007), students can spend their time reading and evaluating the multiple perspectives of a given socioscientific issue instead of "surfing" through a plethora of sometimes misleading information. Of course, this requires that teachers invest the time up front to find both reliable as well as potentially unsound sources of scientific data and perspectives so students may be confronted with mixed evidence and offered scaffolding as they learn to assess the validity of varied claims and data.

The Foresight of Hindsight

> They discuss, argue, and question during each teaching and learning session. The rest of the semester will be dedicated to utilizing their new skills in handling various forms of evidence argumentation skills and learning to become better science students because they know that science is relevant to their lives. I wish we could discuss every question they have... (Teacher's Reflections during the Last Quarter)

While school boards, administrators, and teachers are heatedly debating science curriculum and which science lesson plans make the best medicine, the students have been slipping into a classroom coma. Faculty and department meetings discuss methods of inoffensively introducing contentious topics, such as evolution, while failing to create lesson plans of arguable contemporary scientific issues that are personally relevant to high school students, including issues such as alcohol and drug use, smoking, and obesity. The issues are recognizable, but not as contexts for learning science concepts.

A major problem of education has been the inability of students to identify with the topics they are requested to learn; specifically, the science that does not have easily recognizable relevance. Given that students learn to use cell phones, computers, and iPods without instruction manuals because the content knowledge is useful and meaningful to them, it is not unrealistic to believe that students can also learn science concepts when they meet the same criteria. For individuals to comprehend unfamiliar concepts and materials, they need to create links to personal contexts. The SSI curriculum requires students to formulate claims and conclusions about controversial topics based upon an independent acquisition of information regarding an assortment of socioscientific issues.

What a teacher believes is the reality of their instruction. Creating the mantra of providing rigor and personal relevance in science classrooms is insufficient unless teachers possess pedagogical expertise concerning the investigation of socioscientific issues context to discovering underlying scientific concepts. Encouraging students to examine conflicting evidence, negotiate personal perspective, and challenge their core beliefs about contentious scientific topics is not currently considered standardized curriculum format and lesson planning. However, sociomoral discourse, argumentation, and debate have been clearly established as necessary elements in character development and decision-making ability and therefore should be essential components of science education (Fowler et al., 2009; Sadler, 2006; Zeidler & Sadler, 2008; Zeidler et al., 2009). It is also worth noting that recent academic and educational research has demonstrated the importance of connecting the teaching scientific concepts to contemporary relevance (Applebaum, Zeidler, & Chiodo, in press; Fensham, 2009; Ratcliffe & Millar, 2009; Sadler & Zeidler, 2009; Zohar & Nemet, 2002).

It is important to determine the conditions under which students best grasp the essential concepts of science. The requirements for those conditions, it has been argued and demonstrated, entail that the process of acquiring scientific knowledge should include practices of discovery and learning, where students actively explore socioscientific issues. Current conversations covering the range of issues related to student learning are dramatically different from those of a decade ago. There is a growing national consensus that students should be able to think creatively, work through seemingly ambiguous data, search for novel patterns of thought and tap multidisciplinary expertise (The Council on Competitiveness, 2005; Narum, 2008; Zeidler & Sadler 2011.)

For preservice and practicing teachers, the realization that science education for many (most) students has included years of indoctrination, dogmatism or authoritarianism is a sobering epiphany. However, there is no place in science and, therefore, no place in science education for the protection of concepts and theories from criticism. The challenge for science teachers is to allow students to have personal experiences that do not immediately negate their belief systems; rather, the aim is to provide the conditions necessary to enable the development of a personal epistemology through continued exposure to, and interaction with, the nature of science and SSI. The use of argumentation and relevant SSI as a framework for science curricula is essential for enabling scientific concepts to enter students' individual belief systems.

The customary process of acquiring scientific knowledge should include practices of discovery and learning, where students actively explore socioscientific issues. While this pedagogy requires students to become actively engaged in socially shared activities that "unearth" personal connections and relationships to contentious scientific topics, it is equally important that teachers possess the characteristic leadership and teaching skills necessary for guiding students in their exploration and understanding of science. An aim of socioscientific issues curriculum has been to transform both teachers' and students' epistemological beliefs about the process of learning science by engaging students in a social microcosm where ethical negotiations of "real-world" problems and the use of scientific knowledge in their decision-making is a common occurrence.

Using personal and social issues as context for learning science and acquiring content knowledge is only a novel experience in school, since this is a common method of constructing science knowledge outside of classrooms. Research has demonstrated that SSI instruction can be successfully instituted in classrooms (Walker & Zeidler, 2007; Zeidler et al., 2009); however, this instruction requires teachers to first transform their pedagogical orientation from being purveyors of scientific knowledge to moderators and mediators of a classroom culture that mirrors society and requires individuals to make informed scientific decisions and exercise moral reasoning. Science has to be learned in school very much the same way that it is practiced out of school.

As an issues-based curriculum, SSI instruction requires teachers to provide activities that demonstrate scientific knowledge is not absolute, but forms as a result of social knowledge construction from argumentation and discourse. Curriculum and pedagogy transformation are evident when daily classroom activities require students to discover the personal relevance of science through problem-solving experiences; in particular, the extraction of content knowledge from contextually embedded investigations. It is noteworthy that the success of using SSI-based curriculum is contingent upon redefining the role of the teacher and the responsibilities of the students. Teachers who include socioscientific issue inquiry in their lesson plans will discover their role is transformed from lecturer to mediator and moderator; their focus will be to assist students develop skills in areas of argumentation and evidence evaluation. As part of the transformation process, teachers will become competent in areas of critical thinking, argument quality assessment, and discussing moral dilemmas.

Innovative pedagogy, such as an SSI curriculum, both challenges and compels science teachers to undergo a transformative process that includes, among other things, discarding antiquated teaching methods. The objective is for teachers to transform their pedagogical focus and scientific epistemology so students can better understand how such knowledge is generated and validated (Abd-El-Khalick, 2006). Within the SSI framework, students are exposed to moral problems that involve a number of discrepant scientific, social or moral viewpoints, many of which may conflict with the student's own closely held beliefs. Teachers need to transform their pedagogical orientation away from introducing science concepts through simple lectures and reconfirming laboratory investigations; instead, teachers can create a classroom environment where students can develop a meaningful understanding of scientific concepts in relationship to real-world circumstances. Deforestation, ecojustice, global warming, viral pathogens, and personal fitness are significant topics; nevertheless, students (particularly middle and high school levels) tend to not regard these subjects as personally relevant because they do not instinctively understand that their lives are directly impacted (Mueller & Zeidler, 2010). A SSI framework allows for these personal connections to unfold by way of providing contexts that are organically connected to the students' worldview. Traditionally, science classroom activities rarely include opportunities to make personal decisions regarding contentious topics. Teachers must guide students through SSI activities so that they recognize the importance of informal reasoning in their daily decision-making.

Teachers who include socioscientific issue inquiry in their lesson plans will discover their role is transformed from a more traditional approach to a more progressive stance; their focus will be to assist students develop skills and habits of mind in areas of argumentation and evidence evaluation. As part of the transformation process, teachers will become competent in areas of critical thinking, argument quality assessment, and discussing moral dilemmas. Teacher transformation is further evident as students are directed to discover the personal relevance of science through problem-solving experiences as in the extraction of content knowledge from an academic investigation of SSI context. Perhaps the truest metric of the success of any classroom-based research project is that it survives after the researchers have left the classroom. SSI continues to be the driving pedagogy for this classroom teacher to date.

Appendix 1. SSI Units, Scientific Contexts, Systems and Concept Relationships

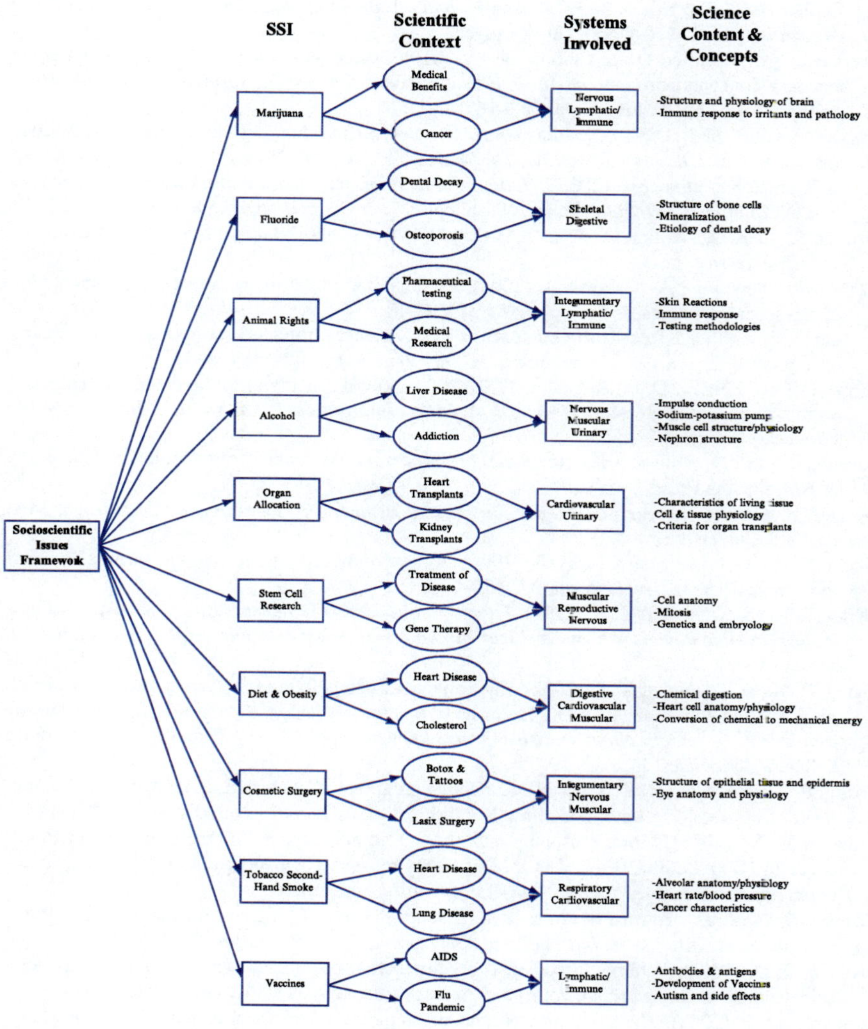

References

Abd-El-Khalick, F. (2006). Socioscientific issues in pre-college science classrooms. In D. L. Zeidler (Ed.), *The role of moral reasoning and discourse on socioscientific issues in science education* (pp. 41–61). Dordrecht: Springer.

Applebaum, S., Zeidler, D., & Chiodo, K. L. (2010). Using socioscientific issues as context for teaching concepts and content. In R. E. Yager (Ed.). *Science for resolving issues/problems*. Arlington, VA: NSTA Press: (pp. 147–163).

Baxter Magolda, M. B. (1999). *Creating contexts for learning and self-author(s)ship: Constructive-developmental pedagogy*. Nashville, TN: Vanderbilt University Press.

Council on Competitiveness. (2005). *National innovation initiative summit and report: Thriving in a world of challenge and change*. Washington, DC: Council on Competitiveness.

Driver, R., Leach, J., Millar, R., & Scott, P. (1996). *Young people's images of science*. Bristol, PA: Open University Press.

Driver, R., Newton, P., & Osborne, J. (2000). Establishing the norms of scientific argumentation in classrooms. *Science & Education, 84*(3), 287–312.

Fensham, P. J. (2009). Real world contexts in PISA science: Implications for context-based science education. *Journal of Research in Science Teaching, 46*(8), 884–896.

Fowler, S. R., Zeidler, D. L., & Sadler, T. D. (2009). Moral sensitivity in the context of socioscientific issues in high school science students. *International Journal of Science Teacher Education, 31*(2), 279–296.

Green, T. F. (1999). *Voices: The educational formation of conscience*. Notre Dame, IN: University of Notre Dame Press.

Kegan, R. (1994). *In over our heads: The mental demands of modern life*. Cambridge, MA: Harvard University Press.

King, P. M., & Baxter Magolda, M. B. (1996). A developmental perspective on learning. *Journal of College Student Development, 37*, 163–173.

King, P. M., & Kitchener, K. S. (1994). *Developing reflective judgment: Understanding and promoting intellectual growth and critical thinking in adolescents and adults*. San Francisco: Jossey-Bass.

King, P. M., & Kitchener, K. S. (2002). The reflective judgment model: Twenty years of research on epistemic cognition. In B. K. Hofer & P. R. Pintrich (Eds.), *Personal epistemology: The psychology of beliefs about knowledge and knowing* (pp. 37–61). Mahwah, NJ: Lawrence Erlbaum Associates, Inc.

King, P. M., & Kitchener, K. S. (2004). Reflective judgment: Theory and research on the development of epistemic assumptions through adulthood. *Educational Psychologist, 39*(1), 5–18.

Keefer, M. W. (2003). Moral reasoning and case-based approaches to ethical instruction in science. In D. L. Zeidler (Ed.), *The role of moral reasoning on socioscientific issues and discourse in science education* (pp. 241–259). Dordrecht: Springer.

Kolstø, S. D. (2001). To trust or not to trust,...'-pupils' ways of judging information encountered in a socio-scientific issue. *International Journal of Science Education, 23*, 877–901.

Kolstø, S. D. (2006). Patterns in students' argumentation confronted with a risk-focused socio-scientific issue. *International Journal of Science Education, 28*(14), 1689–1716.

Labaree, D. F. (2003). The peculiar problems of preparing educational researchers. *Educational Researcher, 32*(4), 13–22.

Mueller, M. P., & Zeidler, D. L. (2010). Moral-ethical character and science education: Ecojustice ethics through socioscientific issues (SSI). In D. Tippins, M. Mueller, M. van Eijck, & J. Adams (Eds.), *Cultural studies and environmentalism: The confluence of ecojustice, place-based (science) education, and indigenous knowledge systems* (pp. 105–128). New York: Springer.

Narum, J. (2008). Transforming undergraduate programs in science, technology, engineering, mathematics: Looking back and looking ahead. *Liberal Education, 94*(2), 12–19.

Pedretti, E. (2003). Teaching science, technology, society and environment (STSE) education: Preservice teachers' philosophical and pedagogical landscapes. In D. L. Zeidler (Ed.), *The role of moral reasoning on socioscientific issues and discourse in science education*. Dordrecht: Kluwer Academic Press.

Ratcliffe, M. (1997). Pupil decision-making about socioscientific issues with the science curriculum. *International Journal of Science Education., 19*(2), 167–182.

Ratcliffe, M., & Millar, R. (2009). Teaching for understanding of science in context: evidence from the pilot trials of the twenty first century science courses. *Journal of Research in Science Teaching, 46*(8), 945–959.

Sadler, T. D. (2004). Informal reasoning regarding socioscientific issues: A critical review of research. *Journal of Research in Science Teaching, 41*(5), 513–536.

Sadler, T. D. (2006). Promoting discourse and argument in science teacher education. *Journal of Science Teacher Education, 17*(4), 323–346.

Sadler, T. D., & Zeidler, D. L. (2005). The significance of content knowledge for informal reasoning regarding socioscientific issues: Applying Genetics knowledge to genetic engineering issues. *Science & Education, 89*(1), 71–93.

Sadler, T. D., & Zeidler, D. L. (2009). Scientific literacy, PISA, and socioscientific discourse: Assessment for progressive aims of science education. *Journal of Research in Science Teaching, 46*(8), 909–921.

Walker, K. A., & Zeidler, D. L. (2007). Promoting discourse about socioscientific issues through scaffolded inquiry. *International Journal of Science Education, 29*(11), 1387–1410.

Zeidler, D. L., & Keefer, M. (2003). The role of moral reasoning and the status of socioscientific issues in science education: Philosophical, psychological and pedagogical considerations. In D. L. Zeidler (Ed.), *The role of moral reasoning and discourse on socioscientific issues in science education* (pp. 7–38). Dordrecht: Springer.

Zeidler, D. L., & Sadler, T. D. (2008). The role of moral reasoning in argumentation: Conscience, character and care. In S. Erduran & M. Pilar Jimenez-Aleixandre (Eds.), *Argumentation in science education: Perspectives from classroom-based research* (pp. 201–216). New York: Springer.

Zeidler, D. L., & Sadler, D. L. (2011). An inclusive view of scientific literacy: Core issues and future directions of socioscientific reasoning (pp. 176–192). In C. Linder, L. Ostman, & P. Wickman (Eds.), *Promoting scientific literacy: Science education research in transaction*. New York: Routledge/Taylor & Francis Group.

Zeidler, D. L., Sadler, T. D., Applebaum, S., & Callahan, B. E. (2009). Advancing reflective judgment through socioscientific issues. *Journal of Research in Science Teaching, 46*(1), 74–101.

Zeidler, D. L., Sadler, T. D., Simmons, M. L., & Howes, E. V. (2005). Beyond STS: A research-based framework for socioscientific issues education. *Science & Education, 89*(3), 357–377.

Zeidler, D. L., Walker, K. A., Ackett, W. A., & Simmons, M. L. (2002). Tangled up in views: Beliefs in the nature of science and responses to socioscientific dilemmas. *Science & Education, 86*(3), 343–367.

Zohar, A., & Nemet, F. (2002). Fostering students' knowledge and argumentation skills through dilemmas in human genetics. *Journal of Research in Science Teaching, 39*, 35–62.

Chapter 17
Metalogue: Balancing Tensions Associated with Extensive Enactment of SSI-Based Teaching

Dana L. Zeidler, Randy L. Bell, Troy D. Sadler, and Jennifer L. Eastwood

Eastwood: Together, the teacher and researcher were able to develop a thorough socio-scientific issues (SSI) curriculum building on science content and nature of science (NOS). The authors mention that the teacher in this study is unique in being willing to commit to such an endeavor. This type of instruction requires a teacher to have a high level of content mastery and confidence in teaching abilities. SSI instruction also requires the teacher to be knowledgeable of content far beyond the science text, reaching into current events, sociological or psychological perspectives, and even religious doctrine. Reading the teacher's reflections gives me a sense of the transformations he went through and the great sense of value he saw in the results for his students. I am already a convert, so I cannot say whether these types of testimonies would encourage me to try an SSI curriculum as a science teacher. My questions relate to teacher recruitment and support. What do you think could encourage more teachers to incorporate SSI in their classrooms? Are there some teachers that simply should not try it? How should teachers learn an SSI teaching approach? Would you recommend professional development sessions, preservice teacher training, or resources for independent study? How important was the close teacher-researcher relationship to enacting the SSI curriculum in this study?

Zeidler: I realize I am preaching to the choir as Jennifer is already a SSI convert. So while I do not need to provide her with additional research to lull her into the plethora of benefits associated with SSI pedagogy, or browbeat her into submission by advancing a deluge of tautologies that begin and end with SSI, I find I do need

D.L. Zeidler (✉)
College of Education, University of South Florida, Tampa, FL, USA
e-mail: Zeidler@usf.edu

R.L. Bell
Curry School of Education, University of Virginia, Charlottesville, VA, USA

T.D. Sadler and J.L. Eastwood
School of Teaching and Learning, University of Florida, Gainesville, FL, USA

T.D. Sadler (ed.), *Socio-scientific Issues in the Classroom: Teaching, Learning and Research*, Contemporary Trends and Issues in Science Education 39,
DOI 10.1007/978-94-007-1159-4_17,

to respond to several "down to earth" questions of practical importance. But I can only speak from my own experiences to best address these issues.

Teachers (particularly science teachers) are generally skeptical of anyone telling them what or how they ought to teach in their own classrooms. And well they should be. I have been fortunate to teach an advanced trends in science education course at the University of South Florida for a Master's program and beginning Ph.D. students. I use this as an opportunity to introduce the research base in SSI and related concepts and have teachers who currently implement SSI in sustained ways in their science classrooms come in and demonstrate various strategies they have found to be successful in engaging their students. I do the same thing on a much more scaled down approach when I implement workshops for the public schools in my area. The bottom line is that the teachers need evidence of theory in terms of providing a justification of the SSI approach, empirical evidence of research as to the effectiveness of SSI on various fronts, and hear about the deeds of other classroom teachers who have successfully implemented this approach. I have found these three aspects to be "key ingredients" in bringing others on board. Once these three factors are in place, the likelihood of teachers moving toward a transformation that is authentically transformative (in the sense discussed in our chapter) is mightily increased.

Are there some teachers who should not try it? Yes. Most likely those teachers who should not be teaching in the first place!

Bell: I agree with Jennifer's concerns about teacher recruitment and professional development. Implementing the SSI curriculum described in the chapter is no easy task. This is clear from the teacher's personal reflections, as well as the description of the intervention and results. Not only is the teacher extraordinary in his background and abilities, the level of support he received is remarkable (and not likely scalable). This teacher had not one, but two experts in SSI working with him as often as on a daily basis. These experts helped with all aspects of the program, including planning, curriculum development, and implementation of the SSI curriculum. This unusual (but powerful) relationship describes what I consider to be an ideal collaboration, and clearly all parties (including the students) benefitted. But it leads to an important question: To what degree can the success of the implementation be attributed to the SSI curriculum versus the level of collaboration/coaching the teacher received? I can imagine other interventions that could result in similar success, should they be accompanied by such high degrees of collaboration and support.

This is not to be taken as a criticism – at the very least the research summarized in the chapter can be taken as a best-case scenario. SSI implemented by an effective teacher with high levels of support from knowledgeable science education faculty can result in measureable improvements in teaching and learning. This is an important finding in itself – if SSI does not work in such an ideal setting, then there is little hope for its success in more typical classroom and professional development contexts. Now that we have evidence of success in this ideal situation, it will be important to see how the authors' curriculum and approach to professional development transfers to other settings.

Zeidler: It is true that the research we report is based on an extensive and intensive academic year experience with an unusually high degree of teacher support and collaboration by the researchers. But we need to be mindful that the purpose of the investigation we undertook was to examine if students could benefit in several important ways from a fully integrated SSI curriculum – by whatever means it took to deliver it. Since this study was the first of its kind, we were interested in addressing certain empirical questions related to reflective judgment, NOS, empathy, etc. Our goal was not to throw the teacher in the classroom with SSI materials and see if he would sink or swim. So, Randy raises an important different empirical question concerning how SSI would fare in a less than ideal-case classroom setting? At this time, I can only report dozens of anecdotal success stories from my own students who have gone on to implement SSI to varying degrees in a variety of classroom contexts and settings – from biology, earth and space science, chemistry, physics, and integrated science to elementary, middle, and high school levels with academically-challenged, average, and honors students. What evidence do I have to back these claims up? Trust me, I'm a doctor! (But be on the lookout for future conference paper presentations and articles that document some of these cases in the years to come.)

Sadler: I would like to shift the discussion to another issue associated with the extensive nature of this project. Most intervention studies in science education in general and certainly those that relate specifically to SSI interventions are based on single units. This study took place over a full academic year. Students were able to explore a series of SSI and build upon ideas and experiences over time. The research conducted under these circumstances documented significant student development in understandings of content, ideas about NOS, reflective judgment, and moral sensitivity. The fact that SSI was a consistent feature of the learning environment over such a long period of time certainly affected these results. However, it seems unlikely that most science teachers will be as willing to fully adopt the SSI framework as the teacher featured in this work. Most teachers like the idea of relating science teaching to socially-relevant issues and many are willing to try implementing a SSI-based unit, but very few are willing or able to restructure their courses such that SSI become a central organizing feature (Sadler, Amirshokoohi, Kazempour & Allspaw 2006). I am interested in hearing about what this group thinks in terms of how SSI ought to be featured in science education under ideal circumstances and what SSI advocates ought to promote given the current realities of schooling. In an ideal world, should SSI be the key feature of science education or should SSI-based instruction be balanced with other approaches and goals? In the real world, should our goal be to encourage the integration of at least some SSI in science classrooms or should we really push for more comprehensive approaches to SSI?

Zeidler: As I suggested earlier, I have found teachers generally accepting and quite enthusiastic about SSI when I supply three "key ingredients" – theory justification for SSI, empirical evidence of research as to the effectiveness of SSI on various fronts, and first-hand accounts from other classroom teachers who have successfully implemented this approach. Just last week, a graduate student stopped me in the hallway and said "I'm sold on the SSI approach – but I am just worried about how I will do this with the administration breathing down my neck about covering the same material at the same time as the other teachers!" These are brute facts of the "real world." Should I still encourage her to transform everything about her science teaching and follow a pure SSI agenda? Should I tell her to just do what she has to do to survive her first years of teaching? Or should I suggest a "blended approach" of integration of SSI where it makes the most sense?

Option two was never a real path for me to take. It could not be considered at all. Option one would have been my response in an ideal world. Of course, in option three, the blended approach retains the best features of what SSI has to offer in what is the "second-best case world." There may be some teachers, who just cannot place themselves in a student-centered classroom where classroom management issues are fully tested. Their own ability to draw out connections from social and ethical issues back to the content at hand, confidence in having a wealth of experiential worldly knowledge to effectively navigate students through a maze of data, misinformation, and passions may be questioned. I am also well aware of the ineffectiveness of simply pointing out connections to social issues while teaching in a more conventional manner. This is why I have my students create SSI units plans that typically will last 3–5 days (sometimes more). This is somewhat of a more sustained approach and the lessons need not follow consecutive classroom meetings. For example, in a 5-day unit plan, the parts of that unit could be extended over the duration of 2–3 weeks, depending on the nature of the SSI investigations and activities. These activities would be blended with other information gained from the more conventional pedagogy taking place in between those 5-day SSI-dedicated lessons.

Given the time it takes to create one effective SSI unit, it may be the only time they implement an SSI approach in the academic year. But as my students gain confidence, they add more units where it makes sense in subsequent years. They have found great success with this more common-sense real-world approach. Eventually, a few, like the teacher in our academic yearlong study, take the transformative plunge to a full-blown SSI classroom!

Eastwood: I fully understand why teachers would be hesitant to take on SSI-based science classes. There are unending challenges as discussed, and integrating SSI throughout their courses can seem like a big risk, especially for new teachers. There are certainly advantages in incorporating SSI units into traditionally taught courses, and as Dana suggested, as teachers become more confident and refine their courses over the years, they can build in more.

Still my position is to encourage a comprehensive approach to SSI. I agree with Troy that the consistency of SSI was important to the success in this particular classroom. Creating a supportive environment where students are willing to express their ideas and make personal commitments takes time. Promoting epistemological development calls for a different type of pedagogy, acknowledging students' capability to take positions on issues, connecting to students' experience, promoting mutual knowledge construction and respect, creating opportunities to explore different perspectives, and encouraging explanations of beliefs

(Baxter Magolda, 1999; King & Kitchener, 1994). SSI is unique in its many opportunities to create this type of learning environment.

Perhaps one possibility to support teachers in a comprehensive SSI approach would be interdisciplinary collaboration and team-teaching. This may be time-intensive, labor-intensive, and difficult to manage, but it may alleviate teachers' concerns about classroom management and being knowledgeable on all aspects of an issue. Collaboration between teachers could allow for deeper understandings of both science and social studies concepts in the context of particular issues. It has been done successfully in college classrooms (Chap. 6). I think there is potential for the approach to also work in kindergarten through twelfth grade (K-12) settings.

Sadler: I like how Dana laid out the basic options one could choose to address the challenge that I initially raised. We can encourage teachers to (1) adopt fully the SSI approach regardless of the challenges they may face, (2) bow to the immediate pressures of schools and classrooms and forget about implementing a SSI based approach, or (3) pursue a blended approach in which teachers try to slowly integrate SSI in their instruction. As Dana suggests, option three makes a lot of sense. However, I would not necessarily advocate option three if teachers employing that option were not pushed to move beyond the development of single units. In my mind, the initial experiences implementing a SSI unit are most significant because they may support growth in the teachers' comfort and confidence such that they become able to adopt more comprehensive strategies over time. My point is that if teachers only implement a single SSI unit over the course of the school year, then the impact on student learning and development will likely be very limited. The general (and ambitious) scientific literacy and citizenship aims of SSI-based education will not be realized if students have opportunities to explore science in context once a year. But if teachers can use these more limited experiences to build expertise such that they become better able to provide more systematic SSI experiences, then the goal of promoting scientific literacy and citizenship may actually be achieved.

Bell: I agree with Troy's point that a single SSI unit is unlikely to have a lasting impact on students' learning and development. However, teachers need to start somewhere, and developing a single unit is a logical place to start. My student teachers and I visited a local high school physics teacher who happens to be one of the most innovative teachers I have ever met. He has incorporated mastery learning, engineering design, self-created flash animations, and rocket science in his curriculum, just to name a few of his projects. His efforts have paid off in that he has made physics one of the most popular courses in the school, and his students do very well, both in class and on standardized tests.

Of course, my students asked this teacher how he finds the time to develop all of these curricular innovations. He responded with some very good advice: "Focus your efforts on one, and only one, new thing every year. Doing so will mean that you are always growing and your curriculum is always improving. Furthermore, focusing on a single innovation will allow you to develop the depth necessary for effective curricular and pedagogical improvement without burning yourself out."

Applying the "one new thing" rule to SSI curriculum development, perhaps interested teachers could work on incorporating one SSI unit each year. Done strategically, it would not take long for SSI to become a theme that extends across the entire year. Even better, individual teachers might share their work with others who are motivated to incorporate SSI into their curricula in deep and meaningful ways. In this manner, SSI could become a major theme of science education across the school, district, and/or state.

Dana: My original "blended" response had to do with building the confidence of new teachers (or teachers wishing to try a new pedagogy) who worry about facing the brute realities of administrative obstacles (both perceived and real) that can seem overwhelming at first blush. I agree that a more sustained effort would obviously be necessary to achieve the kind of functional scientific literacy we have written about in other papers. However, I do think that first steps should be encouraged. Given time, these same teachers can be running SSI marathons!

References

Baxter Magolda, M. (1999). *Creating contexts for learning and self-authorship constructive-developmental pedagogy*. Nashville, TN: Vanderbilt University Press.

King, P. M., & Kitchener, K. S. (Eds.). (1994). *Developing reflective judgment: Understanding and promoting intellectual growth and critical thinking in adolescents and adults*. San Francisco: Jossey-Boss.

Sadler, T. D., Amirshokoohi, A., Kazempour, M., & Allspaw, K. (2006). Socioscience and ethics in science classrooms: Teacher perspectives and strategies. *Journal of Research in Science Teaching, 43*, 353–376.

Chapter 18
A Case Study of the Impact of Introducing Socio-scientific Issues into a Reproduction Unit in a Catholic Girls' School

Vaille M. Dawson

Introduction

Rationale

Internationally, an accepted aim of science education is to enable all students to develop a deeper understanding of the world around them, and to use their understanding of science to contribute to public debate and make informed and balanced decisions about scientific issues that impact their lives (see for example, American Association for the Advancement of Science, 2000; Millar & Osborne, 1998). In Australia, significant emphasis has been placed on the importance of scientific literacy in science education (Rennie, Goodrum, & Hackling, 2001; Tytler, 2007). All Australian State and Territory curriculum documents state that science education should aim to develop students' scientific understandings, problem solving, and critical thinking skills related to science topics of importance in society. A high level of scientific literacy can help young people to question the claims of the scientific community and other stakeholders, weigh up evidence about science issues, and use critical thinking skills and their understanding of science to make informed and balanced decisions. More recently, the newly formed Australian Curriculum, Assessment and Reporting Authority (ACARA) has released guidelines for a national curriculum in science. The guidelines state that the Australian science curriculum must prepare students 'who, as citizens in a global world need to make personal decisions on the basis of a scientific view of the world' (National Curriculum Board, 2009, p. 4).

Thus, it is important that schools equip all young people with the knowledge, skills, and values needed to make informed personal decisions. Young people need to be able to pose questions, weigh up the risks and benefits of alternative solutions,

V.M. Dawson (✉)
Science and Mathematics Education Centre, Curtin University, Perth, WA, Australia
e-mail: v.dawson@curtin.edu.au

T.D. Sadler (ed.), *Socio-scientific Issues in the Classroom: Teaching, Learning and Research*, Contemporary Trends and Issues in Science Education 39,
DOI 10.1007/978-94-007-1159-4_18, © Springer Science+Business Media B.V. 2011

evaluate the integrity of evidence, and distinguish data from opinion. In short, young people need to be scientifically literate to make decisions about socioscientific issues (SSI). Yet, despite these arguments, there have been few published Australian studies that describe or evaluate the implementation of SSI in science classrooms. Indeed, many teachers seem reluctant to include SSI in their teaching programs.

The explicit stated importance of SSI by education leaders coupled with a reluctance by classroom teachers to teach about SSI forms the backdrop for the study described in this chapter. The chapter provides a detailed description and evaluation of the implementation of SSI associated with reproduction and reproductive technology. The research study is important for three reasons. First, the researcher (myself) has first-hand experience as a teacher-researcher who taught students about SSI in the context of human organ and tissue transplantation. Second, the content area for exploring SSI in this study is human reproduction. Third, the context for the research is a Grade 11 class in a Catholic girls' school. Thus, religious beliefs and values are important considerations that impact on this study. Before introducing the teacher, Lillian, and her students, the significance of each of these factors is elaborated.

My Previous Research

I came to this study as a former biological science teacher with a professional background in medical research. My previous experience as a medical researcher and a science teacher had led me to believe that science teachers can play an important role in equipping their students with the skills to make appropriate and well-considered decisions about SSI.

Early in my educational research career I conducted an action research project where as a 'teacher-researcher' I developed, taught, and evaluated a Grade 10 (14–15 year-old students) science unit on the topic of organ and tissue transplantation (Dawson, 1996). The research methodology employed in that study was based on a case study approach (Merriam, 2009). One of the main findings of the study was that the teaching strategies that seemed to be most effective in enabling students to clarify, reflect critically on, and modify their bioethical values were those that were student-centred and underpinned by a constructivist epistemology. These teaching strategies (e.g. role plays and oral presentations) provided students with opportunities to discuss, question, and justify their own values and also listen to the views of others. The students were actively engaged in constructing their own understandings. However, the use of a student-centred teaching strategy does not guarantee success. The style of the teacher, the nature of the students and the dynamics of the classroom environment ultimately determine the learning outcomes (Dawson & Taylor, 1998).

Reproduction and SSI

Previous classroom-based studies on SSI have tended to focus on, obviously, contentious areas in science such as biotechnology (e.g. Dori, Tal, & Tsaushu, 2003), genetics (e.g. Venville & Dawson, 2010; Zohar & Nemet, 2002), and environmental and sustainability issues (e.g. Grace, 2009; Pedretti, 1999).

Human reproduction is an integral and largely uncontroversial part of biology curricula worldwide. In Australia, topics such as the structure and function of the male and female reproductive system, stages of the ovarian and menstrual cycle, gametogenesis, fertilisation, embryonic and foetal development, and birth are typically taught in secondary school science. Although birth control (contraception), sexually transmitted diseases (STDs), and genetic diseases may also be taught, the teaching usually occurs in a factual way with students memorising the advantages and disadvantages of different forms of contraception, the symptoms and treatment of STDs, and common genetic diseases.

If SSI are defined as scientific topics that are 'based on scientific concepts or problems, controversial in nature, discussed in public outlets and frequently subject to political and social influences' (Sadler & Zeidler, 2005, p. 113), then the topic of reproduction is replete with SSI. Modern reproductive technology began with the birth of Louise Brown, the world's first test tube baby. In 1978, her birth caused a sensation (and consternation from religious organisations) when fertility specialists successfully performed in vitro fertilisation by extracting an egg, inserting sperm in a petri dish, and implanting the embryo back into the mother's uterus (Moreton, 2007).

Since that time, a range of reproductive technology procedures have become routinely available in Australia. These procedures are either fully covered or partly covered by Australia's public health system (Medicare) and by private health insurers. Reproductive procedures readily available in Australia include gamete intra-fallopian transfer (GIFT), artificial insemination (AI), intrauterine insemination (IUI), in vitro fertilisation (IVF), zygote intra-fallopian transfer (ZIFT), intracytoplasmic sperm injection (ICSI), ova and sperm donation, and non-commercial surrogacy.

Part of the reason for the increased availability of reproductive technology procedures is increasing need. An increased incidence of STDs such as Chlamydia and increasing incidence of obesity has led to increased female infertility. It is estimated that one in ten couples are affected by infertility. One in thirty babies is born in Australia through IVF. The number of IVF babies is increasing partly because older women are having babies. The number of IVF babies born to mothers in their early thirties is also increasing. The reasons are that women are no longer waiting until they are in their late thirties to seek fertility treatment and the cost is partly covered by the public health system. Recent changes to legislation to allow single women and lesbians to undergo AI and IVF has also increased the pool of women accessing reproductive technology procedures.

Preimplantation genetic testing (PGT) is available to individuals with private health insurance. Australian states and territories follow the National Health and Medical Research Council (NHMRC) guidelines where PGT is used to detect severe genetic conditions (e.g. cystic fibrosis, muscular dystrophy). Healthy embryos can then be implanted through ZIFT. The ZIFT process involves the transfer of an embryo to the mother's fallopian tubes where implantation can occur. PGT can also be used to select an embryo with a similar tissue type to a brother or sister (who may need a tissue or organ donor in the future). This procedure was the focus of the successful novel (and film), *My Sister's Keeper* by Jodie Picoult (2004). Parents accessing PGT must undergo genetic counselling. PGT is not allowed to be used for sex selection (NHMRC, 2007).

With regard to contraception, all forms of contraception are widely available in Australia. Oral hormonal contraceptives are available by prescription. RU-486, which induces abortion, was banned until 2006 when it was reclassified as a 'restricted good' and is available through registered hospitals and doctors. Although it can be used to induce an abortion up to 9 weeks gestation, it is not widely used. Abortion is legal in all states and territories to protect the health (including mental) of the mother. Thus, an unwanted pregnancy is sufficient grounds for an abortion. Early term (less than 14 weeks gestation) abortions are available through clinics and hospitals and costs are covered by Medicare. Consent of the father or parents (if person is a minor) is not required (De Crespigny & Savulesca, 2004).

Regardless of the legal status of the procedures described above, there is by no means universal community support. SSI associated with reproductive technology, contraception, surrogacy, and PGT appear regularly in the print and electronic media. Examples of issues include payment for surrogacy (not currently permitted), the availability of RU-486 and abortion, and availability of reproductive technology procedures to single, lesbian, older, and obese women. The use of abortion for 'minor' abnormalities such as missing limbs and dwarfism, and the high cost of fertility treatment is also contentious.

The Catholic Church and Catholic Schooling in Australia

According to the most recent census, Roman Catholics comprise the most common religious affiliation in Australia with 26% of the population stating that they are Catholic (Australian Bureau of Statistics, 2006). This is partly because a significant proportion of early Australian immigrants were Irish Catholics. Since Europeans settled in Australia, the Catholic Church has provided a low cost, independent (nongovernment) education which includes religious education. The Catholic Church recommends that all Catholic children attend Catholic schools. In 2009, 18% of Australian schools were designated as Catholic schools. It is estimated that more than half of the Catholic children in Australia attend Catholic schools (Dixon, 2005; Potts, 2009). It is accepted (indeed expected by many parents) that children who attend a Catholic school receive a Catholic education. Religious education is compulsory for all years of schooling and it is expected that all teachers uphold Catholic values and remind students of Catholic teachings on various topics. Fees are generally lower than other independent nongovernment schools and the inclusion of education on morals and values is favoured by parents. Although many Catholic high schools do not exclude non-Catholics, priority is given to children who are Catholic and have attended a Catholic primary school.

Catholic Views of Reproductive Technology, Contraception, and Preimplantation Genetic Testing

While human reproduction is typically taught as part of Australian secondary school biology courses, reproductive technology and associated issues would be considered unusual topics in a Catholic school. When such topics are taught, it would be expected that students are also taught the Catholic perspective. Teaching reproductive technology is akin to the teaching of evolution in a conservative Christian school where intelligent design and creation science might also be taught.

With regard to contraception, the Catholic Church is opposed to all methods of artificial contraception (e.g. contraceptive pills, condoms, intrauterine devices, and diaphragms) on the basis that marital sexual intercourse is for reproduction only and using contraception is against 'natural law'. However, natural methods of birth control including the rhythm method, temperature, and mucus patterns are allowed. Also, it is permissible for an unmarried woman to use the contraceptive pill to regulate her menstrual cycle.

The Catholic Church is opposed to abortion, regardless of the reason. Yet, PGT is allowed as long as there is no risk to the child or the mother. This is so that families can prepare themselves and medical treatment may be commenced to early as possible. The Catholic Church permits medical investigations of infertility and surgical and hormonal treatment to restore ovarian or uterine function. Collection of sperm is only allowable through the use of a special permeable condom that is used during sexual intercourse. The only two reproductive technology procedures that are not opposed by the Catholic church are GIFT (a modification of IVF where the mother's ova are placed in the fallopian tube with sperm from the husband) and intrauterine insemination where the husbands' sperm is collected in a condom, washed to extract motile sperm, and injected into the uterus. In both of these procedures, fertilisation occurs in the fallopian tube or uterus.

Introducing Lillian

Lillian, an experienced biological science teacher at a Catholic girls' school, and her 15 students are the focus of this research study. The study was conducted in a medium sized (850 students) girls' Catholic school located in Western Australia.

Enrolment at the school is predominantly children who attended local Catholic primary schools in the same locality. The school caters to girls enrolled in Grades 7–12 (11–17 years of age). About 20% of the students are boarders whose parents live in rural areas of Western Australia. According to national testing in literacy and numeracy, academically, the students attending the school can be considered to be average (ACARA, 2010).

Lillian initially completed a degree in botany and worked in the research field of plant genetics and tissue culture before becoming a teacher. She has taught at the school for 4 years. This study focuses on her teaching of a reproductive technology topic for 16 weeks in a grade 11 science course. The science course is intended for students who do not intend to go to university. In this course, students were introduced to genetic diseases, aspects of reproductive technology (e.g. IVF and GIFT), contraception (e.g. abortion), and SSIs related to these topics. Lillian's teaching goals were to increase her students' understanding of reproductive technology, increase their awareness of ethical issues associated with reproductive technology, help them to appreciate the importance of the issues to themselves and society, and also understand that the issues are complex and not easily resolved.

Purpose of Research and Research Questions

The research reported here was part of a larger study that aimed to examine and evaluate the effectiveness of a range of learning activities utilised by secondary science teachers who included SSI in their teaching programs. Lillian was one of three case study teachers who participated in the study. The findings from the other two teachers have been reported elsewhere (Dawson, 2010).

The research questions were addressed within an instrumental case study research design and underpinned by a constructivist theoretical framework. There were three research questions that guided the research design and three emergent research questions that were unique to Lillian's case.***

Initial Research Question 1
To what extent is a caring and communicative relationship established between the researcher (myself) and the research participants?

Initial Research Question 2
What types of learning activities are selected by science teachers who are incorporating SSI into their teaching programs?

Initial Research Question 3
How effective are these learning activities in enabling students to reflect critically on, articulate, and justify their decisions about SSI?

The first research question relates to the ethics of the research process, an issue that I considered important enough to justify a separate question. The perspectives of Lillian and her students were crucial in data generation and interpretation. By establishing a trusting relationship and encouraging open communication,

I believe that I was better able to understand the research participants' perspectives. In addition to the quality of data generated, I felt that I had a moral responsibility to create a meaningful, trusting, and mutually rewarding relationship. I adopted an ethic of care and empathy in my relationships with Lillian and her students (Noddings, 1984). Empathy is about reception, not projection (of my belief systems). Thus, I attempted to understand the participants' points of view by listening to and valuing their views.

The method of addressing the other two initial research questions was descriptive and based on ethnographic participant observation (Denzin & Lincoln, 2000). A learning activity is defined as any form of teacher-planned interaction that occurs with students in the classroom. Examples of learning activities included oral presentations, portfolios, role plays, debates, and library/Internet research. The implementation of the learning activities depended on the teaching strategies selected by the teachers and included group work, whole-class discussions, cooperative learning, and student-centred learning.

As the study progressed, several factors arose that were unique to this case study. These factors related to the overall purpose of the research (i.e. the impact of introducing SSI) but were not addressed by the initial research questions. Emergent research questions were developed to address these factors.

Emergent Research Question 1
To what extent did the teacher achieve her teaching goals related to the teaching of SSI?

Emergent Research Question 2
What effect does the reproduction course have on students' attitudes toward science?

Emergent Research Question 3
From the multiple perspectives of the students, the teacher, and myself, what impact did the learning activities have on student learning?

Research Method

The research design was an instrumental case study (Merriam, 2009; Stake, 2000). A case study is an intense examination of a specific issue; in this case, the issue is the teaching of reproduction, reproductive technology, and associated SSI by a science teacher in a Catholic girls' school. There are many different types of case study designs used in qualitative research. Stake (2005) describes three types of case studies: intrinsic, instrumental, and collective. Intrinsic case studies are those in which the specific case is unique and it is not intended that the research findings be generalised or transferred to other environments. Collective or multiple case studies are those in which a number of cases about a particular situation or phenomena are studied. The research presented here is an instrumental case study because the case 'serves the purpose of illuminating a particular issue' (Creswell, 2008, p. 476). Stake (2005) asserts that the selection of cases should be those that provide the greatest opportunity to learn about the issue. Briefly, the researcher gathers as much data as possible about the research problem (the teaching of SSI) and then categorises, analyses, and develops tentative hypotheses. This case study is descriptive and interpretive. The use of multiple data sources enabled triangulation and contributed to the trustworthiness of the findings.

Data Generation Methods

Data generation methods included participant observation, multiple interviews, open ended questionnaires, and reflective journal writing. These methods enhanced my understanding of the multiple perspectives of the research participants. Data generation and interpretation occurred in a cyclical fashion. Interpretation was informed by grounded theory methods (Strauss & Corbin, 1990) as I identified categories and themes within the data.

Participant Observation

Participant observation is an ethnographic method that is commonly used when attempting to understand what is happening in a particular environment, in this case, Lillian's classroom. As a participant observer, I visited and observed Lillian, her students and the classroom environment. When observing in the classroom, I observed and recorded my perceptions of the classroom environment and the responses and emotions of the participants. Data collection was prolonged, and occurred over 16 weeks; I was unable to attend every class due to my own teaching commitments but participated in the classes twice a week (32 visits). Although the SSI component was explicitly taught in the last 5 weeks of the course, I felt I would gain a deeper understanding of the learning environment if I observed Lillian's class throughout the entire reproduction topic (see Table 18.1 for an outline of the structure of the course). I believed that this would enable me to develop a caring and trustful relationship with Lillian and her students. It would also enable me to share with Lillian my ongoing perceptions and interpretations of the classroom environment.

Interviews

During the 16 weeks, I interviewed Lillian about her perceptions of the learning activities and students' learning at the end of each lesson. The length of interviews varied from 20 to 90 min. The interviews were usually audio-taped and transcribed as soon as possible. Sometimes though, our interviews took place in an open staff room where the background noise was not conducive to audio taping. Thus, I took brief notes which I wrote up and added to the reflections recorded in my journal.

Table 18.1 Outline of science course

Week	Topic	Activities
1–2	Sexual and asexual reproduction	DVD
	Mitosis	Growing and preparing onion root tips
	Meiosis	Lilium anther slides
3	Plant vegetative propagation Techniques (cuttings, runners, bulbs, grafting)	Practical activities using different plants and techniques
		Growing carrots and African Violets on agar
4	Plant sexual reproduction	Dissection of flowers
	Flower structure and reproduction	Pollen tube growth in sugar solutions
		DVDs
	Monocotyledon and dicotyledon seed structure and germination	Seed dissection and germination
5	Reproduction of vertebrates and invertebrates	DVD
		Rat dissections
		Raising tadpoles
	Reproductive anatomy and life cycles	Morphology of mice
6–10	Simple Mendelian genetics	Breeding mice – F1, F2 and back crosses
11–13	Human reproduction	DVD
	Anatomy/physiology	Worksheets
	Contraception	
	STDs	
	Genetic diseases	
14–16	SSIs	DVD
	Embryo testing	Defining the terms
	IVF	Oral presentation
	Genetic engineering	

The interviews were unstructured. I initiated teacher interviews with open-ended questions like 'how did you feel about the lesson?', 'what do you think the students learnt?', 'would you do anything differently next time?' I paraphrased Lillian's responses to determine whether my understanding of what she said approximated what she intended. I encouraged Lillian to talk not only about what happened in the classroom but how she felt about what happened.

While observing Lillian's class, I spoke informally with her students. I also conducted formal interviews with students on two separate occasions. Toward the middle of the reproductive technology topic, I interviewed all 15 students in the class. The purpose of this initial interview was to determine the students' perceptions of the importance of the topic and also to ascertain what they expected to learn. A second interview of four students was conducted at the end of the course where I asked them to elaborate and explain their questionnaire responses (described below). In relation to the selection of these four students, I used a purposeful sampling method (Patton, 1990) that allowed for maximum variation so as to perceive a wide range of students' views. The student interviews were conducted individually.

Questionnaires

All 15 students completed two questionnaires at the end of their course to determine their perceptions of what they had learnt. One of the questionnaires contained open-ended questions and was designed to elicit information about the students' perceptions of the learning activities, teaching style, course content, and learning outcomes. In the second questionnaire, students were asked open-ended questions about what they had learnt in the course and how the course had influenced their views about issues. For example, students were asked, 'what are the three most important things that you have learnt in this unit?', and 'Is this course different from your other science classes? Please explain.'

Personal Journal

A personal journal has been shown to be a useful tool in qualitative research to encourage critical reflection (Connelly & Clandinin, 1988; Holly, 1992). I used a personal journal throughout the research study. I used it for developing ideas and reflecting on the actions of myself and others. I also wrote my journal when analysing data, posing questions, and developing arguments for and against emergent hypotheses. My use of journaling was consistent with Cooper's (1991) suggestion that 'Journals allow us to examine our own experiences, to gain a fresh perspective, and by that means begin to transform the experience themselves' (p. 99). The journal also helped me to hear and listen to my own voice. By seeing the written words on the page, the process of journal writing also assisted me in making explicit the personal values and beliefs that influenced my reflections and interpretations.

Constructivist Standards

In adopting a constructivist theoretical framework, the criteria used to guide the research process was based partly on those described by Guba and Lincoln (1989). They described a set of 'parallel' criteria for judging the quality of constructivist research. The criteria are based on *trustworthiness* and mirror the traditional standards of judging quantitative research (i.e. validity, reliability, and objectivity). The parallel trustworthiness criteria are *credibility*, *transferability*, *dependability*, and *confirmability*.

Credibility is a measure of the extent to which the participants' realities are faithfully portrayed. Credible reporting of the experiential realities of Lillian and her students was ensured by prolonged engagement and persistent observation over a 16 week period. By spending an extended period of time in the research environment, coupled with sensitive and careful observations, I had an increased opportunity to become aware of and follow-up on patterns, trends, and relationships as they emerged.

Data generated by different techniques allowed triangulation and cross-checking of emergent hypotheses. This was coupled with negative case analysis as I searched for disconfirming evidence of these hypotheses. The practice of peer debriefing, or talking to others unrelated to the research assisted in developing and discarding emergent hypotheses. I also maintained a record of my prior and existing constructions about the environment and attempted to ensure that undue weight was not given to these a priori constructions.

The practice of member checking by seeking continual feedback from Lillian demonstrated not only that I valued and respected her interpretation but also that it was an effective means of checking, clarifying, and refining working hypotheses. This process of member checking occurred during the data generation phase and through the initial data analysis.

Transferability is the degree to which the research findings are applicable in other situations. In constructivist research, the onus is on the reader to make that decision to their context. In this study, transferability was optimised by the use of rich descriptions of the participants, events, and context. Thus, where relevant, comprehensive details are provided about the school environment, classroom environment, characteristics of students, and curriculum content. Dependability and confirmability were maximised by extensive reporting of the data generation methods and interpretation so that the reader can link my interpretation to the original data, thus reducing the amount of personal bias.

Ethics in This Research Study

The issue of ethics is important in all qualitative research studies. In research underpinned by a constructivist approach, a consideration of values, and more specifically ethics, is paramount to ensure that the participants are treated with care and respect. Thus, attention to ethics was an especially crucial element in this research study. Throughout the study, I endeavoured to act in a way that respected the rights of all involved. That included myself, Lillian, her students and others who contributed to this study.

In formulating a guiding theoretical framework for this study, I was profoundly influenced by the work of Nel Noddings. I adopted Noddings' (1984) caring ethics as a means of guiding and evaluating my actions in this study. Caring ethics is empathic, responsive, and concerned with relationships. I believed that I had an ethical responsibility to be receptive to the needs of the participants. I needed to empathise with Lillian and her students if we were to develop a relationship based on trust, care, honesty, and respect, that is, to listen actively, and offer support, encouragement, affirmation, and resources when required. Only then can a mutually beneficial, collaborative relationship develop. Noddings (1984) states that, 'Caring involves stepping out of one's personal frame of reference into the other's' (p. 24). Erikson (1986) agrees also that 'a non-coercive, mutually rewarding relationship with key informants is essential if the researcher is to gain valid insights into the informant's point of view' (p. 142). The adoption of a caring ethic helped me to remain constantly alert to the feelings of the participants during the process of data generation and interpretation.

Results

The first part of this section describes the context and structure of the science course, including Lillian's teaching goals. Then, I outline how I developed a trusting and caring relationship with Lillian and her students (initial research question 1). Two vignettes that illustrate the type of learning activities in which students engaged are presented (initial research questions 2 and 3). I also address the three emergent research questions. After presenting the two vignettes, I describe the extent to which the learning activities enabled Lillian to achieve her teaching goals. Finally, the students' perceptions about what they learnt are presented.

Context of Science Course

I first met Lillian when I ran a professional development workshop (on SSI in science) for the Science Teachers' Association of Western Australia. Lillian attended the session. She shared with the group her recent experience teaching Grade 11 science students about SSI associated with reproductive technology. She felt she had not been overly successful partly because of the lack of suitable resources and also because of the limited academic ability of her students.

At the end of the workshop, I approached Lillian and asked her if she would be willing to participate in my research study. She agreed and the following year I organised regular visits to her class. Because Lillian taught in an all-girls Catholic school, I wanted to explore whether the teaching of SSI associated with reproductive technology would conflict with, or complement the religious teachings of the school. I asked Lillian how she approached reproductive technology procedures that are not permitted by the Catholic Church. For example, I asked her what stance she took on informing students about PGT and genetic counselling. Typically, one of the options discussed by genetic counsellors would be the possibility of an abortion. She told me that, even though she is not a Catholic, she is a teacher in a Catholic school and therefore has a duty to ensure that the students are aware of the Catholic Church's views on these issues. The Catholic perspective is that the purpose of PGT is to enable the parents to prepare for the birth of their child and abortion is not an option. In relation to methods of contraception, even though their use is not permitted by the Church, she believes that her students still need to understand the procedures. Thus, when Lillian taught the students about various methods of contraception, including abortion, she stated clearly the Catholic perspective on abortion and other forms of contraception.

As the only teacher of the grade 11 science course at her school, Lillian had relative freedom to develop a teaching program to suit her students' needs. In an early interview, I asked Lillian how she had decided on the scope and sequence of the course. She explained that she had endeavoured to teach a logical sequence. Rather than a linear sequence, she used a 'spiral approach, gradually introducing more information' on the same topic. For example, at the start of the unit, Lillian obtained a pure bred male black mouse and a pure bred female white mouse. The students were very excited by the arrival of the mice and willingly participated in a cleaning and feeding roster. Lillian used the mice to help students begin to understand the structure and function of sexual reproductive systems. The students observed and drew the external genitals. Sexual reproduction was reintroduced during the dissections of male and female rats. More information on sexual reproduction was supplied later, regarding the human reproductive organs. An understanding of reproduction led into inheritance and genetics, gene and reproductive technologies and, finally, SSI.

Lillian deliberately introduced SSI at the conclusion of the course. She believed that students needed a thorough understanding of the topic of reproduction before they could examine SSI. When I asked Lillian whether she felt there were any advantages in leaving SSI until the end, she told me that the SSI were the 'culmination of the course'. She believed that the students would not understand the issues if they were taught from a 'position of ignorance'. Lillian stated that 'otherwise their views will be naive and simplistic'.

Lillian's Teaching Goals

Lillian believed that this course may be the last (and only opportunity) for her students to become aware of issues associated with reproductive technology. She believed that some of her students may need to deal with fertility problems or genetic diseases either directly or indirectly. Thus, Lillian wanted to equip her students with the knowledge and skills to understand how these issues impact themselves and society. She explained that:

> My students will be voting in a year or two and it is important that they understand issues. They will not be continuing with science at university, so they need to be exposed to issues in science now. Explicitly teaching bioethics is important in the area of reproduction because the students ask ethical and moral questions. You can't avoid the issues.

Lillian hoped that by introducing students to SSI, students could understand, for example, how it would feel to be pregnant with a child with a genetic disease. She explains:

> I want students to understand that genetic diseases are more than words on paper. I want them to appreciate the anguish faced by a parent with a child who has a severe genetic disease. They can only address the ethical issues if they have some appreciation of the trauma involved.

Otherwise, she asserted, they would be unable to appreciate the seriousness of the situation. In her experience, students with little understanding tended to be naive or to adopt an inflexible stance. Lillian also told me that she wanted her students to appreciate the complexity of issues. She explained:

> I wanted them to see there is no such thing as black and white. These issues, even though there might be morally right and morally wrong viewpoints, real life comes in shades of grey. If we are to live in a compassionate society, even though maybe this is not a morally right thing in some ethics book, it's happened and what are we going to do? Are we going

to forgive them? Help them? What sort of choices do we have? They have to learn to cope with the grey. I want them to see that moral issues are a complex set of relationships and you cannot just say this is right and this is wrong. They have to learn to cope with that to see their way through it and not be simplistic.

Structure of Course

The reproduction part of the science course was taught over 16 weeks with 4×45 min periods a week. The topics studied included plant, animal and human reproduction, contraception, genetics, genetic diseases, and reproductive technology. Table 18.1 summarises the teaching program.

The final part of the course addressed SSI associated with reproductive technology. Lillian began the section by showing the students a DVD about families that were affected by genetic disorders. In some instances, the parents were aware prior to the birth of an affected child and the DVD emphasised that it is the parents' choice to decide. The importance of genetic counselling was discussed. The DVD made the point that a genetic disease was not always a catastrophe. At the end of the DVD, Lillian discussed with students how they might cope if they were the parents of such a child.

Students also watched and discussed a DVD about embryonic stem cell research. The DVD explained how embryos (14 days post fertilisation) are harvested and frozen for future fertility treatment and research purposes. The scientists believed it was ethical to experiment with embryos of one to four cells. The DVD raised a number of ethical questions which students debated in a whole-class discussion. For example, if an embryo has a genetic disease and is subsequently not implanted, is this abortion? If the parents die, who owns the embryos? What should happen to frozen embryos that are not used by parents? Is it ethical to harvest and freeze ova taken from the ovaries of a 16-week-old foetus.

Students also individually completed a research project. First, students constructed a glossary of terms including IVF, amniocentesis, genetic engineering, bioethics, muscular dystrophy, and eugenics. Students were required to research the definitions of the terms. The students then selected a reproduction issue from a predetermined list to examine in-depth. Each student prepared a 10 min talk which they presented to the rest of the class. Examples included diagnosis and treatment of sex-linked diseases such as haemophilia, IVF, and GIFT (including the Catholic perspective) and eugenics.

Establishing a Caring Relationship

In this section, I reflect on the extent to which I established a research relationship with Lillian that was based on care and respect (i.e. first initial research question).

In addition to a moral obligation, I believed that it was essential to establish a trusting relationship so that Lillian would be able to raise problematic issues with me as I visited her classroom and also as I shared my interpretations with her (i.e. member checks to enhance credibility). I have used excerpts from classroom observations and my personal journal to illustrate how our relationship developed and evolved.

The First Interview

Because I spent an extended period of time (4 months) with Lillian and her students, I had the opportunity to establish and maintain a trusting and caring relationship. I found it relatively easy to establish rapport with Lillian, partly because we shared common interests and backgrounds. I had been a science teacher in a girls' school with an interest in teaching SSI in science education and I empathised with the nature of the innovatory teaching that she was attempting to undertake in her class.

Like Lillian, I had also taught students with limited academic ability and was thus aware of the difficulties faced by those students when they are required to read and comprehend scientific information. Lillian and I also discovered that our employment history overlapped in that we had both tutored in the same subject (Cell Biology) in adjacent departments at the same university. We had also both worked in the same hospital laboratory, although at different times. The fact that we shared these common experiences meant that it was easier to begin to establish a trusting, communicative relationship. Overall, I found Lillian to be a warm and easy going person.

At the time I wrote in my journal:

> I feel that Lillian and I have communicated well. We helped each other in our respective teaching by swapping resources and ideas. For example, Lillian has given me her drawings of the male and female reproductive systems and a copy of an assignment on sexual reproduction while I gave her an Internet assignment, a cloning, and some reproductive technology resources. This mutual exchange of resources and sharing of ideas helps me, I believe, to demonstrate to Lillian that I value her. I was not there just to 'get data'. I am willing to learn from Lillian.

Enhancing Credibility

In order to enhance credibility, I endeavoured to stay behind after each lesson not only to discuss with Lillian her perception of the lesson, partly for debriefing, but also to clarify my perceptions. At regular intervals, I gave Lillian my case record which contained interview details, journal extracts, and classroom observations. When I gave Lillian the notes that I had written about my first visit, I explained to her that there were two reasons for giving her what I had written. Firstly, it was important that she knew what I had written as I may have misinterpreted or not explained fully the issues we had spoken about. Secondly, from an ethical perspective, it was important that Lillian had an opportunity to express her view point.

By seeking feedback, I was also able to reassure Lillian about her teaching. After 2 weeks I gave Lillian a copy of the case record to read and comment on. The following week, I asked Lillian what she thought of the case record. She replied, 'Yes, it was good, but strange to read. A mixture of verbatim conversation between us and also what the students had said'. I said that I was trying to record as complete an account as possible of what was happening.

Raising Problematic Issues

I felt that by establishing a trusting and caring relationship, Lillian would be able to raise problematic issues with me. This did seem to be the case. The following extract from my journal relates to an incident where Lillian perceived that I wanted her to change the structure of the course.

> Today, Lillian asked me what sort of time line I was envisaging in the study. "How flexible is your research?" she inquired. She explained that normally she would have taught the SSI component at the end of the course when the students understand reproduction, principles of genetics, and genetic diseases. She said that she was a little concerned as to how she was going to structure the course with so much happening.
>
> She thought I was only interested in the SSI component of the course and that I wanted her to teach it at the start of the course. I reassured her it was not so. I said that I was interested in the context of the SSI section and that she should teach the course as she had planned. "Please," I said. "Don't adjust the course to suit me! Teach it the way you normally would. I am flexible. I would, however, like to visit your class regularly so that I can understand what the students have learnt during the whole course."
>
> "Oh, good." she said. "I was worried that you wanted me to teach about SSI now."
>
> "No, no." I replied. "Don't change the course just to suit me."

I think Lillian was reassured by our conversation. I was relieved that she had felt able to raise her concern with me because I certainly did not intend for her to feel pressured to change the course to suit me.

Reciprocal Assistance

After 2 weeks of sitting in Lillian's class observing and writing notes, I asked her if she would mind me helping the students. This was partly to offer something in return for her extended invitation and also to engage the students in dialogue about their understanding of the topics covered. She replied that it would be 'great to have an extra pair of hands'. I was glad that Lillian did not feel threatened about me assisting in her classroom. Even though the class is small (15 students) the students seemed to benefit from the extra one-to-one assistance. Absenteeism was high and students found it difficult to stay on-task in student centred activities.

Thus, when students were engaged in practical work, I frequently helped in class as the following classroom observation extract demonstrates.

> Students are staining and observing root tips where some of the cells are undergoing mitosis.
>
> Lillian holds up the microscope and shows the students the position of the coarse and fine focussing knobs and the objective lens. She asks the students to prepare a slide with the acid-softened root tips, add the dye, and look for chromosomes ("like worms") in the center of the cell. The students, working in groups of two, quickly and efficiently prepare a slide.
>
> I approach a pair of students and ask if they can see anything. 'I don't know.' one replies. I ask if I can look. Under the microscope, I see many unstained rectangular shaped cells. I change the objective lens to high power and move the slide so that the edge of the root tips can be seen. I know from experience that this is where a single layer is most likely to be found and also that the stain will penetrate these outer cells first. I can see square cells with an orange nucleus and I suggest to the girls that they look around this area.
>
> Lillian is occupied with students on the other side of the room. I move to an adjacent group. "Is it focused?" I ask. "Yes, I can see cells." "May I look?" The student moves aside and through the lens I see five small air bubbles. I explain what they are and focus under high power. I find a group of four cells. I can see the chromosomes in one of them. I ask them both to look at the four square cells. I ask them both if they can see the round structure in the middle. "What is it?" I ask. "A nucleus." "Are all four cells the same?" "No." replies the student. "One of them has little lines in it." "They are the chromosomes," I explain. "They separate to each end of the nucleus." I demonstrate it with my fingers. One of the student's information sheets depicts the stages of mitosis. I show the students what stage the cell is at. Over the next 10 minutes, I help several more groups to focus on cells where chromosomes are visible.

Learning Activities

As described earlier, the final part of the science course was concerned with issues associated with reproductive technology. This section presents two vignettes compiled from classroom observations during the final part of the course. The purpose of the two vignettes, *Defining the Terms* and *Oral Presentations*, is to illustrate the types of learning activities that students engaged in.

Vignette One: Defining the Terms

For the previous four lessons, the students have been working in the library searching for information to define and describe a range of terms associated with reproductive technology and genetic diseases. Some of the terms provide background information (e.g. genome, foetus) while others raise ethical issues (e.g. eugenics, Human Genome Project).

During this activity, Lillian uses a whole-class question and answer discussion to go through the definitions. As each term is defined by a student, Lillian, if necessary, expands on the definition. She also asks questions to draw her students' attention to issues associated with each procedure. For example, in an early part of the lesson, a student defines eugenics as 'using genetics to improve the human race'. Lillian informs the students that eugenics was used in Europe during the first part of this century in an attempt to rid Europe of Jews, gypsies, and mentally incapacitated individuals. Lillian asks the students, 'what are the disadvantages of eugenics?' Students reply. 'Variation is needed for the survival of our species.' 'Unethical.' 'Goes against human rights.' 'It's not natural.'

Lillian continues with the list of terms asking individual students to read their definitions and asking questions related to the term. For much of this part of the activity, the students were very quiet with only four students responding with brief answers to Lillian's questions about issues. However, toward the end of the second lesson, the term surrogacy is defined as 'where another bears and then gives up the child who may or may not be genetically related'. Lillian shows the students a newspaper article where a woman had a child for her sister. The woman had tried using her sister's ova and the husband's sperm, but it was not successful so they used her own ova and the husband's sperm. Thus, the child is genetically related to the surrogate mother. Lillian asks if there are any problems with this.

A number of students answer.

The mother may not want to give up the child.
The baby might find out.
You must tell the baby when it's young.
They should involve the genetic mother in its upbringing.
It might be okay for related people, but may not be realistic.
There was a case in America where the woman was paid
The child needs to know in case they fall in love with a relative. They need to know or incest may occur and the chance of recessive diseases may increase.

The students are for the first time participating in a whole-class discussion. They appear to be animated and interested. At the end of this brief interchange, Lillian tells them that they have been talking about issues, about the ethics. They have been having an ethical discussion.

Vignette Two: Oral Presentations

After the students had defined the terms (in the vignette above), they were assigned a topic related to an aspect of reproductive technology or to genetic diseases. They were required to prepare a 10 min oral presentation, which they presented to their peers. Lillian assigned each student a topic from a prepared list of topics. The students were provided with four lessons to search for information and prepare their oral presentation. The students used the library and the Internet. Lillian provided assistance as needed. In this vignette, several of the students are presenting their oral presentations. Jacquie was the first student in the class to give her presentation, while Rachel and Maria presented in the middle and Tanya was the final presenter. The descriptions of the oral presentations illustrate how the students in the class gradually changed from a passive role to a more active one through asking and answering questions about SSI. The presentations occurred over six lessons.

Jacquie: Sex Linked Diseases

Jacquie appears to be nervous and reads her talk rapidly from written notes. She places an overhead transparency on the overhead projector. She states that Duchenne muscular dystrophy is a sex-linked disease, explains what this means, and why males inherit the disease while females are carriers. On the overhead transparency, she has a pedigree that shows how the recessive gene is inherited. The students are trying to copy the notes on the overhead transparency as she speaks. Jacquie realises this and pauses to let them write.

As the students were writing, Lillian initiated the following whole-class discussion:

Lillian: Are there any ethical issues?
Jacquie: One issue is that scientists are spending time and money trying to understand the abnormal gene and finding a cure.
Lillian: If you knew you had a history of muscular dystrophy, what issues would that raise?
Jacqui: Whether to get married or abort the baby.
Lillian: What would you do?
Jacquie: Not have children.
Lillian: What about the rest of the class?
Student 1: Get counselling.
Student 2: Get information so you know what you're getting into.
Lillian: What would a counsellor do?
Student 1: Look at the family history and give advice.
Lillian: What if the mother was already pregnant and had an antenatal test. How do you make up your mind about whether to have a baby?
(Students do not respond.)

In this first oral presentation, there is little interactive discussion. The students are copying down notes. Although Lillian uses questioning to attempt to engage the students in discussion, they are unwilling. No student asks the speaker a question or responds to another student's comments.

Rachel: Surrogacy

On an overhead is the question 'what is surrogacy? – an arrangement under which a woman agrees to bear a child for another person who cannot have a child of their own'. Rachel explains the difference between altruistic surrogacy between sisters or close friends and commercial surrogacy where the surrogate mother is paid.

Rachel concludes her presentation by outlining arguments in favour of surrogacy (i.e. the only method if other reproductive technology methods fail, child is really wanted, and women should be free to use bodies how they wish) and against (i.e. deliberate creation of life, not natural, planned separation of child and birth mother at early stage of life is unnatural, and not all participants (e.g. relatives of surrogate) may agree).

When Rachel is asked by Lillian what she believes, she says that it is a personal decision and must be based on a person's moral beliefs. However, she recognises that it offends some groups, as it puts women in the role of being reproductive incubators. A student asks, 'what's your personal opinion?' Rachel replies, 'It's fine'. Lillian asks Rachel whether she would be a surrogate for a sister. Rachel responds that it would depend if she had her own children and on how old she was. Lillian asks her if she is aware of the Catholic Church's view. Rachel says, 'This is like playing God so they probably don't agree'. Lillian tells her that she is correct. Lillian asks the class if they have any questions. No one does.

During this presentation, students are still unwilling or unable to engage in discussion although one student does ask Rachel for her opinion to which she answers, 'it's fine'. Although provided with opportunities, students are still reluctant to ask questions or make comments. Although Lillian's aim was for students to raise and discuss issues (which she stated explicitly), the students seem to perceive that the purpose of this activity is to listen and copy down notes.

Maria-Amniocentesis and Chorion Villus Sampling (CVS)

Maria begins her presentation by outlining some of the problems associated with chromosome abnormalities. She draws students' attention to the overhead indicating the increasing risk of Down's syndrome with age. She defines amniocentesis and CVS. 'If children have a disability', Maria asks, 'is it right to terminate? It reduces our tolerance to disabilities. Are these tests of benefit or do they assist "perfect baby syndrome". What do you think?' Maria pauses to allow students to respond; some of their responses are captured below:

Student 1:	I don't know if I want to say in public (long pause).
Lillian:	Go on.
Student 1:	I think everyone over 43 should have an amnio., not for abortion but to prepare them.
Lillian:	Should doctors tell their patients the sex of the foetus?
Student 2:	The doctor has a responsibility to tell.
Student 3:	I reckon it's the parent's choice [to decide whether to have an abortion].
Lillian:	Should the father have a say?
Student 4:	The husband has as much right as the wife.
Student 5:	The mother should have the final say.
Student 6:	No, both.
Student 7:	I don't think anyone should be born with abnormalities.
Lillian:	How do you decide if it is major problem?
Student 7:	It's a big responsibility. What if the baby were dumped in a home?

Lillian: What if it was a Catholic hospital?
Student 8: Nothing would happen.

For the first time during the oral presentations, the students are talking and responding to each other's comments. They are very animated, asking and answering each others' questions. It seems that it has taken time for Lillian to help students realise that it is alright to ask questions and that they must be able to defend their views.

Tanya: Abortion

Tanya begins by defining abortion as 'the ending of the foetus or embryo'. It may be 'natural or induced'. She describes how abortion laws vary throughout the world. For example, India and China include abortion as part of their population control programme whereas in Ireland, an abortion may be performed only if the mother's or baby's life is in grave danger.

She outlines the different methods of abortion; curettage/vacuum aspiration in the first trimester, injection of a salt solution to kill the foetus followed by prostaglandins to induce premature labour in the second trimester, and RU-486, which may be used up to 9 weeks gestation. Tanya shows graphic photographs from websites to illustrate the different procedures. She then initiates what becomes a lively exchange among the students and teacher.

Tanya: In Australia, the average abortion is performed at 10 weeks. One day you may have to go against everything you believe, all that your family and friends believe, and make a decision that could be murder.
Lillian: What would a victim of rape in Ireland do? [Students do not respond.] Go to the United Kingdom. Who should decide the law?
Student 1: Government people.
Student 2: Why is it legal?
Lillian: Because women were dying of 'backyard' abortions. Thus, it was legalised so it can be controlled and also to provide counselling.
Student 3: It should be allowed because the mother might not care for it. I am personally against it, but it should be up to the person.
Student 4: It is not the same as killing a child.
Lillian: What is the difference? [Student do not respond.] If it is done at seven months, the baby will feel pain.
Student 5: [Presents a story about a baby who was born with a blocked oesophagus and could not be fed, so gradually died of starvation.]
Lillian: Which is kinder?
Student 6: Abortion. Not letting the baby suffer.
Student 7: If a baby has a major defect, it can be put up for adoption.
Lillian: It might be difficult if it has a major defect.
Student 7: It should still be offered.
Student 8: I agree if the mother is under 16 or the baby is deformed.
Student 7: The baby could have been put up for adoption.

In this final presentation, the students freely engage in discussion and express a range of views. Lillian guides the discussion by asking questions and playing devil's advocate by presenting students with alternative scenarios.

Impact of the Learning Activities

The discussion in this section relates to the learning activities illustrated in the two vignettes and addresses the emergent research questions. That is, to what extent did Lillian achieve her teaching goals of increasing her students' awareness of the complex ethical issues associated with reproductive technology and encouraging them to appreciate the impact of the technology on themselves and society? From the multiple perspectives of the students, the teacher and myself, what impact did the learning activities have on student learning and their attitude to science?

Lillian's Perceptions of the Learning Activities

After Lillian had read the vignettes, I interviewed her about the learning activities described. I asked her why she selected the activities, and also the extent to which each activity contributed to her achieving her teaching goals. Lillian told me:

> The first activity, [defining the terms], was an introductory activity, leading down the path to the issues. I wanted them to have an idea of what the words meant so that when we got into it they would realise what it was about.

By requiring the students to research the terms, she hoped that they would read beyond the definition. Even though Lillian could easily have told students the definitions, she believed that they would learn more if they found the meanings through their own reading. Thus, this activity indirectly addressed her articulated goals. The students were developing an understanding of the language of reproductive technology that would allow them to engage in debate at a later stage. As Lillian explained:

> I wanted them to have the background knowledge of what those terms meant so they could make a more informed choice about what areas to do research on. Otherwise they would tend to choose the words they knew already and avoid the words that looked harder because they did not know what they were.

Lillian believed that all of the students did learn what the terms meant. She based her belief on their ability to define and use the terms appropriately in their final written test and also during discussions.

Lillian told me that the second activity, the oral presentation was 'the culmination, when they actually showed their understanding and what they got out of it'. She felt that investigating one topic in depth and giving an oral presentation on the issues made them think about their own values. Lillian felt she had begun to achieve her teaching goals during the oral presentation when 'they all started talking over each other'. She continued. 'That's what I wanted to happen. I wanted them to get interested and start bubbling over. Wanting to say, "I've got an opinion. I've got to tell you about it." Rather than sitting and thinking, I don't have an opinion. I don't care. I don't want to know'. My classroom observation notes (e.g. the second vignette above) agree with Lillian's perception that the students became increasingly engaged toward the end of the oral presentation activity. They asked more questions of the presenter, responded to each other's questions, and answered Lillian's questions.

Lillian used the discussion after each oral presentation to increase the students' awareness of the complexity of issues. She explained:

> In the questions after each talk, I tried to give them scenarios to try and make them see that maybe this is not the best way to look at this case. Sometimes I played devil's advocate and said the opposite of what I believed to challenge their point of view. They need to realise there are others out there with different points of views and they have a right to them.

Overall, Lillian felt she had achieved her teaching goals because of the way the students answered her questions. Also in the final written test, Lillian included some of the same questions as those raised and discussed during the oral presentations. Lillian felt that when she marked the tests, most of the students 'showed that they had listened and learnt. Not this is right or wrong, but that there are several answers'.

Students' Perceptions of the Learning Activities

I interviewed the 15 students individually while they were researching the topic of their oral presentation (prior to the second vignette). I asked the students why their topic was important and what they hoped to learn. I was interested in whether their learning goals were similar to Lillian's teaching goals. The three interview extracts from Amelia, Maria, and Jacquie were representative of the range of students' responses.

Amelia was investigating the topic of bioethics. She told me:

> It is about moral decisions and finding out about religious and other people's views. I want to expand on what I know so that when it comes to making a decision I know what options I have. As a Catholic, I need to know what can and cannot be done. Reproduction has helped a bit because I can understand about some of the issues. Like abortion is killing. I know what abortion is, but I can think more about my attitude rather than just what it is.

Amelia seemed to want to know more about ethical issues so that if she needed to make a decision she would be aware of her choices. She also wanted to be aware of the views of others, including the Catholic Church.

Maria (whose talk was in the second vignette) investigated the screening tests of CVS and amniocentesis. She told me that it was important to learn about her topic because:

> It is about what is happening right now. It is used by a lot of people. It is an issue today and I need to work out whether it is right or wrong. Abortion might be involved if the baby is abnormal. Other issues are producing the perfect child to produce a perfect race. And also fertile people may be picky or selfish. People have to think about whether to test or

> not. I want to see different points of view, get stories from people who have had it. We need to learn about these issues because they come up in the news and we need to understand them. A lot of adults don't know about these things.

Maria seemed to believe that her topic was important because amniocentesis and CVS are procedures that are available to at-risk mother during pregnancy. They raise issues that are relevant in our society. She felt, therefore, that she needed to be able to express her view point. She also wanted to be aware of the views of others.

Jacquie had been assigned the topic of sex linked diseases. She explained that her topic was important because:

> One day, I will be a mother. I need to know what is happening and to be more aware. These topics are issues in today's society. It is important to understand. We will be the future generation, politicians, etc. So, we will be making the decisions.

Jacquie felt that the topic was personally relevant because she may be affected herself. She also seemed to realise that reproductive technology issues are important in our society now and in the future.

Overall, the students' reasons for studying these topics seemed to relate either to being able to cope better if they experienced problems associated with reproduction (e.g. Amelia) or to a need to be aware of current advances in those areas so they can make decisions in the future (e.g. Maria and Jacquie). The students' interview responses suggested that most of the students felt that they needed to know more about the procedures while a smaller proportion were concerned with the issues. The Catholic Church and religious beliefs were mentioned by several students. With these students, the emphasis seemed to be on doing what is 'natural'. Some of the students expressed their own ethical values in those terms.

Students' Perceptions of Their Learning

At the end of the course, students completed two questionnaires with open-ended questions where they were asked to comment on what they had learnt about science, ethics, and reproductive technology. In addition, four students were interviewed and asked to elaborate on their questionnaire responses. The questionnaire responses were combined into a single word file. The students' responses were read and recurring themes were noted. There were four areas or categories that were mentioned by most of the students. Using this grounded theory approach, the students' written responses were grouped into four categories. The four categories were:

1. An understanding of the topic of reproductive technology
2. Impact of science on society
3. Personal relevance of issues to themselves
4. An awareness of ethical issues

The categories of 'impact of science on society' and 'an awareness of ethical issues' were classified into subgroups. The subgroups are described later in this section.

In this section, exemplars are provided to illustrate the range of students' views.

An Understanding of the Topic of Reproductive Technology

- I learnt what happens if you want a baby and you can't have one the normal way.
- We have learnt about a lot of issues to do with human reproduction such as diseases and sex linked disorders. Also methods of helping pregnancies and saving or ending lives.
- The issues that affect everyone such as genetic diseases, types of diseases that may affect people. I have also learnt what genetic engineering is.
- I know a lot more about sex linked diseases and different forms of creating pregnancies. I also know that genetic counselling is available that will go into detail about these issues and help people make up their mind.
- I learnt all about surrogacy and data about abortion and other topics. All about genetic engineering and bioethics.

All of the students mentioned at least one of the topics presented in the oral presentation. The topic most frequently mentioned was abortion. Many students also outlined their views about a topic and supplied reasons. In order to understand the associated issues, students need to be familiar with the procedures used in reproductive technology. The learning activities used to promote an understanding of SSI also provided students with the opportunity to increase their knowledge about reproductive technology.

Impact of Science on Society

This category (impact of science on society) was further subdivided into those comments which were considered to be positive, negative or neutral. Comments classified as positive related to the benefits of science or reproductive technology on peoples' lives, while those classified as negative related to the risks and dangers of science or reproductive technology. Comments that recognised both the risks and benefits of science were classified as neutral.

The topic of reproductive technology seemed to challenge the students' beliefs and attitudes about science. There were a wide range of views expressed by students about the impact of science on society. The number of positive, negative, and neutral comments expressed by students was about one-third each. However, negative comments were more vehement. The neutral comments indicated that this group of students were aware of the benefits of science while also cautioning against the potential harm of scientific advances.

Positive comments included:

- I now realise that science covers many different avenues and that science is a major factor in our world.
- Science is a major part of today's society. Through science, our world is developing hugely. Science is bringing us forward in the world.
- People trust science to make major decisions e.g. on abortion.

- It is very helpful because you can work out cures for diseases and prevent them from showing up in the first place.
- I know that science is a major part of life that the society relies on everyday to bring new theories.
- I know that science can be used to help people suffering from genetic diseases.

Neutral comments included:

- There's a right and wrong side in science and it's so much more complicated than I thought.
- Science is developing ways to help infertile couples, but it may also be wrong in ways such as eugenics.
- We can use it to better the state of things, but we can unfortunately abuse it as well.
- That it can have advantages and disadvantages depending on one's personal and moral thoughts of the topic.

Negative comments included:

- Messing with nature can be disastrous.
- What the scientists are doing is wrong.
- I realise that scientists really can get carried away with things. What happens when things get out of hand?
- I have learnt that science in our society is moving very fast, but I think that before they start to make changes to the world they should consider every factor and I don't think they are.
- I believe we are playing God when it comes to genetic engineering and similar because we are altering humans.
- I believe that they are destroying nature and that they should maybe stop and think about what they are doing before they go and stuff things up (e.g. genetic engineering).
- I believe that most of the issues shouldn't be done on humans. e.g. genetic engineering is wrong to do on humans. It is not natural and we can't all be perfect.

During classroom observations, Lillian did not explicitly state that reproductive technology was good or bad. However, during discussions, Lillian did inform students about some of the problems associated with the use of reproductive technology. Students' religious beliefs might have contributed to their attitude. As mentioned previously, during interviews, several of the students stated that some of the procedures were not 'natural'. Regardless of the students' attitude to science (positive, negative or neutral) they were able to state a position, indicating that Lillian did achieve her goal of helping students to appreciate the impact of science on society.

Personal Relevance of Issues to Themselves

Lillian had stated that one of her teaching goals was to help students reflect on how they would respond if faced with any of the reproductive technology procedures discussed in class. The written responses indicate that some students seemed to be aware that they may be personally affected.

The students' responses included:

- I have learnt about how all of the topics discussed could be appropriate to me or the people around me, the options available and what I think about the issues so it can help me make up my mind. e.g. on abortion.
- I have learnt about how to cope if we happen to have a Down's syndrome baby and what the symptoms of other diseases are.
- It was more personal and relates more to personal experiences we might have in later life.
- By finding my own ethical and moral decisions to certain topics. Like with abortion. I don't think I would personally be able to have one, but I know people who have and now I don't feel any different towards them.

An Awareness of Ethical Issues

Students' comments related to aspects of ethics were numerous. Thus, the comments from this category were regrouped into four subcategories. The subcategories were

1. An awareness that ethical issues exist
2. The importance of using knowledge in decision making
3. To be tolerant to views of others
4. How to resolve ethical issues

The subcategories and examples of comments were

1. An awareness that ethical issues exist
    - I have become aware of all the issues that society is faced with today and what peoples' beliefs and moral points of view are.
    - There are many issues on one simple topic.
    - I don't know about any ethical issue before.
    - I think all students should be taught about this topic so they can come to a decision if they are faced with an issue.
    - I researched bioethics and discovered the issues are so controversial.
2. The importance of using knowledge in decision making
    - Using the knowledge I now have, I think it will be easier to make a decision.

- I can decide better in what I believe since discussing these issues because I now know the facts behind each issue and exactly what the issue is.
- I can say or back up what I believe now concerning issues better with all the facts influencing my beliefs.
- I know a lot more about the issues that relate to genetics. Because I know more information I find it easier to come to a decision on what I believe.
- Now I know most of the information about a topic, I would be able to put forward my opinion easier and more persuasively.
- With these facts, it has helped me to change some viewpoints I once had and to express what they are and how they differ.
- I can express my view point because I know all the facts and situations about these issues.

3. To be tolerant to views of others

- I believe everyone has a right to their opinion and other people should be open-minded to these opinions and not turn against that person because of their view point.
- Some people are narrow-minded – they should allow people to speak even if it is not what they believe.
- I think some classmates should learn to keep an open mind and not force their beliefs on others. We listen to them. They should listen to us.
- By this I learnt other people's point of view and this made me think that in a situation I will consider other's beliefs.

4. How to resolve ethical issues

- Before, I wouldn't have thought much, just said no. Now, I still would not have an abortion, but I would think about it more. I would talk to my family. If I was raped, I would talk to as many people as possible to get information. I would think about the long term effects. It's hard to make a decision.
- I would deal with these issues by getting all the information possible and doing some research. If necessary I would get genetic counselling and have peoples' opinions open to me, but overall it would be my decision.
- Listen to people's views. Think what my view is about the issues. Then I'll make a true decision about the issues.
- It has taught me to look at the advantages and disadvantages.
- I'd find out all the facts about the issues, then find out if the Catholic Church is against it or for it and then I'd find out people who have dealt with this issue and from them their opinion.
- Well, if it was abortion, I would first look at the situation and weigh up other issues involved.

There was a wide range of student comments related to reproductive technology issues. The types of comments suggested that most students were aware that issues exist and that their views may differ from their peers. They also seemed to be aware that to resolve ethical issues, they needed to collect and weigh up information and listen to the views of others.

Conclusion

Discussion of Results

Lillian taught a reproduction topic in a Catholic girls' school. The topic formed the second semester of a Grade 11 science course. Lillian is an experienced teacher, although this was only the second time that she had taught the reproductive technology and SSI component of the course. The course comprised four lessons a week for 16 weeks with the SSI component covered in the final 4 weeks. The 15 students studying the course were academically weak with varying levels of motivation. I observed Lillian's class over an extended period (4 months) which enabled me to develop a caring relationship with Lillian and her students. In relation to my presence in the classroom, when students were discussing their beliefs about emotive issues (e.g. abortion) they seemed very open and uninhibited. This was one of the benefits of visiting the class over an extended period of time.

Lillian was committed to the teaching of SSI. Her teaching goals were to prepare her students for the future when they may be faced with SSI associated with reproductive technology. She wanted them to comprehend the complexity of issues and also to appreciate the impact of science, in particular reproductive technology, on society. She considered that the teaching of SSI was a central part of her science teaching role. She had deliberately chosen to modify her existing teaching programs to include SSI. Lillian persevered with an increased workload as she sought out or designed suitable learning activities for her students.

Lillian introduced SSI toward the end of the course. She considered that it was important for her students to have an understanding of the theory associated with the topic before they could fully appreciate the related issues. Reproductive technology has specific, complex, and technical aspects that students may not have been exposed to previously.

Lillian planned and implemented a series of learning activities that she felt would best address her goals. The SSI learning activities were student centred, requiring students to choose and research one reproductive technology topic in-depth to present to their peers in an oral presentation. Whole-class discussion formed a major part of all learning activities. However, despite a student-centred approach, Lillian maintained control of the discussion; that is, she initiated and guided classroom discussion. Lillian assessed ongoing student learning primarily through questioning during whole-class discussion.

In the first vignette, the students participated in a learning activity where they defined and discussed terms associated with reproductive technology. Thus, the students were introduced to the language of reproductive technology. Through questioning, Lillian introduced her students to some of the issues associated with reproductive technology. Early in this activity, the students seemed reluctant to participate. Only when asked directly would a student offer a definition. They answered Lillian's questions about issues with brief responses. At this stage, the students did not augment their classmates' definitions. Also, they did not engage in debate about the issues. My perception from observing the students' behaviour during these lessons was that they seemed to perceive that the purpose of the learning activity was to check that their written definitions were correct, whereas Lillian's aim was to use the definitions as a springboard to introduce issues. It was only toward the end of this activity that students started to become more actively involved through discussion.

In the second vignette, the students presented a talk on a topic that they had investigated. They had been informed that they would need to present and defend their views and also contribute to discussion following other students' presentations. When students expressed extreme views, Lillian did not refute their comments. Rather, she asked questions to encourage them to consider alternative viewpoints. Lillian also challenged students who offered simplistic answers by providing information to help them become more aware of the complexity of the situation. Through careful questioning, Lillian also encouraged students to reflect on how they would respond if they were in a problematic situation (e.g. abortion, surrogacy).

During the discussions that followed the student presentations, I noted that Lillian did not directly express her ethical values. On occasion, she would appear to be supporting a particular ethical stance, but when a student supported her comment, she would adopt an alternative viewpoint and challenge that student's views. Lillian always ensured that students were aware of the Catholic Church's viewpoint.

Lillian selected learning activities that provided students with the opportunity to be active and interact collaboratively with their peers and the teacher. Most of the learning activities incorporated whole-class discussion. The teacher used resources, such as DVDs and newspaper articles to inform whole-class discussion. Although Lillian could have chosen to use small-group discussion as a teaching strategy she did not. I concur with her pedagogical decision. My observation of the students when they were engaged in practical group work early in the course was they were easily distracted when they were allowed to talk freely.

Evidence from this research suggests that a consideration of issues in science may influence students' attitudes to the subject of science. In the other two cases in this research study (not reported here), many of the students indicated that examining SSI in science had modified their views of science and society and influenced positively their choice to continue studying science (Dawson, 2010). In contrast, Lillian's students displayed a range of viewpoints about the impact of science on society. Their written comments suggested that science could be both harmful and beneficial. They seemed to recognise that the use of science has an impact on their lives and society. However, many of the students expressed their concerns about science 'going too far'. In her teaching, Lillian continually adopted the role of devil's advocate, alerting her students to the risks as well as the benefits of reproductive technology.

The research findings suggested that through participation in the learning activities, Lillian's students developed an increased understanding of reproductive technology procedures and associated issues. Most of the students seemed to be aware that in order to make a decision they needed to obtain information and then weigh up the advantages and disadvantages. The students also realised that others may have different views and that they need to consider multiple perspectives. Finally, the students seemed to be aware that science has an impact on our society that may be beneficial and/or harmful.

Implications for Teaching and Research about SSI

From the research findings, it appears that the teacher's beliefs about the purpose of science education and their resultant teaching goals are crucial elements in the quality of SSI education. When introducing SSI, teachers need to believe that SSI are an important and worthwhile topic to teach. Teachers need to have clear pedagogical goals. In this case study, Lillian clearly expressed her teaching goals that students need to understand the science and ethical issues associated with reproductive technology and how science impacts on society. All of the students stated that they had learnt about the topic and their questionnaire comments suggested that they did become aware of ethical issues and the impact of science on themselves and society. Clear teaching goals will influence the selection, timing, and implementation of learning activities. When selecting SSI learning activities, teachers need to be aware and tolerant of the variable cognitive ability and ethical maturity of their students. Lillian understood that her students were unlikely to continue with science. Thus, she patiently implemented learning activities where students could explore their beliefs and values. Lillian recognised that it takes time and substantial effort to negotiate and encourage reluctant learners to engage with SSI especially those students who do not like science or school.

The teacher should also endeavour to establish and maintain a safe and caring learning environment. Students need to be aware that they will not be ridiculed about their beliefs and values although they may be called upon to justify them. The articulation of values which are strongly held or different from those of their peers and society can be a threatening experience for students. Lillian adopted a caring approach throughout her teaching of the topic. Nevertheless, at times, she did play devil's advocate and challenged her students so that they became aware of different perspectives. Students need sufficient time and appropriate resources to investigate issues associated with a topic. They need time to understand the scientific content as well as the ethical issues surrounding a topic. Lillian spent a total of 16 weeks teaching the topic of reproduction which included 3 weeks focussing explicitly on issues associated with reproductive technology.

In selecting learning activities, it is recommended that teachers should explicitly provide students with opportunities to understand that

(a) decision making is a process;
(b) SSI in the science curriculum do exist;
(c) a group of individuals will hold a wide range of beliefs and values about;

(d) they need to listen to and respect the bioethical values of others;
(e) to resolve issues, they need to seek information;
(f) they need to provide sound reasons to justify their decisions.

Students need frequent opportunities to engage in learning activities where solutions to complex ethical problems are produced collaboratively with their peers. This model approximates how decisions are made in our democratic society. One rarely makes an important ethical decision in isolation, relying solely on rules, without first seeking information and the views of those whom we respect and trust, be it family, friends, experts or church leaders.

In relation to qualitative case study research, when conducting prolonged classroom-based research the researcher must make the time and effort to establish a rapport with the teacher and students. Classroom-based research requires a significant commitment on the part of the teacher and students. It is important to consider (in advance, if possible) how one can contribute back to the school, teacher, and students. In this case study, I assisted the teacher and her students with practical work. This was possible because I was a science teacher who had taught the topic of reproduction to similar aged and ability students in a similar school type.

References

American Association for the Advancement of Science. (2000). *Designs for science literacy*. Washington, DC: American Association for the Advancement of Science.

Australian Bureau of Statistics. (2006). 2006 census tables [Data set]. Retrieved from Australian Bureau of Statistics website http://www.abs.gov.au

Australian Curriculum Assessment and Reporting Authority (ACARA) (2010). My School. Retrieved April 20, 2010, from http://www.myschool.edu.au

Connelly, F. M., & Clandinin, D. J. (1988). *Teachers as curriculum planners: Narratives of experience*. New York: Teachers College Press.

Cooper, J. E. (1991). Telling our own stories: The reading and writing of journals or diaries. In C. Witherell & N. Noddings (Eds.), *Stories lives tell: Narrative and dialogue in education* (pp. 96–112). New York: Teachers College Press.

Creswell, J. W. (2008). *Educational research: Planning, conducting, and evaluating qualitative and quantitative research* (3rd ed.). Upper Saddle River, NJ: Pearson Education.

Dawson, V. M. (1996). A constructivist approach to teaching transplantation technology in science. *Australian Science Teachers Journal, 42*(4), 15–20.

Dawson, V. M. (2010). Outcomes of bioethics education in secondary school science: Two Australian case studies. In A. Jones, A. McKim, & M. Reiss (Eds.), *Ethics in the classroom* (pp. 69–86). Rotterdam, the Netherlands: Sense Publishers.

Dawson, V. M., & Taylor, P. C. (1998). Establishing open and critical discourses in the science classroom: Reflecting on initial difficulties. *Research in Science Education, 28*(3), 259–278.

De Crespigny, L. J. & Savulesca, J. (2004). Abortion: Time to clarify Australia's confusing laws. *The Medical Journal of Australia*. Retrieved March 18, 2010, from http://www.mja.com.au/public/issues/181_04_160804/dec10242_fm.html

Denzin, N., & Lincoln, Y. (Eds.) (2000). *The Sage handbook of qualitative research*. Newberry Park, CA: Sage.

Dixon, R. (2005). The Catholic community in Australia. Retrieved March 31, 2010, from http://www.catholicaustralia.com.au/page.php?pg=austchurch-history

Dori, Y. J., Tal, R., & Tsaushu, M. (2003). Teaching biotechnology through case studies-Can we improve higher order thinking skills of nonscience majors? *Science Education, 87*, 767–793.

Erikson, F. (1986). Qualitative methods in research on teaching. In M. C. Wittrock (Ed.), *Handbook of research on teaching* (3rd ed., pp. 119–161). New York: Macmillan.

Grace, M. (2009). Developing high quality decision-making discussions about biological conservation in a normal classroom setting. *International Journal of Science Education, 31*, 1464–1489.

Guba, E. G., & Lincoln, Y. S. (1989). *Fourth generation evaluation*. Newbury Park, CA: Sage.

Holly, M. (1992). *Keeping a personal – Professional journal*. Geelong, Australia: Deakin University Press.

Merriam, S. B. (2009). *Qualitative research: A guide to design and implementation*. San Francisco: Jossey-Bass.

Millar, R., & Osborne, J. (Eds.). (1998). *Beyond 2000: Science education for the future*. London: School of Education, King's College.

Moreton, C. M. (2007). World's first test-tube baby Louise Brown has a child of her own. Retrieved March 9, 2010, from http://www.independent.co.uk/life-style/health-and-families/health-news/worlds-first-testtube-baby-louise-brown-has-a-child-of-her-own-432080.html

National Curriculum Board. (2009). *Shape of the Australian Curriculum: Science*. Carlton: National Curriculum Board.

National Health and Medical Research Council. (2007). *Ethical guidelines on the use of assisted reproductive technology in clinical practice and research*. Canberra, Australia: Australian Government Publishing Service.

Noddings, N. (1984). *Caring: A feminine approach to ethics and moral education*. Berkeley, CA: University of California Press.

Patton, M. Q. (1990). *Qualitative evaluation and research methods* (2nd ed.). Newbury Park, CA: Sage.

Pedretti, E. (1999). Decision making and STS education: Exploring scientific knowledge and social responsibility in schools and science centers through an issues-based approach. *School Science and Mathematics, 99*, 174–181.

Picoult, J. (2004). *My sister's keeper*. Crows Nest, Australia: Allen and Unwin.

Potts, A. (2009). Public and private schooling in Australia: Historical and contemporary considerations. Retrieved March 15, 2010, from http://www.history.ac.uk/resources/e-seminars/potts-paper

Rennie, L., Goodrum, D., & Hackling, M. (2001). Science teaching and learning in Australian schools: Results of a national study. *Research in Science Education, 31*, 455–498.

Sadler, T. D., & Zeidler, D. (2005). Patterns of informal reasoning in the context of socioscientific decision-making. *Journal of Research in Science Teaching, 42*(1), 112–138.

Stake, R. E. (2000). Qualitative case studies. In N. Denzin & Y. Lincoln (Eds.), *The Sage handbook of qualitative research* (3rd ed., pp. 443–466). Thousand Oaks, CA: Sage.

Strauss, A. L., & Corbin, J. (1990). *Basics of qualitative research: Grounded theory procedures and techniques*. Newbury Park, CA: Sage.

Tytler, R. (2007). Re-imagining science education: Engaging students in science for Australia's future. Camberwell, Vic: Australian Council for Educational Research Press. Retrieved March 15, 2010, from http://www.acer.edu.au

Venville, G., & Dawson, V. (2010). The impact of an argumentation intervention on grade 10 students' conceptual understanding of genetics. *Journal of Research in Science Teaching, 47*(8), 952–977.

Zohar, A., & Nemet, F. (2002). Fostering students' knowledge and argumentation skills through dilemmas in human genetics. *Journal of Research in Science Teaching, 39*, 35–62.

Chapter 19
Metalogue: Critical Issues in Teaching Socio-scientific Issues

Jennie S. Brotman, Vaille M. Dawson, and Felicia Moore Mensah

Pedagogy for SSI

Brotman: Despite numerous links to complex SSI, the topic of reproduction has not been a frequent focus of SSI research. This chapter vividly illustrates the potential for topics related to reproduction to engage students in the kind of thinking promoted by the SSI movement. It was refreshing to see that students responded positively to the reproduction unit, and that it prompted them to become more aware of their own and others' beliefs, to more deeply understand these issues, to see these issues as personally relevant, and to reflect upon the interplay between science, society, and ethics. I would like to further explore the question of what aspects of this reproduction unit made it largely successful, as well as what recommendations for improving the unit the author and others might suggest. More specifically, I would like to raise questions about two aspects of the curriculum: the learning activities and the approach of the teacher.

The specific learning activities chosen to address reproduction issues (defining terms, oral presentations, whole-class discussion) raised several thoughts and questions for me. It was interesting to see how students' engagement and involvement in the discussions around the oral presentations improved over time. It seemed that this shift was partially related to an adjustment to a new kind of classroom culture, where open-ended discussion was expected as opposed to diligent note-taking, as well as to students' need for some time to get comfortable sharing beliefs and ideas about personal, contentious issues. I think these are both important issues for

J.S. Brotman (✉)
Teaching Matters, Inc, New York, NY, USA
e-mail: jsb2137@columbia.edu

V.M. Dawson
Science and Mathematics Education Centre, Curtin University, Perth, WA, Australia

F.M. Mensah
Department of Mathematics, Science and Technology, Teachers College, Columbia University, New York, NY, USA

T.D. Sadler (ed.), *Socio-scientific Issues in the Classroom: Teaching, Learning and Research*, Contemporary Trends and Issues in Science Education 39,
DOI 10.1007/978-94-007-1159-4_19, © Springer Science+Business Media B.V. 2011

teachers of SSI to consider. However, I also wondered whether additional structures might have been built into the activities themselves to trigger more debate and discussion from the outset. For instance, could the specific requirements/directions for the oral presentations have been articulated so as to trigger debate and discussion more explicitly? Were there ways in which the teacher might have prepped students for the shift in classroom culture that addressing SSI often demands? What have we learned in relation to the benefits and drawbacks of these particular learning activities as ways to address SSI?

In terms of the approach of the teacher, I appreciated Vaille's highlighting of the importance of Lillian's "caring approach" as well as the need to establish a "safe and caring learning environment" when discussing sensitive issues such as these. What strategies or insights can we explicitly articulate from this chapter about how to go about doing this? How can we as a research and teacher education community better support teachers in developing this kind of caring approach and safe classroom environment?

Dawson: Jennie has raised some very good (and difficult) questions that have made me ponder what I learnt from this case study and what I have learnt since. In some ways I think that I am probably less certain about some aspects of teaching SSI than I was when I commenced research in this area. I have wondered whether more structure might have prompted students to engage in more debate and discussion. This would have required explicit intervention from me which was not the intent of the research. In conducting professional development more recently I have provided explicit lesson plans to teachers. Many teachers in these sessions have gone on to use the materials but for some the lessons became contrived and had limited effectiveness. Alternatively, some of the experienced teachers with whom I have worked chose not to use the materials and facilitated effective discussions.

In the case of this particular research, Lillian was very sensitive to the needs of her students. Her 16-year-old students were attending a conservative Catholic girls' school, and the topic of reproduction needed to be handled carefully. By keeping her activities relatively unstructured she was able to be responsive to their needs. The teacher was also finding her way. I think that if teachers need to take away anything from professional development on SSI it is to realize that it may be difficult and scary to change your teaching. Also, it may take longer to teach content. Students need to be receptive to change and a wide range of strategies; they need to be able to participate in group work and whole-class discussions. The type of SSI used needs to be in context. Finally, I think that the most important factor is the teacher!

Brotman: Vaille raises an important point about the possibility that lessons can become contrived and thus less effective with more structured plans, particularly those designed by others. And, it is interesting that Vaille found the less structured approach to make it possible for Lillian to respond to her students' needs. In my work, I have also observed the powerful impact and critical role of a good teacher, particularly one that develops trusting, caring relationships with students, as it seems Lillian did.

Mensah: I agree with the importance of teachers having some flexibility and less structure to teach, and how a "good teacher" is able to teach in ways that address the range of learning needs of her students. I also agree that good teachers should develop trusting, caring relationships with students. We cannot underestimate the significance of teacher-student relationships to promote an environment conducive to learning. However, I want to push the conversation in order to explore the role of the good teacher as social advocate. I mention this in the context of work that I do in preservice teacher education. I am also flexible in my teaching, accounting for the diversity of teacher candidates that enroll in my courses, and I develop trusting and caring relationships with them, yet I also assume the role of a social advocate trying to push my students' thinking beyond their current perspectives (Mensah 2009). I think Lillian attempted to do the same acting as "devil's advocate", yet the girls were slow to speak but quick to listen. As teachers, how do you move students from comfortable places of listening—where the girls were completely engaged in writing notes as they listened to oral presentations—to engage in difficult dialogues? I know these are high school girls, but I am also challenged to engage teacher candidates in difficult dialogues pertaining to education, and in my secondary biology methods course, the same for addressing SSI issues in biology. I know this is challenging for teachers, but I also believe that difficult dialogues within classrooms will mirror and prepare students and teachers for "adult" and real-world conversations and debates that are sure to arise regarding SSI. It seems that SSI issues and understanding the complexity that comes from learning about multiple perspectives warrants more in the role of the teacher to be a social advocate or devil's advocate. How might the teacher take up this role in the classroom and structure these moments in critical points in classroom conversation or debate?

Contexts for Teaching and Learning

Mensah: Context was also an integral part of this study and how the teacher and students learned about the topic of reproduction in an all girls Catholic school setting. Context for so many researchers is often assumed, yet in this study, context was relevant in a myriad of ways. For example, I find it interesting that in an all girls Catholic school, the issue that was discussed most often in the oral presentations and on the questionnaire responses was abortion, and that gender issues did not seem to have as strong a focus during classroom oral presentations and questioning. Vaille stressed that the teacher highlighted the Catholic church's viewpoint; yet, I was curious to know more about how the specific context of an all girls Catholic school setting influenced conversations about reproduction and abortion and values, beliefs and context. Lillian is a female, and I wondered why conversations about reproduction and gender issues were not addressed more explicitly. Therefore, as I consider context and the specifics of this study, I began to wonder about the role of our own values and beliefs as teachers (and researchers) and if/how/why we neglect to include values in classroom discussions, particularly in our discussions of SSI? And what do we neglect in the development of lessons and curriculum, whether SSI or not, that closes avenues for students and teachers to engage in pedagogical practices that allow for deeper conversations of the content and contexts that we are teaching and learning?

Decision-Making

Brotman: This chapter also prompted me to think about what we want students to understand about decision-making processes. In the chapter, decision-making is largely described as a process of using facts and information, including the multiple perspectives of others, to weigh the advantages and disadvantages regarding particular SSI in order to come to a decision. Students come to recognize the "importance of using knowledge in decision-making" (p. 339) and their questionnaire responses related to this topic highlight the significance of "know[ing] the facts" and having all the "information" about the issue. Their responses to "how to resolve ethical issues" also call attention to "getting all the information possible" and "look[ing] at the advantages and disadvantages" (p. 340).

Certainly, gathering information and multiple perspectives in order to weigh different sides of an issue are critical aspects of decision-making, and ones that we want to relay to students. But are there other, perhaps messier, aspects of decision-making processes, or alternative models of thinking about decision-making that we might want students to also understand? For instance, people in the fields of cognitive psychology and health behavior have questioned whether individuals making choices do in fact weigh pros and cons in a rational way in order to make decisions in real-life situations (Reyna & Farley, 2006). My own research has attempted to complicate how we think about decision-making by illustrating how sexual health decision-making is tied to issues of identity for high school students (Brotman, Mensah, & Lesko, 2010). I think Sadler and Zeidler's (2005) study illustrating "rationalistic, emotive, and intuitive forms of informal reasoning" (p. 112) and how these forms interact in nuanced ways in relation to socioscientific decision-making provides further support for this perspective.

Vaille's recommendations for teachers include teaching students about "decision-making processes" and that "to resolve issues, they need to seek information" (p. 344). In addition to this important point, what else, if anything, do we want students to understand about decision-making processes?

Dawson: I would like to emphasize that this case study was a naturalistic study where the teacher set the agenda and my role was as an observer and supportive critical friend. Thus I was not imposing an external intervention on Lillian and her students. This teacher and others I have worked with do prioritize the importance of having information and knowing the facts before a decision can be made. There are other factors related to decision-making which in professional development with teachers I would now emphasize. These factors include listening to and tolerating, if not respecting, the views of others (peers, teacher, and community); being aware of multiple perspectives; suspending judgment; being prepared to change one's mind if presented with evidence; questioning the veracity of evidence; distinguishing evidence from opinion; realizing that making a personal decision may be difficult and time consuming; and being prepared to defend a decision with evidence.

Mensah: Vaille, these are very important factors in promoting decision-making and I would add also in promoting more inclusive learning environments. One major area that researchers and teachers fail to connect SSI with is multiculturalism, and I offer this connection to multiculturalism broadly to include culturally relevant teaching (Ladson-Billings, 1995). Though culturally relevant teaching is emphasized as a method of educating all students, I find similarities between decision-making and culturally relevant teaching. Ladson-Billings presents three criteria for culturally relevant pedagogy: (a) Students must experience academic success; (b) Students must develop and/or maintain cultural competence; and (c) Students must develop a critical consciousness through which they challenge the status quo of the current social order. Vaille, I see a direct connection between issues you have raised in your chapter and the third criterion for culturally relevant teaching. I believe the idea of critical consciousness was evolving from Lillian and her students' understanding of reproductive issues.

Ladson-Billings (1995) contends that culturally relevant teachers "engage in the world and others critically," and in order to do this, "students must develop a broader sociopolitical consciousness that allows them to critique the cultural norms, values, mores, and institutions that produce and maintain social inequities" (p. 162). Here I see a strong and useful framework to assist teachers in becoming "good teachers" of students and helping students in the process of decision-making. If students come to understand that the process of schooling is not simply learning facts, but learning allows them to take facts and other information and situate it all within a larger sociopolitical framework, this will equip them to make adult decisions—decisions that are value-laden, emotional, political, and debatable.

Brotman: Decision-making often involves a strong emotional component, one that cannot always be supported with evidence. I would also argue that decision-making can be tied to how people see themselves, and how they want others to see them. I have been trying to understand the ways in which our relationships and interactions with others and the world around us, including societal discourses around SSI, factor into our decision-making processes. These aspects of decision-making seem to me to be important for teachers to understand, as they highlight the potential complexity of decision-making as it may happen in the real-life scenarios we are aiming to prepare our students for.

Mensah: And the real-life scenarios for which we prepare our students to make decisions definitely require critical consciousness. What may be even more challenging is that teachers will have to be selective of topics and take a certain level of risk in their topic selection.

SSI Topics

Dawson: There are many different SSI that may be selected by a teacher. Choosing a particular SSI topic will depend on many factors including the mandated curriculum, interest and expertise of the teacher, students' interests and age. In the case study described in this chapter, the teacher chose the topic of reproductive technology, an area that is controversial in the context of a Catholic girls' school where students must be made aware of the Catholic perspective. The content area of reproduction is of interest to 16-year-old girls. Also the teacher had a biology background and expert knowledge in the area. Reproductive technology is available in Australia but is expensive and is the subject of debate in the media. Issues around reproductive technology raise strong emotional responses from those affected by infertility. However, in some contexts, for example conservative Christian schools, boys' schools or schools located in countries with less health care, the topic may not be considered appropriate. Thus, what criteria should be used when deciding on a topic? Are there some SSIs (perhaps climate change) that are universal?

Mensah: Teachers will select and teach topics that they have confidence in teaching. Their decision-making regarding topic selection is very much related to self-efficacy, content knowledge, and pedagogical content knowledge. However, should we treat all topics equally? How do we decide which topics to teach? Do we pay particular attention to those topics that have local interest yet global relevance? Because our world is increasingly becoming more accessible to issues of local, state, national, and global significance, how does a teacher, or a community, truly decide which SSI to teach in schools? For example, there are some issues that are under-addressed in the school curriculum, such as clean water or organic food production. I recognize that the school curriculum is stressed—teachers are held accountable for teaching the "mandated" curriculum, yet how many learning opportunities are lost when topics that potentially have more far-reaching consequences on our understanding (i.e., local and global stability) are not considered to be part of the mandated curriculum?

Brotman: Vaille and Felicia highlight many of the criteria that I see as being critical for selecting appropriate SSI: alignment with the curriculum, teachers' interests and comfort level, students' interests, school and societal context, and prevalence of ongoing debate in the current media. In a related vein, some SSI might be particularly relevant locally, and may provide opportunities to reach out to local informal institutions and resources with the potential to enrich SSI curriculum through real-world connections. While certain issues might often fit these criteria and thus be frequently addressed in classrooms, like climate change, the above factors will always need to be considered in deciding on a topic. In particular, I think students' interests and perceptions of the relevance of the issue to their lives are critical (as we saw in this chapter), and those interests will vary depending on the context. I would also argue that providing teachers with strategies for choosing SSI based on students' interests would be a useful approach to explore.

References

Brotman, J. S., Mensah, F. M., & Lesko, N. (2010). Exploring identities to deepen understanding of urban high school students' sexual health decision-making. *Journal of Research in Science Teaching, 47*(6), 742–762.

Ladson-Billings, G. (1995). But that's just good teaching! The case for culturally relevant pedagogy. *Theory into Practice, 34*(3), 159–165.

Mensah, F. M. (2009). Confronting assumptions, biases, and stereotypes in preservice teachers' conceptualizations of science teaching and diversity through the use of book club. *Journal of Research in Science Teaching, 46*(9), 1041–1066.

Reyna, V. F., & Farley, F. (2006). Risk and rationality in adolescent decision making: Implications for theory, practice, and public policy. *Psychological Science in the Public Interest, 7*(1), 1–44.

Sadler, T. D., & Zeidler, D. L. (2005). Patterns of informal reasoning in the context of socioscientific decision making. *Journal of Research in Science Teaching, 42*(1), 112–138.

Chapter 20
Socio-scientific Issues-Based Education: What We Know About Science Education in the Context of SSI

Troy D. Sadler

Over the last decade, interest in socio-scientific issues (SSI) as research themes and instructional contexts for science education has grown dramatically. During this period, science educators have presented several articles and chapters that outline the conceptual and theoretical dimensions of this emerging field (e.g., Hodson, 2003; Sadler & Zeidler, 2009; Zeidler & Keefer, 2003; Zeidler, Sadler, Simmons & Howes, 2005). Scholars working in this area have also conducted a fair amount of research addressing a wide range of questions making use of various methodologies including correlational analyses, quasi-experimentally designed quantitative studies, case studies, and grounded theory. Much of this research has been reviewed and synthesized elsewhere (Bennett, Lubben & Hogarth, 2007; Sadler, 2004; 2009). These contributions provide an important backdrop for the development of this volume.

In this volume, the chapter authors, metalogue contributors, and I have attempted to draw from established frameworks for SSI-based education and build from the base of empirical literature to push the field forward and offer new insights. In framing the overall book project, we decided to focus explicitly on classroom-based projects featuring SSI. While research into other aspects of SSI is certainly important, many of the most pressing questions for advancing the SSI agenda are best explored in classrooms. In creating this volume, we have drawn together a diverse array of classroom-based SSI investigations. This assemblage of projects and the conversations associated with their presentation has potential to inform the SSI research base, generate new questions for advancing the SSI agenda, and offer new insights for educators looking to implement SSI-based education.

T.D. Sadler (✉)
School of Teaching and Learning, University of Florida, Gainesville, FL, USA
e-mail: tsadler@coe.ufl.edu

T.D. Sadler (ed.), *Socio-scientific Issues in the Classroom: Teaching, Learning and Research*, Contemporary Trends and Issues in Science Education 39,
DOI 10.1007/978-94-007-1159-4_20, © Springer Science+Business Media B.V. 2011

Synthesis of Research

Detailed accounts of the research questions explored, the methods employed, and the findings produced are embedded throughout the chapters. My intent here, in this final chapter, is to briefly highlight some of the themes that emerged across chapters.

Varying Contexts

The projects featured in this volume provide evidence of the potential of SSI-based education across a wide range of settings including diverse national contexts (Australia, Cyprus, Hong Kong, Israel, Spain, the United Kingdom, and the United States). These projects also showcase an extensive range of learner audience in terms of age. Evagorou reports on work done with upper elementary learners, and Eastwood, Schegel, and Cook describe a program for university undergraduates. The seven other projects are situated in various secondary contexts. This tells us that SSI can be used productively with learners spanning broad age ranges, but it does not answer the question of what is the lower age boundary for productive use of SSI education. In the Foreword, Aikenhead suggests sixth grade as a lower limit, and Evagorou, who worked with students as young as 10 years old expressed some concerns regarding the effectiveness of SSI approaches with younger audiences. In one of his metalogue posts, Zeidler mentions having worked with elementary educators, so it may be possible to extend SSI audiences, but we lack strong evidence for where the boundaries should be set and what criteria should be used in establishing these boundaries.

Issues and Approaches

The projects discussed throughout the volume feature a wide variety of SSI. Topics include issues related to using science and technology to enhance human performance, public policy issues such as the fluoridation of water supplies, reproductive issues, and various diseases such as cystic fibrosis, SARS, and AIDS. Some projects focused on scientific controversies with important social implications such as biological determinism and the cause of AIDS. Finally, several studies featured learner exploration of local and/or global environmental issues including water quality, air quality, and climate change. This list of issues may not be particularly useful in and of itself, but some of the discussions that emerged around issue selection and affordances are important in terms of identifying what makes a good SSI for instructional purposes.

One of the most critical issue attributes is relevance and interest to the students. It is essential to engage students in issues that they find inherently interesting or that the students may come to see as relevant and interesting. Potential connections to science and the ability to draw these connections out in classroom settings are also very important. An issue may be technically classified as a SSI-based on some operationalized definition, but if teachers and students cannot locate the science behind the issue, then it cannot serve as an effective context for science education. It is also very helpful if students and teachers can access the ethical tensions underlying a particular SSI. It tends to be these ethically contentious aspects of SSI that are most interesting and can generate the most meaningful engagement among learners. These three points are not new to the field; other authors have made these points. But there was one novel consideration related to issue attributes that was raised in one of the metalogue discussions (Wong, Zeidler, & Klosterman). In this discussion, the contributors considered the use of actual SSI cases and how information about the historical outcomes of these issues necessarily shapes the use of those issues. This line of thought presents interesting challenges for the use of issues that are constantly evolving. It is generally agreed among the SSI community that SSI are open-ended and unresolved, but it is entirely possible that some issues become effectively resolved because of scientific or technological advancement or policy enactment. Therefore, the use of a particular issue in educational contexts will necessarily shift as the issue itself evolves.

Another dimension of variance observed across the chapters relates to instructional approaches. The teachers and researchers showcased in this work used a wide range of pedagogical strategies in order to engage their learners in SSI. These strategies included innovative simulation activities, role plays, discussions, and a variety of group activities. Technology was used in the projects in a variety of ways including comprehensive platforms for structuring content delivery (Tal, Kali, Magid, & Madhok), to support student collection of scientific data (Evagorou), and to provide student access to varying perspectives on the issue under consideration (Sadler, Klosterman, Topcu). Teachers and curriculum designers also used strategies more typical of traditional science classrooms like lecture, laboratory exercises, and guided inquiry activities. The take home message is that there is no one way to deliver SSI instruction; educators need to carefully consider their contexts, their learners, and the nature of the issues they want to present in order to optimize SSI education.

Teachers

An important theme that emerged in multiple chapters relates to the challenging nature of creating and delivering SSI education. We see several examples of teachers struggling with aspects of SSI education enactment (Dawson; Puig & Jiménez-Aleixandre; Simon & Amos; Wong, Wan & Cheng; Zeidler, Applebaum & Sadler). Teaching science through SSI is challenging work that requires commitment and a willingness to struggle with uncertainties. Teaching science such that the representation of science content is the exclusive or at least primary focus is much easier than assuming the challenge and "messiness" of SSI.

Teachers struggle with SSI instruction in all of these cases (cited above) but they also show signs of development as professionals. We see evidence of teachers skeptical of an SSI approach come to embrace the approach (Simon & Amos), teachers who initially struggle come to design their own SSI units (Wong et al.), and teachers who declare their students incapable of sophisticated reasoning transform to become passionate advocates of SSI approaches (Zeidler et al.). All of the cases that document teachers' struggles also document teacher and student successes. SSI-based education is not a magic bullet for science education; it is a difficult strategy to use but the work presented in this volume documents that teachers from many different backgrounds working in very different situations can succeed with appropriate supports.

Student Interest

Several of the chapters document student interest in learning science through SSI (Dawson; Eastwood et al.; Evagorou; Simon & Amos; Tal et al.; Zeidler et al.). As I mentioned earlier in this chapter, student interest in the issues under consideration is important for successful implementation of SSI-based education. The chapters presented here offer evidence of the potential of embedding science instruction in SSI as a means of generating interest and enthusiasm among learners. Although there is one chapter (Sadler et al.) that suggests that students were indeed engaged in learning activities, it questions the extent to which the focus on SSI was the most important factor in generating engagement and interest.

Learning Outcomes

The research presented in these chapters also documents several student learning outcomes associated with SSI-based education. Several projects document student learning of science content through explorations of SSI. Content areas ranged from genetics (Puig & Jiménez-Aleixandre; Tal et al.) and reproductive biology (Dawson) to environmental science (Evagorou; Simon & Amos) and chemistry (Sadler et al.). Two of the studies focused on student understandings of nature of science (NOS) and offered examples of how SSI-based education can support development of more sophisticated ideas of NOS aspects (Wong et al.; Zeidler et al.). A few chapters also present evidence suggesting that SSI-based education can raise ethical awareness and sensitivity among students (Dawson; Eastwood et al.; Zeidler et al.)

Many of the projects focus on student engagement in higher order reasoning, argumentation and/or decision-making processes. Most of the projects suggest that students in SSI interventions showed improvements in these higher order cognitive processes. Eastwood and colleagues found that undergraduates engaged in long-term (4 years) explorations of SSI demonstrated higher quality reasoning than their peers who had less exposure to SSI-based learning opportunities. Evagorou documents ways in which student use of evidence for argumentation improves over the course of a SSI unit. In the study by Zeidler and colleagues, students learning science through SSI show significant gains in reflective judgment. Finally, the students in Dawson's study become better able to reason through complex issues and appreciate the varied perspectives and views pertinent to particular SSI.

In looking across the chapters, there is a sense that students engaged in SSI learning experiences can become better prepared to deal with science-related issues in their lives. The learners featured across the projects tended to become more aware of issues, better equipped with relevant scientific information, and more experienced with the practices necessary for negotiating complex SSI. Overall, students became more scientifically literate as a result of learning science through SSI.

Open Questions

Most good research answers some questions while raising others. Each of the chapters featured in the volume discuss implications of the individual projects including some of the unanswered questions. In this section of the volume's final chapter, I revisit some of the questions that emerge across the chapters and that offer the science education community new ideas for advancing the SSI agenda.

Assessment is an issue that emerges frequently in the chapters as well as the supporting metalogues. Assessment questions relate to how best to assess student practices associated with SSI both in classroom teaching contexts and research contexts. The practitioners and researchers whose work is featured in this volume use a variety of argumentation (e.g., Evagorou; Simon & Amos) and reasoning (e.g., Eastwood; Sadler et al.) constructs and associated rubrics and scales to document student practices and growth in those practices. However, as pointed out in multiple metalogue discussions, the assessment tools currently in use could be significantly improved. As a community, we need better measures of student engagement in SSI learning experiences and more valid assessments of growth. In essence, we are highlighting the long-standing problem of assessing scientific literacy. While the contributions of this volume offer some suggestions for moving forward (e.g., particular ways of using argumentation frameworks and the presentation of "socio-scientific reasoning"), much work will be needed to address the conceptual and pragmatic challenges associated with assessing student learning and practice in the context of SSI.

Another important question relates to the nature of the partnerships featured across these chapters. In almost all of the projects presented, teachers and university-based researchers form strong partnerships to facilitate the work done. In some cases, these collaborations resulted in customized curricula uniquely suited to individual classrooms. In other cases, university-based educators helped classroom-based educators translate and/or implement innovative curricular materials. The extensive nature of these relationships and the positive project results provides evidence that SSI can serve as effective contexts for education under ideal or at least very favorable conditions. Collectively, the results highlight the need to explore SSI-based education in less-than-ideal conditions. How can SSI-based education efforts be replicated in situations where classroom teachers do not have the benefit of collaborations as extensive as those featured in this volume? What kinds of systems can be created to support teachers and the facilitation of classroom explorations of SSI? As discussed earlier, implementing SSI-based education is challenging work that demands a lot from teachers. So it stands to reason that new models for teacher education to support these kinds of classroom practices are certainly warranted. Understanding how different models of professional development for in-service teachers can be used to support teacher practices around SSI-based education would be very valuable. Furthermore, research to address whether or not the science teacher education community should present SSI-based education as a standard aspect of preservice science teacher education and if so, what kind of emphasis it should receive would be equally informative.

One point that emerged in multiple chapters was the importance of modeling desired practices for students whether those practices were framed in terms of argumentation, decision-making or reasoning. Authors consistently highlighted the significance of teachers and curricula providing learners with access to examples and models of what it means to engage with SSI in informed ways. Additional work aimed at understanding how to optimize learner experiences with these kinds of models and how best to create these models would be very useful.

The final question I wish to direct attention to in this section relates to how much SSI is enough to foster scientific literacy. This is an issue that I alluded to in the introductory chapter, resurfaced implicitly in a few other chapters and was discussed directly in one of the metalogues (Zeidler, Bell, Sadler, & Eastwood). The question of how much attention ought to be focused on SSI in science classrooms is an important one, and one for which the community has limited empirically based answers. As advocates of SSI-based education, most of the contributors to this volume would likely argue that SSI does not receive the attention and classroom time that they warrant in most schools. But we really do not have good answers for questions related to how much classroom time and effort ought to be devoted to SSI in order to achieve the progressive goals of the SSI movement. Research that could help to address these issues would help the field move forward particularly in terms of affecting practice in science classrooms.

Models for SSI-Based Instruction

When I began the work of planning this volume, I decided that including a concluding chapter would help tie the project together. My initial thoughts were that I could synthesize some of the main research findings and highlight some important questions that remain unanswered (not unlike the beginning of this chapter).

But then two things happened that lead me to think that this chapter could offer something more than simply tying together the main themes of the book. First, I had an opportunity to see a presentation by Ingo Eilks (2010) in which he described a series of classroom-based SSI research projects that he and his students had been working on in Germany. Eilks' group had worked with several different teachers as they used a series of controversial, socially-relevant issues as contexts for science education and the promotion of scientific literacy. Eilks used these experiences across varying classroom settings to propose an instructional model for SSI-based education. Eilks' model offered five steps for instruction:

1. Problem analysis. In this step, students are presented with an issue of interest through media reports or other strategies that highlight the reality and relevance of the issue.
2. Clarification of the science. Teachers help students understand the basic science underlying the issue.
3. Refocus on the socio-scientific dilemma. Students refocus their attention on the issue and the associated social problems or controversies.
4. Role-playing task. Students assume roles for engaging in the negotiation of SSI. These roles may include parties to the issue debate or creators of media related to the issue.
5. Meta-reflective activity. Students are encouraged to reflect on their overall experiences with the issue and the underlying science.

Eilks' model offered a relatively simple model for operationalizing what it means to teach with SSI. Teachers and curriculum designers could easily apply this model as they worked to create new SSI-related instructional units. Ultimately, the model makes the process of planning and implementing SSI instruction more manageable and therefore is very helpful. However, in thinking about how the model might be applied across broader educational contexts, I found it to be too prescriptive. For me, the fixed progression of steps could help inform how to approach SSI instruction in some settings but could also be overly restrictive. So while I certainly liked the idea of offering a model of SSI instruction as a means of making the integration of SSI in science classrooms more manageable, I thought that a more flexible model was needed. (It should be noted that Eilks was presenting the model that he and his group had used for guiding their work; he did not intend this to be a model for all SSI instruction.)

Earlier in this section, I mentioned that two things lead me to think about this concluding chapter as more than just a summary and synthesis of research findings. The first was Eilks' presentation; the second was reading through the full collection of chapters submitted for the volume. As the editor, I had read multiple iterations of all of the chapters, but as I prepared to write this final chapter I read through the full complement of contributions in sequence. It became clear in this final read through that the authors offered several models for SSI instruction, many of which shared features with Eilks' work and others deviated because of different contexts, purposes or foci. It also became clear to me that this diverse collection of projects could be used as the basis for generating a more general framework for SSI-based education. (I am using the term "model" to reference specific guidelines for instruction such as those offered by Eilks. I use the term "framework" to reference less specific ideas upon which different models could be built).

An Emergent Framework for SSI-Based Education

Based on my experiences considering Eilk's model and reflecting on the chapters prepared for this volume, I decided to use the various models offered within these chapters to inductively derive a framework that could be used to inform design and implementation of SSI-based education in various contexts. In order to achieve this goal, I reexamined the specific models presented in the chapters. Some of these had been clearly formalized in their original presentation (e.g., Tal et al.). In other cases (e.g., Puig & Jiménez-Aleixandre) I made inferences about the underlying instructional model based on the presentation of the project. After isolating the individual models, I looked across the models for recurrent themes and generated an extensive list of shared characteristics. Next, I condensed the categories and grouped the themes in a conceptually coherent fashion ultimately resulting in a hierarchical organization of emergent ideas. These sequential iterations of analysis and representation were very similar to open and axial coding as described in grounded theory research (Strauss & Corbin, 1998). With the remainder of this chapter, I present the results of this work: an emergent framework for SSI-based education.

The emergent framework (graphically presented in Fig. 20.1) is made up by four primary aspects: *design elements*, *learner experiences*, *classroom environment*, and *teacher attributes*. *Design elements* refer to considerations important for the design of successful SSI-based education. *Learner experiences* represent the kinds of opportunities that learners ought to have access to during SSI-based education. *Classroom environment* refers to contextual features of the learning environments

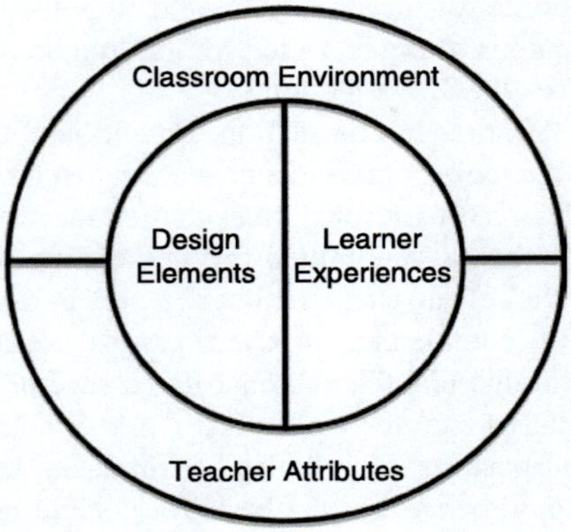

Fig. 20.1 Graphic representation of the emergent framework for SSI-based education

that are necessary for successful design and enactment of experiences. Similarly, *teacher attributes* reference characteristics and practices that teachers should assume for successful implementation of SSI-based education. *Design elements* and *learner experiences* are situated centrally in the graphic representation to show these as core features of SSI education. *Classroom environment* and *teacher attributes* are positioned more peripherally to indicate the role both of these elements play in shaping implementation of *design elements* and *learner experiences*. All of these aspects of the framework elements are described in greater detail in the sections that follow.

Design Elements

The first aspect of the framework relates to design considerations that practitioners and curriculum authors should incorporate in their efforts to create units of instruction based on SSI. Essential and recommended *design elements* are represented in Fig. 20.2. Each of these elements will be discussed here in sequence. Fundamental to the idea of SSI-based education is the notion that teaching and learning will relate to a compelling social issue with connections to science. The second essential design element relates to the positioning of that issue within an instructional sequence. It is important to feature the issue early in the instructional sequence so that it can serve as a true context for science learning. It may be useful to present issues and cases following science instruction as a demonstration of how science content may be applied, but if the goal is to truly contextualize the learning experience, as is the case for SSI-based education, then the issue needs to be highlighted at the outset of the experience. Engaging students in sophisticated argumentation, reasoning, and/or decision-making practices is a challenging goal, and it cannot be assumed that students will intuitively improve their practices in these areas. The projects featured in this volume that were most

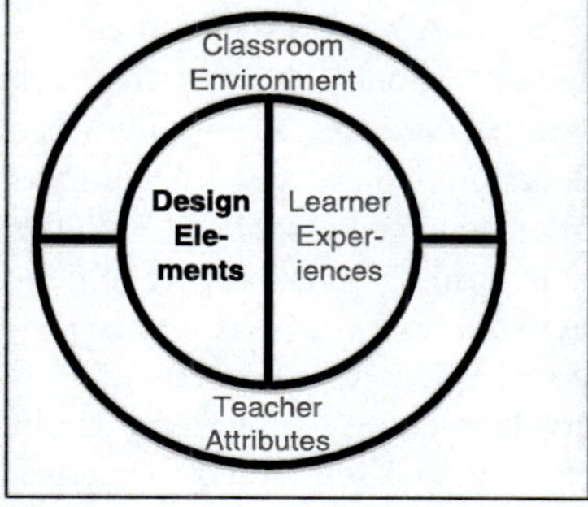

Essential *Design Elements*
1. Build instruction around a compelling issue. 2. Present the issue first. 3. Provide scaffolding for higher-order practices (e.g. argumentation, reasoning and decision-making). 4. Provide a culminating experience.
Recommended*Design Elements*
5. Use media to connect classroom activities to the "real world." 6. Use technology to facilitate student learning experiences.

Fig. 20.2 Essential and recommended *design elements* for SSI-based education

successful in supporting higher-order practices among students offered scaffolding and/or models from which to work. Providing targeted supports for and exemplars of desired practices such as argumentation and reasoning are necessary for the design of SSI-based education if affecting scientific literacy is an instructional goal. The final essential *design element* is the need for a culminating experience that provides an opportunity for learners to tie together what they have learned and relate those understandings and practices to the issue that has been explored. The culminating experiences featured in these chapters ranged from role-play and debate activities to the creation of formalized recommendations and service learning projects. The culminating activity may take multiple forms as long as the chosen strategy provides opportunities for students to reflect on their experiences and use what they have learned in making and representing decisions, arguments, and/or other products.

In addition to these essential *design elements*, two other considerations emerged in some but not all of the chapter projects. These elements may be very useful for teachers and designers to consider, but their incorporation into the design of SSI-based education is not absolutely necessary. The first point is that media characterizations of the SSI under investigation can provide useful resources to help generate interest among students and to help draw connections between classroom representations of the issue and representations of the issue in the "real world." Similarly, educators should consider leveraging technology to help learners access and explore SSI. Technology was used in a variety of ways in the chapter projects including as a teaching and learning platform, as a tool for collecting scientific data related to issues, and as a means of student exploration of information. Teachers cannot rely solely on textbooks or other traditional forms of content representation when featuring SSI in classrooms. SSI are, by definition, current and evolving issues and technologies can be very helpful in providing student access to these issues.

Learner Experiences

The *learning experiences* aspect of the emergent framework for SSI-based education describes the kinds of experiences that students should have as they are engaged in SSI learning. Essential and recommended *learning experiences* are represented in Fig. 20.3. In order for SSI-based education to be successful, students have to have opportunities to engage in higher order practices associated with the negotiation of SSI. These activities may be framed in terms of reasoning, argumentation, decision making, or position taking. The actual activity label is less important than the fact that learners have opportunities to participate in one or more of these activities. In other words, it is not enough for learners to simply learn about SSI; they need to be engaged in higher-order practices associated with those activities. Students also need to have opportunities to confront the scientific ideas and theories that relate to the issue under consideration. Students may confront the

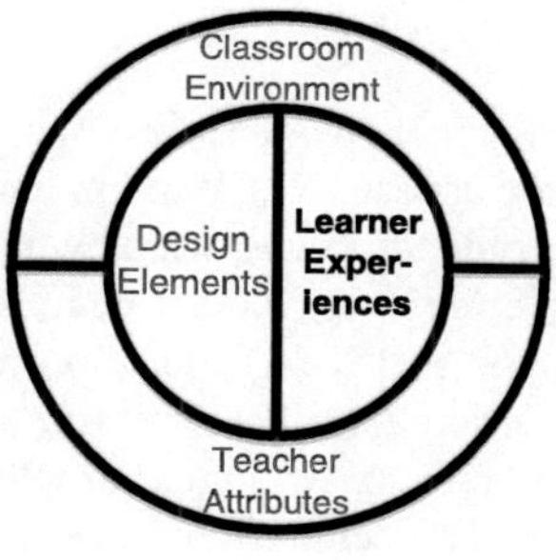

Essential ***Learning Experiences***
1. Engage in reasoning, argumentation, decision-making and or position taking. 2. Confront the scientific ideas and theories related to the issue being considered. 3. Collect and/or analyze scientific data related to the issue being considered. 4. Negotiate the social dimensions of the issue being considered.
Recommended ***Learning Experiences***
5. Confront the ethical dimensions of the issue being considered. 6. Consider nature of science themes associated with the issue.

Fig. 20.3 Essential and recommended *learning experiences* for SSI-based education

science of issues in a variety of ways as long as there are explicit opportunities to do so. In addition to exploring the science content of issues, students need opportunities to collect and/or analyze scientific data related to the issue. Scientific data are usually not enough to provide solutions to complex SSI, but data are essential for the informed consideration of these issues. Collecting relevant data may be difficult to orchestrate in classroom explorations of SSI, but technologies can provide students with access to data sets that can be analyzed and used in their decision-making processes. The final essential learning experience relates to the social dimensions of SSI. Social considerations such as politics and economics are a defining element of SSI, and students need opportunities to explore these considerations in SSI-based education.

In addition to the essential *learning experiences*, the emergent framework identifies two recommended *learning experiences*. The first recommended experience is actually an extension of the final essential experience. Ethical issues are one of the social considerations inherent to SSI. There is some debate (see the Wong, Zeidler, & Klosterman metalogue discussion) regarding the extent to which featuring the ethical tensions of SSI is a necessary element of SSI-based education. While exploring at least some of the social dimensions of SSI is essential for SSI-based education, an explicit focus on the ethics associated with an issue may not be absolutely necessary. Some of the contributors to this volume likely disagree with my placement of ethics in the recommended category, rather than the essential category. However, the volume does provide evidence of SSI-based education that does not highlight ethical tensions. So, encouraging students to confront the ethics of an issue is certainly a recommended element of SSI-based education, but it may not be absolutely necessary. Likewise, providing opportunities for students to consider nature of science (NOS) themes in the context of SSI is certainly recommended for quality SSI science instruction but not necessarily essential.

Classroom Environment

Features of the *classroom environment* are one of the two framework aspects that are necessary for successful implementation of the two core aspects. Without a supportive *classroom environment* and a teacher exercising certain attributes, which will be discussed next, desired *design elements* and *learning experiences* cannot be implemented effectively. Classroom environments that support SSI-based education will have established norms and expectations for student participation. These norms and expectations set the stage for a collaborative and interactive environment in which students and teachers can and do engage in discourse and scientific practices. In order for these interactions to be productive and engaging, it is important for all of the participants to demonstrate healthy respect for one another and the positions they advocate even if there are legitimate disagreements about those positions. High levels of mutual respect tend to create environments in which students and teachers can feel safe in expressing their views and perspectives. SSI are controversial in nature and can create opportunities for heated disputes. Students and teachers need an environment in which they can share their views and not fear ridicule or alienation for the expression of unpopular perspectives. These features are essential for classroom environments that support SSI-based education (Fig. 20.4).

Teacher Attributes

The final aspect of the framework relates to characteristics of teachers who are able to successfully facilitate SSI-based education. These essential *teacher attributes* are presented in Fig. 20.5. First and foremost, teachers need to be familiar with the SSI around which they build instruction. Beyond just simple familiarity, teachers need to be knowledgeable about the science content and aware of the social considerations associated with the issue. Because we are talking about embedding science

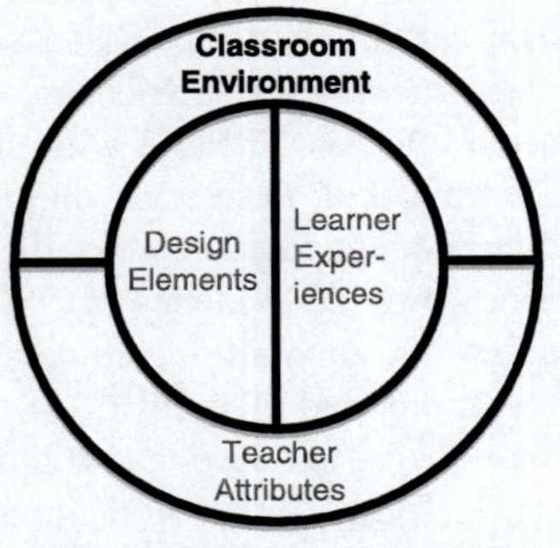

Essential *Classroom Environment* features

1. High expectations for student participation.
2. Collaborative and interactive.
3. Students and teachers demonstrate respect for one another.
4. Students and teachers feel safe within the environment.

Fig. 20.4 Essential *classroom environment* features for supporting SSI-based education

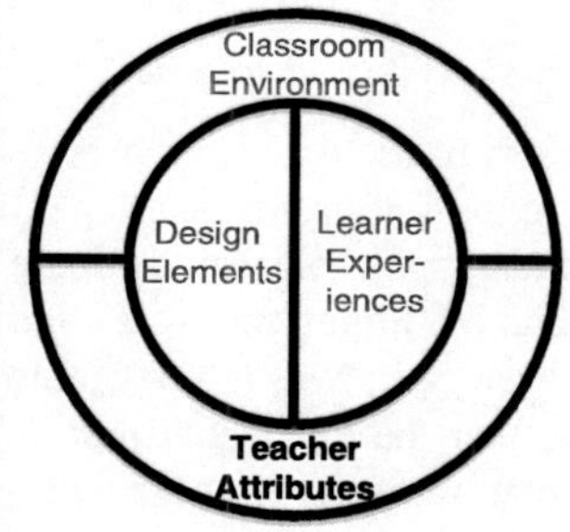

Essential *Teacher Attributes*
1. Familiar with issues being considered. a. Knowledgeable about science content related to the issue. b. Aware of the social considerations associated with the issue.
2. Honest about knowledge limitations.
3. Willing to deal with uncertainties in the classroom.
4. Willing to position self as a knowledge contributor rather than sole authority.

Fig. 20.5 Essential *teacher attributes* for supporting SSI based education

instruction in SSI, it is necessary for teachers to understand the related science. There is less of a burden of expertise in the area of the social considerations associated with a particular SSI. While it is certainly important for a teacher to be aware of the social considerations, it is not necessary, nor is it realistic to expect, for teachers to be experts on all of the political, economic, and ethical dimensions of SSI under consideration. Again, having these understandings would certainly help, but a teacher may not be expertly fluent in all of these areas, and yet, may still be able to present SSI learning experiences in productive and successful ways. It can be very helpful to develop collaborations with other educators who possess expertise in these other areas.

Although it is essential for a teacher to be familiar with an issue being featured in SSI-based education, being knowledgeable in the related science, and at least aware of the associated social concerns, no teacher (or anyone else for that matter) can possibly know everything that could be known about a particular issue. Therefore, it is important for teachers to acknowledge their own knowledge limitations. Students need to understand that neither they nor their teachers can possibly have all the information or answers for SSI but that an important part of the process of thoughtfully negotiating SSI is seeking new information and understandings. The fact that SSI are open-ended problems that can be addressed in multiple ways leads to inevitable uncertainties. To be successful using SSI, teachers must be willing to deal with some uncertainties in their classrooms. Unlike some traditional forms of science education where teachers know the answers to the questions they pose, SSI present questions for which teachers do not have answers and classroom situations that they may not be able to predict. Given these uncertainties, there are times within SSI-based educational experiences that teachers find themselves in nonauthoritarian roles. Sometimes teachers have to position themselves as inquirers and knowledge contributors on par with their students as opposed to the primary or sole authority of the classroom. For some science teachers, such as those who embrace inquiry-based approaches in their classrooms, this is a relatively easy transition to make. For others, who view their position as a disseminator of knowledge, this transition can be more challenging.

Conclusions

At the outset of this book project, my primary goal was to bring together a diverse group of researchers working in the area of SSI to share the findings they are generating in classroom-based research projects. A secondary goal was to facilitate conversations among these researchers to help problematize important issues and challenges for the field. My intent relative to both of these goals was to contribute to the empirical knowledge base of the field. I believe that the other contributors and I have accomplished these goals through the presentation of the research chapters and associated metalogues. I leave it to the audience to determine how well we accomplished these goals and whether the advances made are substantive enough to prove valuable.

Using the classroom-based research projects as a basis for deriving a framework for SSI-based education was not an a priori goal for the book. However, the collection of projects and the common themes that were addressed in all of the chapters like detailed descriptions of the nature of the interventions studied and development of relationships between teachers and researchers (see Chap. 1) provided the kinds of information necessary for deriving an emergent framework. Although developing a framework for SSI-based instruction may not have been an initial aim of this project, such a framework is desperately needed if the science education community intends to advance the SSI agenda beyond the realm of theory, academic research, and very isolated pockets of practice. If we are serious about extending this work such that it impacts science teaching and learning more generally, we must work to develop new tools for helping teachers and curriculum designers situate SSI in classrooms (Sadler, Amirshokoohi, Kazempour, & Allspaw, 2006; Sadler & Zeidler, 2009). The emergent framework presented offers design and implementation principles that could be of use to educators interested in transforming school science through SSI.

The framework does not offer simple prescriptions for using SSI; it does not provide a lock-step sequence of teacher and student activities. Contextualizing science education in SSI is too complicated an endeavor to be reduced to a list of simple steps. In order to be useful across multiple contexts, a framework has to be general and flexible enough to account for variance in the issues themselves, learning goals, classroom situations, etc. Because the framework described above was generated from very different teaching and research contexts, it naturally possesses a degree of flexibility. Whether that flexibility is sufficient enough while not being too ambiguous to provide value to practitioners remains to be seen. New research will be necessary to explore the extent to which the framework can be of use to science educators and ways in which the framework can be improved. I believe that the framework provides a new tool with potential for advancing the SSI agenda but here again, the audience will be the ultimate arbiters of its utility.

In Chap. 1, I highlighted two fundamental questions for the science education community: (1) What should the goals of science education be? and (2) How can these goals best be achieved? The first question is normative in nature and the volume's contributors and I have offered some perspectives on this issue. We view science education as an opportunity to support student development as competent and informed citizens capable of engaging in the negotiation and resolution of challenging societal problems particularly when those problems intersect with science. Given a particular perspective on the first question, the second question can be addressed through research. The individual studies featured in this volume provide evidence pertinent to this question. The synthesis of individual findings and derivation of a framework for SSI-based education yields additional evidence that informs ways in which science educators may use SSI to support progressive aims of science education.

References

Bennett, J., Lubben, F., & Hogarth, S. (2007). Bringing science to life: A synthesis of the research evidence on the effects of STS and context based approaches to science teaching. *Science & Education, 91*, 347–370.

Eilks, I. (2010, September). *Making chemistry teaching relevant and promoting scientific literacy by focusing on authentic and controversial socio-scientific issues*. Presentation at the annual meeting of the society for didactics in chemistry and physics, Potsdam, Germany.

Hodson, D. (2003). Time for action: Science education for an alternative future. *International Journal of Science Education, 25*, 645–670.

Sadler, T. D. (2004). Informal reasoning regarding socioscientific issues: A critical review of research. *Journal of Research in Science Teaching, 41*, 513–536.

Sadler, T. D. (2009). Situated learning in science education: Socioscientific issues as contexts for practice. *Cultural Studies of Science Education, 45*, 1–42.

Sadler, T. D., Amirshokoohi, A., Kazempour, M., & Allspaw, K. (2006). Socioscience and ethics in science classrooms: Teacher perspectives and strategies. *Journal of Research in Science Teaching, 43*, 353–376.

Sadler, T. D., & Zeidler, D. L. (2009). Scientific literacy, PISA, and socioscientific discourse: Assessment for progressive aims of science education. *Journal of Research in Science Teaching, 46*, 909–921.

Strauss, A., & Corbin, J. (1998). *Basics of qualitative research: Techniques and procedures for developing grounded theory*. Thousand Oaks, CA: Sage.

Zeidler, D. L., & Keefer, M. (2003). The role of moral reasoning and the status of socioscientific issues in science education. In D. L. Zeidler (Ed.), *The role of moral reasoning on socioscientific issues and discourse in science education*. Dordrecht, the Netherlands: Kluwer Academic Publishers.

Zeidler, D. L., Sadler, T. D., Simmons, M. L., & Howes, E. V. (2005). Beyond STS: A research-based framework for socioscientific issues education. *Science & Education, 89*, 357–377.

Index

T.D. Sadler (ed.), *Socio-scientific Issues in the Classroom: Teaching, Learning and Research*, Contemporary Trends and Issues in Science Education 39,
DOI 10.1007/978-94-007-1159-4, © Springer Science+Business Media B.V. 2011